Topics in Safety, Risk, Reliability and Quality

Volume 33

More information about this series at http://www.springer.com/series/6653

Patrick T. Hester · Kevin MacG. Adams

Systemic Decision Making

Fundamentals for Addressing Problems and Messes

Second Edition

Springer

Patrick T. Hester
Engineering Management and Systems Engineering
Old Dominion University
Norfolk, VA
USA

Kevin MacG. Adams
Information Technology and Computer Science
University of Maryland University College
Adelphi, MD
USA

ISSN 1566-0443 ISSN 2215-0285 (electronic)
Topics in Safety, Risk, Reliability and Quality
ISBN 978-3-319-54671-1 ISBN 978-3-319-54672-8 (eBook)
DOI 10.1007/978-3-319-54672-8

Library of Congress Control Number: 2017932922

Printed on acid-free paper

This Springer imprint is published by Springer Nature
The registered company is Springer International Publishing AG
The registered company address is: Gewerbestrasse 11, 6330 Cham, Switzerland

To my wife for her love and partnership, my children for their comic relief, and my parents for their encouragement. All of your support through the years has been invaluable.

—Patrick T. Hester

To my wife for her love and companionship; to my parents, sisters, children, and grandchildren for their unconditional love and support; and to my many colleagues and friends for their help and forbearance. All of you added to this endeavor in ways you do not know.

—Kevin MacG. Adams

Preface

Quick, think about a problem that vexes you. Too easy, right? The only difficulty you'd likely face is narrowing it down to *a* singular problem. Now think of another one. But this time, dig deep into your brain. Think of a problem that keeps you up at night, one that bothers you day in and day out, one that is seemingly intractable. Got one? Good, now think about what it is that characterizes this problem. What makes it hard? Why haven't you solved it yet?

Lyons (2004) offers the following barriers to solving what he calls *systemic problems*:

- Lack of incentives
- Limited resources
- Limited levers to change
- Limited power/authority
- Uncertain outcomes

We may summarize this list as saying that your problem is *complex*. But what, exactly, does that mean? What makes a problem complex? Is complexity a binary characteristic of a problem? That is, is a problem definitively complex or not? Does the complexity of a problem change throughout its development? These and more issues lead to perhaps the most fundamental introductory question for us, that is, how do we define complexity in a manner that is meaningful to us as practitioners and researchers.

Well, complexity is a loaded term. In fact, the notion of complexity is one that has been debated for decades in the scientific community and yet, no consensus on its definition has been reached (Gershenson, 2007; Lloyd, 2001; McShea, 1996; Mitchell, 2009). Precisely defining what is intended by the term complexity evokes former US Supreme Court Justice Potter Stewart's [1915–1985] famous description of obscenity, *I know it when I see it*; we know something is complex when we see it. Of course, from a scientific perspective, this is imprecise and problematic.

Literature abounds with measures proposed for evaluating complexity. We can measure the complexity of a system using a number of metrics such as Shannon's information entropy (Shannon & Weaver, 1949), algorithmic information content (Chaitin, 1966; Kolmogorov, 1965; Solomonoff, 1964), effective complexity (Gell-Mann, 1995), logical depth (Bennett, 1986), thermodynamic depth (Lloyd & Pagels, 1988), statistical complexity (Crutchfield & Young, 1989), hierarchy (Boulding, 1956; Simon, 1962), a set of predefined characteristics (Cilliers, 1998; Funke, 1991, pp. 186–187), and a number of other measures (Lloyd, 2001). Criticisms of these measures range from a lack of intuitive results when using some measures (information entropy, statistical complexity, and algorithmic information content) to the lack of a practical means for consistently utilizing other measures (logical depth, effective complexity, and thermodynamic depth). Mitchell (2009) discusses the drawbacks of many of these measures and suggests that none have obtained universal appeal as a practical and intuitive means of measuring the complexity of a system. McShea (1996) agrees, stating, "…no broad definition has been offered that is both operational, in the sense that it indicates unambiguously how to measure complexity in real systems, and universal, in the sense that it can be applied to all systems" (p. 479). In the absence of a universal measure of complexity, we will investigate two perspectives for defining complexity, namely characteristic complexity and hierarchical complexity, in an effort to provide some structure to the concept.

Characteristic Complexity

We may conceive of complexity as being measured by the extent to which a situation or problem exhibits a number of predefined characteristics. One such set of characteristics was posed by noted psychologist Joachim Funke (1991, pp. 186–187) as characterizing complex problem-solving situations:

- *Intransparency*: Intransparency refers to the lack of availability of information in our problem. An intransparent problem represents a situation in which all variables cannot be directly observed. In this case, we may have to infer information about the underlying state of the system, or too many variables exist, leading to our selection of only a handful for observation and analysis.
- *Polytely*: From the Greek words *poly* and *telos* meaning *many goals*. This set of goals can be thought in many forms. We may have many individuals associated with our problem, and each harbors their own needs and wants. These interests are likely not to be directly aligned; thus, they compete for our attention, requiring trade-offs. Similarly, objectives within our problem are not typically straightforward. Complex problems involve multiple, conflicting objectives. Finally, our problem will likely require competition for resources. We do not have unlimited resources; thus, we are limited in our ability to address our problem in the most straightforward and effective manner.

- *Complexity*: Here, Funke is referring to the number of variables, the connectivity between these variables, and the nature of their relationship (i.e., linear vs. nonlinear). Funke (1991) summarizes complexity as:

> A complex problem-solving situation is not only characterized by a large number of variables that have to be considered, but also by their complex connectivity pattern, by the possibilities to control the system, and by the dynamic aspects of the system. The growing complexity of situational demands may conflict with the limited capacity of the problem solver. (pp. 186–187)

- *Variable connectivity*: A change in one variable is likely to affect the status of many other variables. Given this high connectivity, consequences are difficult to predict. That is, there is substantial unpredictability in the behavior of the problem. Even the most tried-and-true of modeling techniques fail to capture the behavior of modern problems—events such as Hurricanes Katrina or Sandy, the housing market crash, and other so-called *Black Swans* (Talib, 2007). These unpredictable phenomena go beyond the bounds of our uncertainty analysis techniques and require us to consider the robustness of our institutions, organizations, and supporting systems. Considering these phenomena in concert with shrinking resources, we have a quandary. More resources are required to plan for unpredictability, yet we lack sufficient resources to address these concerns completely. Thus, we must make compromises to account for this inherent contradiction.
- *Dynamic developments*: There is often considerable time pressure to address problems before they worsen. Positive changes also occur, but these changes could lead to further unpredictability. This is complicated by humans' bias for action. Most people are uncomfortable with situations that are unresolved. We want an answer and we want it now. One must simply look at the increase in information availability over the last decade to understand how the world has transformed into one demanding instant gratification. No longer are we content to pull an encyclopedia off our book shelf (that is, if we even own an encyclopedia anymore) and look up the answer to a question. Instead, we pull out our smart phone and *Google it*, expecting an instant answer, and grumbling when our Internet connection hits a snag. This behavior is problematic when the problems of substantial complexity are considered. Choosing to act, to get an answer *right now*, rather than obtaining additional information, may lead to an inferior choice based on insufficient information. We must carefully weigh the desire to obtain more information with our potential for loss and what may have been. To put it another way, we must choose between *getting it right* and *getting it right now*.
- *Time-delayed effects*: Effects often occur with a time delay. This requires patience on the part of the individual concerned with the problem. This is in direct contrast to the need for near-term action discussed in the previous element.

To this list, we add two characteristics:

- *Significant uncertainty*: Complex problems have substantial uncertainty. That is, there are unknown elements which plague our problem. Some are so-called *known unknowns* such as the fact that market demand for a new product is unknown. These uncertainties come from the variables that are known to exist in a problem (but that have some level of random behavior associated with them that can be expressed by probability distributions). These types of uncertainties are present in any real-world problem due to the inherent variability of the natural world. So we use probabilistic information to reason about and predict these phenomena. More difficult to deal with are *unknown unknowns* such as the fact that we do not know what our competitors will do. This type of uncertainty comes from lack of knowledge of the larger system of problems (which we will later classify as a mess) of which our problem is a part. Will we be instantly outclassed by our competitors the day our new product is introduced to the market (or worse, before we even release our product)? To estimate these uncertainties, we typically turn to experts for their insight. Both sources of uncertainty, known and unknown unknowns, complicate our problem landscape but cannot be ignored.
- *Humans-in-the-loop*: Designing a mechanical system given a set of specifications may be straightforward, but designing the same system while incorporating human factors, including elements such as ergonomics, fatigue, and operator error prevention, is substantially more complex. Once we insert humans into our problem system, all bets are off, so to speak. In many ways, humans are the ultimate trump card. They represent the one factor that seemingly ignores all the hard work, all the calculations, all the effort, that has gone into the development of a solution to our problem. They exploit the one weakness or vulnerability in our problem system that no amount of simulations, trial runs, mock-ups, or counter-factuals could have accounted for. They are intransparent, uncertain, competitive, unpredictable, and have a bias for action, all factors that we've indicated make a problem hard. To boot, they are not mechanistic; they have feelings and emotions, and difficult problems are often especially emotional issues. Think about some of the most difficult problems facing our current society, e.g., health care or higher education; they are highly emotional topics likely to elicit an emotionally charged response from even the most level-headed of individuals. Thus, even when we think we have it all figured out, humans enter the equation and blow it all apart.

Hierarchical Complexity

Conversely, it may be advantageous for us to think of complexity as existing in a hierarchical fashion. Jackson (2009) summarizes the work of Boulding (1956) in creating a nine-level hierarchy for real-world complexity, as shown in Table 1 and in keeping with the *principle of hierarchy* (Pattee, 1973).

Table 1 A summary of Boulding (1956) hierarchy of complexity (Jackson, 2009, p. S25)

Level	Description	Example
1	Structures and frameworks which exhibit static behavior and are studied by verbal or pictorial description in any discipline	Crystal structures
2	Clockworks which exhibit predetermined motion and are studied by classical natural science	The solar system
3	Control mechanisms which exhibit closed-loop control and are studied by cybernetics	A thermostat
4	Open systems which exhibit structural self-maintenance and are studied by theories of metabolism	A biological cell
5	Lower organisms which have functional parts exhibit blue-printed growth and reproduction, and are studied by botany	A plant
6	Animals which have a brain to guide behavior are capable of learning, and are studied by zoology	An elephant
7	People who possess self-consciousness know that they know, employ symbolic language, and are studied by biology and psychology	Any human being
8	Sociocultural systems which are typified by the existence of roles, communications and the transmission of values, and are studied by history, sociology, anthropology, and behavioral science	A nation
9	Transcendental systems, the home of 'inescapable unknowables', and which no scientific discipline can capture	God

Each of these levels is of increasing complexity, and each contains emergent properties not found in the levels below. Thus, in seeking to understand a given level, we must also understand those levels beneath it, invoking the *principle of recursion* (Beer, 1979). Boulding (1956) comments on the maturity of our knowledge about the levels in his hierarchy:

> One advantage of exhibiting a hierarchy of systems in this way is that it gives us some idea of the present gaps in both theoretical and empirical knowledge. Adequate theoretical models extend up to about the fourth level, and not much beyond. Empirical knowledge is deficient at practically all levels. Thus, at the level of the static structure, fairly adequate descriptive models are available for geography, chemistry, geology, anatomy, and descriptive social science. Even at this simplest level, however, the problem of the adequate description of complex structures is still far from solved. (p. 205)

Despite our relative naïveté about the higher levels of the hierarchy, Boulding (1956) notes that all hope is not lost:

> Nevertheless as we move towards the human and societal level a curious thing happens: the fact that we have, as it were, an inside track, and that we ourselves are the systems which we are studying, enables us to utilize systems which we do not really understand. (pp. 206-207)

Thus, even though we may not *understand* systems at the higher levels of this hierarchy in the theoretical sense, we can work with, utilize, and make sense of them. This is absolutely necessary as we attempt to determine the appropriate opportunity to intervene in a problem system.

So, what is one to do? Well, we could avoid all problems exhibiting one or all of the characteristics of complexity, existing within Boulding's hierarchy, or fundamentally identified as complex by us as researchers and practitioners. This leaves a very small, uninteresting subset of the world to deal with. Alternatively, we suggest that all hope is not lost. We simply need a new way to reason about these problems that goes beyond the traditional methods we employ. Full disclosure—the authors of this book are engineers by education. But we've worked in industry and the military for many years and we've come to understand that no single discipline can solve truly complex problems. Problems of real interest, those vexing ones that keep you up at night, require a discipline-agnostic approach. They require us to get out of our comfort zone a little bit, to reach across the aisle and embrace those fundamental concepts of other disciplines that may be advantageous to our effort. Simply, they require us to think *systemically* about our problem.

Fundamentally, we need a novel way to *address* these problems, and more specifically, to do so *systemically*, hence the title of this book. It is the hope of the authors that, after reading this book, readers will gain an appreciation for a novel way of thinking and reasoning about complex problems that encourages increased understanding and deliberate intervention. We set out to provide this in a manner that is not predicated on the reader being either an engineer or a scientist. Indeed, most of the complex problems vexing us are not engineering or scientific problems, at least in the strictest sense. Complex problems such as climate change, world hunger, poverty, and global conflict know no disciplinary boundaries. So, you'll see us draw from engineering and science to be sure, but we'll also draw from psychology, mathematics, sociology, management, and many other fields in an effort to develop a robust approach to thinking about and addressing problems. To support this approach, this book is divided into four major sections: (1) A Frame of Reference for Systemic Decision Making; (2) Thinking Systemically; (3) Acting Systemically; and (4) Observing Systemically.

This book is intended for use by practitioners tasked with addressing complex problems or individuals enrolled in a graduate or advanced undergraduate class. Given its discipline-agnostic nature, it is just as appropriate for use in a business, sociology, or psychology course as it is in an engineering or scientific course. Regarding its instruction, the chapters should be taught in order. Part I provides the proper theoretical foundation necessary for Parts II–III. Part II provides a multimethodology for thinking systemically about complex problems and problem systems. Part III provides an approach for acting on the complex problems and problem systems investigated in Part II. Finally, Part IV discusses observation of actions undertaken in Part III, and it provides a comprehensive case study demonstrating the material discussed throughout the text.

References

Beer, S. (1979). *The heart of the enterprise*. New York: Wiley.

Bennett, C. H. (1986). On the nature and origin of complexity in discrete, homogenous, locally-interacting systems. *Foundations of Physics*, *16*, 585–592.

Boulding, K. E. (1956). General systems theory—The skeleton of science. *Management Science*, *2*(3), 197–208.

Chaitin, G. J. (1966). On the length of programs for computing finite binary sequences. *Journal of the Association of Computing Machinery*, *13*, 547–569.

Cilliers, P. (1998). *Complexity and postmodernism: Understand complex systems*. New York: Routledge.

Crutchfield, J. P., & Young, K. (1989). Inferring statistical complexity. *Physical Review Letters*, *63*, 105–109.

Funke, J. (1991). Solving complex problems: Exploration and control of complex systems. In R. J. Sternberg & P. A. Frensch (Eds.), *Complex problem solving: Principles and mechanisms* (pp. 185–222). Hillsdale, NJ: Lawrence Erlbaum Associates.

Gell-Mann, M. (1995). What is complexity? *Complexity*, *1*, 16–19.

Gershenson, C. (2007). *Design and control of self-organizing systems*. Mexico City, Mexico: CopIt ArXives.

Jackson, M. C. (2009). Fifty years of systems thinking for management. *Journal of the Operational Research Society*, *60*(S1), S24–S32. 10.2307/40206723

Kolmogorov, A. N. (1965). Three approaches to the quantitative definition of information. *Problems of Information Transmission*, *1*, 1–17.

Lloyd, S. (2001). Measures of complexity: A nonexhaustive list. *IEEE Control Systems Magazine*, August, 7–8.

Lloyd, S., & Pagels, H. (1988). Complexity as thermodynamic depth. *Annals of Physics*, *188*, 186–213.

Lyons, M. (2004). Insights from complexity: Organizational change and systems modelling. In M. Pidd (Ed.), *Systems modelling: Theory and practice* (pp. 21–44). West Sussex, UK: Wiley.

McShea, D. W. (1996). Metazoan complexity and evolution: Is there a trend? *Evolution, 50*(2), 477–492.

Mitchell, M. (2009). *Complexity: A guided tour*. New York: Oxford University Press.

Shannon, C. E., & Weaver, W. (1949). *The mathematical theory of communication*. Champaign, IL: University of Illinois Press.

Simon, H. A. (1962). The architecture of complexity. *In Proceedings of the American Philosophical Society*, *106*(6), 467–482.

Solomonoff, R. J. (1964). A formal theory of inductive inference. *Information and Control*, *7*, 224–254.

Talib, N. N. (2007). *The black swan: The impact of the highly improbable*. New York: Random House.

Acknowledgements

Patrick T. Hester

I would like to thank the numerous organizations, and all the individuals, that I have worked with over the past 15 years and that have shaped my perspectives on life and work. Thanks to all my colleagues and classmates from Webb Institute, National Steel and Shipbuilding Company, Vanderbilt University, Sandia National Laboratories, Naval Surface Warfare Center Dahlgren Division, and Old Dominion University. Each stop along my journey has led to new insights and even more questions.

To my students during my time at Old Dominion University. You never lack the courage to ask the tough questions that have filled up my white board with chaos that boggles my mind and has helped me to lose what little hair I had left. Thanks especially to those students from my *Systemic Decision Making* class who beta-tested this text with me. Your insights greatly improved this book and the thinking behind it.

To my co-author Kevin, who has always approached everything in his life with 100% enthusiasm. Your willingness to jump into writing another book, no questions asked, is admirable. To my wife Kasey, who is my partner and equal on this wild ride through life. To my kids, Maryn and Gavin, who cheer me up even on the most dreary of days. To my parents, Kevin and Peggy, for your unwavering support through the years. Thanks to all of you for inspiring me in ways you can never know.

Kevin MacG. Adams

I would like to thank the many real-world systems thinkers who have mentored me over the past 46 years. There are too many to list, but special recognition goes to Dr. Steve Krahn, Dave Herbein, Jeff Reed, Dr. Tim Arcano, Mark "Einstein" Welsh, and Dr. Bob Charette, all of whom provided continuous counsel in keeping me from committing the *Type VII error*.

To the many students I have been privileged to teach at both the University of Maryland University College and Virginia Wesleyan College. Your quest for knowledge has challenged me to constantly renew and improve my own understanding as part of the learning process.

To my co-author Patrick, for making the ideas in this book come to life. As colleagues and friends, we have had the opportunity to discuss and debate just about everything under the sun. Our mutual interest in problem-solving, through *systemic thinking* is what has motivated us to share the thoughts in this book. Our remarkably similar views on the transdisciplinary nature of problem-solving methodologies focused on the effort and permitted us to collaborate in an exciting and energizing way. Thanks for making this fun!

Once again, to my parents, for providing me with the educational foundation that fostered both learning and the love to share this with others. To my children and grandchildren, for many hours of challenges, joy, and amazement. Finally, to my wife, for her thorough editing, companionship, and love throughout the process of completing this book and our journey through life together.

Contents

Part II Thinking Systemically

Part IV Observing Systemically

About the Authors

Dr. Patrick T. Hester is an associate professor of Engineering Management and Systems Engineering at Old Dominion University. Dr. Hester received a Ph.D. in Risk and Reliability Engineering from Vanderbilt University and a B.S. in Naval Architecture and Marine Engineering from the Webb Institute. He has been involved in research and consulting activities focused on systems thinking and decision analysis for diverse agencies, including NIST, Naval Surface Warfare Center Dahlgren Division, NASA Langley Research Center, DHS, Sandia National Laboratories, NOAA, TRADOC, and General Dynamics National Steel and Shipbuilding Company. The results of his research have led to over one hundred publications in books, journals, and conferences. His research has been published in *Systems Engineering*, *International Journal of System of Systems Engineering*, *International Journal of Operations Research*, *International Journal of Critical Infrastructures*, and *Journal of Defense Modeling and Simulation*, among others. Dr. Hester is a senior member of the Institute of Industrial and Systems Engineers (IISE), the Performance Management Association, and the International Society on Multiple Criteria Decision Making, a board member of IISE's Society of Engineering and Management Systems, and an editorial board member of *Systemic Practice and Action Research*. He is a two-time Outstanding Author Contribution Award Winner from the Emerald Literati Network, and in 2016, he was named one of the 20 "Engineering Management Professors You Should Know" by OnlineEngineeringPrograms.com.

Dr. Kevin MacG. Adams is a much sought after speaker and lecturer who specializes in providing system-based solutions for complex real-world problems. He is the author of books on systemic thinking and nonfunctional requirements and has published over 50 articles in peer-reviewed journals and conference proceedings.

His presentations merge both philosophical and theoretical concepts from academia with the everyday problems being faced by businesses operating in the highly connected, fast-paced, global economy—providing effective and lasting solutions. Kevin's system-based solutions emphasize the importance of organizational stakeholders and the use of tailored methodologies for each and every solution.

His unique foundation for problem-solving is based upon over 35 years of practical management experience. He has been able to integrate his first career into the United States Navy, where he served aboard three submarines and as a maintenance officer with second and third careers as a private-sector information technology consultant specializing in enterprise solutions and as an academic teaching in both engineering and management curriculums.

He is currently an adjunct associate professor at the University of Maryland University College and Virginia Wesleyan College. He volunteers on the Editorial Board of the Journal *Systemic Practice and Action Research* and as a program evaluator for the Accreditation Board for Engineering and Technology (ABET). Dr. Adams is a retired Navy submarine engineering duty officer, and a senior member of the Institute of Electrical and Electronics Engineers (IEEE).

Dr. Adams is a member of *Tau Beta Bi* (NJ Beta)—the National Engineering Honor Society, *Keramos*—the National Ceramic Engineering Honor Society, and *Phi Kappa Phi*, the nation's oldest, largest, and most selective collegiate honor society for all academic disciplines.

Dr. Adams hold engineering degrees from Rutgers University (B.S.), the Massachusetts Institute of Technology (dual M.S. degrees), and Old Dominion University (Ph.D.).

Part I
A Frame of Reference for Systemic Decision Making

Chapter 1
Introduction

Abstract The first step in addressing a problem is recognizing you have one. It is with this notion in mind that the authors begin their discussion. This chapter begins with the fundamental tenet of systemic decision making, which we term the *TAO approach*, a general approach for increasing our understanding about problems that is invoked throughout the text. Then, a discussion of systems errors is presented.

1.1 The TAO Approach

As we said before, we have all got problems. Some are big; some are small. Some are fleeting, while some are nagging and persistent. All could benefit from a structured way of reasoning about them. To that end, we provide a general approach for improved understanding that we call the TAO approach, for **t**hink, **a**ct, and **o**bserve. The idealized relationship between these elements is pictured in Fig. 1.1. Ideally, these steps would progress in a linear fashion in a manner that maximized understanding and minimized wasted effort due to rework. The reality is, however, that real-world decision making is rarely this smooth, as we will see as the topics in this book unfold. Our aim throughout this text is to provide information to assist the reader in completing each of the think, act, and observe stages. Chapters 6 through 11 in Part II will detail those steps necessary for systemic thinking, Chaps. 12–14 in Part III discuss systemic action, and Chaps. 15 and 16 in Part IV address systemic observation.

Knowing that we have problems and more importantly, knowing that we need approaches to deal with these problems, requires us to first understand what systematic mistakes we make that may be avoided. To this end, we turn to a discussion of systems errors.

P.T. Hester and K.M. Adams, *Systemic Decision Making*, Topics in Safety, Risk, Reliability and Quality 33, DOI 10.1007/978-3-319-54672-8_1

Fig. 1.1 Idealized TAO approach to increased understanding

1.2 Systems Errors

Reasoning about a complex problem routinely employs the use of one of a number of systems-based approaches (Jackson, 1991, 2000, 2003). Independent of the method used to address a complex problem is the opportunity to commit a number of errors. Analytical and interpretational errors are common while thinking about, acting on, and observing problems; however, none of these systems approaches explicitly addresses these potential errors. Further, despite their prominence, there is not an agreed-upon taxonomy for errors in problem solving approaches. Thus, the authors have worked to establish an initial taxonomy for error classification (Adams & Hester, 2012, 2013), which is expanded upon in this text. This taxonomy has drawn from research performed by researchers representing four of the 42 fields of science (OECD, 2007), as depicted in Table 1.1.

Based on our review of the literature in Table 1.1, we were able to develop a taxonomy of eight common errors that individuals are prone to encounter while thinking about, acting on, and observing problems. We will not discuss the errors in numerical order; rather, we begin with discussion of the Type III error and proceed by discussing errors in the chronological order in which they are most often encountered while attempting to address a complex problem.

Table 1.1 Science sector and field of science that have conducted inquiry on errors (adapted from Adams & Hester, 2013, p. 319)

Science sector	Field of science	References
Social sciences	Educational sciences	Betz and Gabriel (1978), Kaufman, Dudley-Marling, and Serlin (1986), Marascuilo and Levin (1970, 1976), Onwuegbuzie and Daniel (2003), Rosnow and Rosenthal (1989, 1991)
	Psychology	Games (1973), Kaiser (1960), Leventhal and Huynh (1996), Levin and Marascuilo (1972, 1973), Meyer (1991), Mitroff (1998), Mitroff and Featheringham (1974), Reason (1990)
	Economics and business	Boal and Meckler (2010), Umesh, Peterson, McCann-Nelson, and Vaidyanathan (1996)
Natural sciences	Mathematics	Holland (1986), Kimball (1957), Mosteller (1948), Neyman and Pearson (1928a, b, 1933), Tracz, Nelson, Newman, and Beltran (2005)

1.2.1 Type III Error

The extant literature on the Type III (γ) error originated in statistics. Frederick Mosteller [1916–2006], one of the most eminent statisticians of the twentieth century, reported:

> In other words it is possible for the null hypothesis to be false. It is also possible to reject the null hypothesis because some sample O_i has too many observations which are greater than all observations in the other samples. But the population from which some other sample say O_j is drawn is in fact the right-most population. In this case we have committed an error of the third kind. (p. 61)

This is commonly referred to as "the error associated with solving the wrong problem precisely" (Mitroff, 1998, p. 15). Type III errors normally occur during the formulation of problems, the phase in which the actual details surrounding the reported problem are exposed, validated, and verified as part of the process of problem reformulation (reformulation is where the initial *reported* problem statement is validated by relevant stakeholders). We denote this revised problem statement the *real* (or *formulated*) problem, to differentiate it from the reported problem. Mitroff (1998) identifies the five most common causes of a Type III error:

1. Picking the wrong stakeholders
2. Selecting too narrow a set of options
3. Phrasing a problem incorrectly
4. Setting the boundaries/scope of a problem too narrowly
5. Failing to think systemically.

Each of these issues is addressed in this text, with the fifth cause (and its avoidance) being the ultimate driver in writing this text.

Adams and Hester (2012) devise a medical analogy to explain the Type III error:

> The systems practitioner faced with a reported problem needs to act much like a physician. The physician listens to the symptoms reported by a patient, but does not accept the diagnosis of the patient. The physician cannot rely solely on the patient's story and symptoms, but must gather empirical data by conducting tests, taking physiological measurements, and conducting a physical examination. The systems practitioner is in a similar professional relationship with the client that has a systems problem. Problem reformulation ensures that the scope of the problem is properly abstracted from the real-world and defined. The problem system must be adequately bounded, include empirical data of both the quantitative and qualitative types, and include an understanding of both the environment and relevant stakeholders. (p. 28)

Mitroff and Featheringham (1974) elaborate on the importance of proper problem formulation:

> The initial representation or conceptualization of a problem is so crucial to its subsequent treatment that one is tempted to say that the most important as well as most difficult issue underlying the subject of problem solving is precisely 'the problem of how to represent problems.' (p. 383)

Failure to properly define the scope of the problem results in inadequate problem statements and is commonly referred to as "the error committed by giving the right answer to the wrong problem" (Kaiser, 1960, p. 134). Once we have appropriately formulated our problem (i.e., thought about it), we must decide what to do about this problem (i.e., act on it). In acting (or abstaining from action), we may encounter a number of errors, to which we now turn.

1.2.2 Type IV Error

A review of the extant literature on Type IV (δ) errors shows that this type of error has been discussed principally in the psychology and the educational sciences. To the authors' knowledge, the first mention of the Type IV error in the literature was by Marascuilo and Levin (1970). They define the Type IV (δ) error as:

> A Type IV error is said to occur whenever a correct statistical test has been performed, but is then followed by analyses and explanations that are not related to the statistical test used to decide whether the hypothesis should or should not have been rejected. (Marascuilo & Levin, 1976, p. 368)

The primary discussion related to Type IV errors has been associated with statistical testing, most notably ANOVA models (Kaufman et al., 1986; Rosnow & Rosenthal, 1989, 1991; Umesh et al., 1996). We prefer, however, to endorse the Type IV error as one concerned with a higher level of abstraction, most notably as "the incorrect interpretation of a correctly rejected hypothesis" (Marascuilo & Levin, 1970, p. 398).

Boal and Meckler (2010) elaborate on the problems caused by a Type IV error, introducing the concept of iatrogenic solutions:

> Acting to solve a problem, be it the right problem or the wrong problem, can create other difficulties. Sometimes solutions are 'iatrogenic,' meaning that they create more, or bigger problems than they solve. Faced with such a possibility the decision maker should thoroughly examine all the potential system effects, and perhaps refrain from action. In the case that it was an attempted solution to the right initial problem, one important problem is now replaced by another, perhaps worse problem. (p. 333)

Thus, even though the problem has been correctly identified (i.e., thought about), the action identified to resolve the problem is incorrect. Systems and management expert Russell Ackoff [1919–2009] (1994a, b) referred to this simply as an error of commission, or "doing something that should not have been done" (p. 3).

Further, there is potential in this situation for the identified actions to actually exacerbate the problem.

Adams and Hester (2013) continue their medical analogy:

> This could be the case where the physician commits a Type IV (δ) error by correctly diagnosing the problem and prescribes the right medication. However, the medication side-effects for a particular patient are worse than the original symptoms. The systems

> practitioner is prone to committing this error. The most typical instance is when the practitioner has properly reformulated and defined the client's problem and then applies an improper solution approach (i.e., methodology, method, or technique) in an attempt to resolve this problem. Failure to match the solution method to appropriate solution of a problem has been an important subject in the systems literature (Adams & Mun, 2005; Jackson, 1984; Jackson & Keys, 1984). (pp. 320–321)

1.2.3 Type V Error

The Type V Error, like the Type IV Error, concerns actions taken in support of problem resolution. The field of cybernetics and the systems principles of *homeostasis* (Cannon, 1929) and *homeorhesis* (Waddington, 1957) inform individuals that systems have the ability to self-regulate to maintain a stable condition. Thus, some problems may resolve themselves by simply allowing a natural order to restore itself. The converse of this is that many problems require intervention in order to be addressed and simply wishing for a problem to disappear on its own will not make it go away. There is a substantial risk in not acting when action is called for. Boal and Meckler (2010) discuss this sentiment as the Type V (ε) error:

> Deciding to take no action, when no action is called for, is the correct solution. However, falsely believing that the problem will either solve itself or simply go away is an error of the 5th kind. Such errors allow the situation to linger, at best, or to fester and worsen requiring greater resources to solve. (p. 334)

Ackoff (1994a, b) described such an error as an error of omission, or "not doing something that should have been done" (p. 3). Errors of omission are more difficult to identify as they seldom are recorded due to their implicit nature, i.e., we don't usually record what we don't do; rather, we simply do not do it. The lack of accountability afforded to errors of omission has a curious effect in that it actually exacerbates the likelihood of their occurrence. Ackoff (1994a, b) addressed this phenomenon directly:

> Because errors of commission are easier to identify than errors of omission, many decision makers try to avoid making errors of commission by doing nothing. Although this increases their chances of making an error of omission, these errors are harder to detect. (p. 4)

In the medical analogy of this error, the physician commits a Type V error when he or she correctly diagnoses an ailment (i.e., thinks about the problem properly) yet fails to take corrective action to resolve the problem. The reason for the failure to act in this case may reside in the physician's belief that the ailment will simply resolve itself (or the desire to avoid a poor decision and thus commit a Type IV error).

Causes for the Type V error are many. Lack of stakeholder consensus (e.g., the doctor, insurance company, and patient do not agree on treatment options) may lead to inaction due to the lack of a singular prevailing option, or due to a predominant stakeholder forcing an inaction strategy (e.g., the insurance company denies a request for an MRI, leading to a wait-and-see approach). Further, there may be a fundamental lack of understanding which permeates the analysis of the problem. This may lead to the stakeholders being unable to generate a plausible scenario for resolving the problem. Finally, stakeholders may fear worsening the problem by interfering. While this is a valid concern, we must weigh the balance between the Type IV and Type V errors, that is, between taking the wrong action and taking no action at all.

1.2.4 Type VIII Error

The Type VIII error refers to the phenomena where the correctly decided action has been incorrectly implemented. While it is coined in this text by the authors, it has its roots in the study of human error by psychologist and human factors researcher James Reason. In his seminal text on the subject, Reason (1990) discusses a number of different causes for accidents involving humans, and Type VIII errors exist under the general category known as *unsafe acts*. In order to understand the Type VIII error, it is useful to distinguish between errors and violations. Errors are defined as "mental or physical activities of individuals that fail to achieve their intended outcome" (Shappell and Wiegmann, 2000, p. 3), while a violation is a "willful disregard for the rules and regulations…" (Shappell and Wiegmann, 2000, p. 3). The Type VIII error specifically addresses this notion of error and not a violation. That is to say, unsafe acts that are committed on purpose, i.e., acts of sabotage, are not errors and are not addressed for the purposes of our discussion.

An example of the Type VIII error is one that many of us have no doubt experienced in our daily lives. A distracted driver is talking on his cell phone, adjusting the radio and generally, not devoting 100% of his attention to the task of driving. As a result, he misses his exit on the freeway. Having traveled the route numerous times before, there was no confusion regarding the way to his destination (i.e., the correct action), rather he incorrectly implemented it (i.e., he deviated from his plan). His actions represented a Type VIII error and not a willful violation. Within the medical world, we can consider an analogy in which a patient is responsible for taking a dose of medication at a prescribed time every day, say before bed. Failure to do so, in spite of knowledge of the directions, constitutes a Type VIII error. The patient knew that the medicine was to be taken each night before bed and simply committed an error. Perhaps he forgot, or some other circumstance prevented him from correctly implementing the correct action, i.e., taking his medicine according to the directions.

Once we have acted and hopefully avoided the Type IV, V, and VIII errors, we must now observe the effects of our actions. During observation, there are also opportunities for committing errors.

1.2.5 Type I and Type II Errors

The extant literature on the Type I and Type II errors is founded in the mathematics (i.e., statistics) field of science, originating with Neyman and Pearson (1928a, b, 1933). The Type I and Type II errors have been explored extensively in the literature associated with these fields. They are driven by discussions of statistical inference; specifically, they are motivated by the traditional two-sided hypothesis test. In such a test, there are only two possible error conditions: (1) deciding that a difference exists when, in fact, there is none (i.e., committing a Type I (α) error), and (2) deciding there is no difference when, in fact, there is a difference (i.e., committing a Type II (β) error) (Kaiser, 1960). Table 1.2 contains a representation of and definitions for the Type I and Type II errors framed in terms of the testing of a null hypothesis, H_0.

To continue our medical analogy, there are two classic examples from the medical world of the Type I (α) and Type II (β) error, based on the premise of H_0 being the hypothesis that a person does not have a disease:

- *Type I (α) Error*: A medical test indicates a person has a disease that they do not actually have.
- *Type II (β) Error*: A medical test indicates a person does not have a disease that they do actually have.

Both of these errors typically occur after the problem has been thought about and acted on (and after practitioners hopefully have avoided committing a Type III, IV, V, or VIII error). Thus, this phase is considered to be the observation phase (observation, as we intend it, will be elaborated on later in this book). Another potential error of observation is the Type VI error.

1.2.6 Type VI Error

Here, we introduce a Type VI (θ) error as one that is well known yet not characterized in error terms traditionally. This error originates in statistics and is that of unsubstantiated inference. Succinctly, Holland (1986) states famously, "Correlation does not imply causation…" (p. 945). Given two variables, A and B, we can

Table 1.2 Type I and Type II errors

Test result	Actual condition	
	H_0 true	H_0 false
Reject H_0	Type I error (α) False positive	Correct inference True positive
Fail to reject H_0	Correct inference True negative	Type II error (β) False negative

measure the strength of the relationship between these variables, known as their correlation. If we continue our medical analogy, denoting *A* as the number of tests taken to diagnose an illness and *B* as money spent on treatment, then we see what is termed a positive correlation between these two variables, meaning that the more tests that are performed, the more money that is spent. We can now change *B* to money remaining in your bank account. As additional tests are run, assuming they are being paid for by you, your bank account balance decreases, indicating a negative correlation. The correlation coefficient measures the strength of the relationship between these two variables.

Causation is not as straightforward, however, and it is often erroneously taken as a given when correlation is present. For example, if we have two additional events, (1) a man receives a positive test for a given disease (*A*) and (2) his brother receives a positive test for the same disease (*B*), we may be able to establish correlation. However, inferring that *A* caused *B* or *B* caused *A* is faulty, unless we have information (more specifically, observations) that corroborates this assumption, e.g., the disease in question is a blood-borne disease and the brothers admit to sharing needles during drug use. In this case, we might be able to establish causality. More often than not, however, our notion of causality is simply conjecture. This behavior represents the Type VI error. In fact, there are four possible outcomes for any two correlated variables, *A* and *B*:

1. *A* could cause *B*.
2. *B* could cause *A*.
3. An additional third variable, *C*, could be contributing to the change in both *A* and *B*.
4. It may simply be a coincidence that the two events have a correlation.

We must be careful not to infer causality regarding *A* and *B* in an effort to explain unknown phenomena. Establishing causality requires a significant number of observations and should not be done erroneously.

1.2.7 Type VII Error

Complex problems are further exacerbated by committing a Type VII (ζ) error, "a system of errors" (Adams & Hester, 2012, p. 30) to complement Ackoff's characterization of "messes as systems of problems" (Ackoff, 1979, p. 100). A Type VII error occurs when all other error types compound to create a larger, more complex problem than originally encountered. Boal and Meckler (2010) elaborate on the nature of Type VII errors:

> …the resulting problem may no longer be recognizable in its original form. The problems are not easily diagnosable, the resources and choices available become less sufficient or desirable, the solution is not readily apparent, and the solution not so attainable. (p. 336)

Adams and Hester (2012) complete their medical analogy by discussing this error:

> …a Type [VII] error can be conceived as one that first involves a physician diagnosing an incorrect problem for a patient, perhaps due to incorrect information provided by the patient (thus committing a Type III error). Let's suppose for the sake of argument that the patient is uninterested in receiving a true diagnosis of his symptoms as he fears grave news from the physician, so he downplays his symptoms. Given this incorrect (and underemphasized) problem, the physician decides to take no action to a problem otherwise requiring action (thereby committing a Type V error). His reasoning, based on the information he's received, is that the problem will go away on its own. The problem, untreated, worsens, thereby resulting in an inoperable condition, such as the progression of a benign cancer to a stage at which treatment is unavailable. Clearly, this system of errors has exacerbated the original in a form unimaginable by the original stakeholders (i.e., the patient and physician). (p. 30)

It is the Type VII error that we must truly be concerned about.

1.2.8 Analysis of Errors

We have discussed eight classifications of errors that may be experienced while thinking about, acting on, or observing a problem. A taxonomy of these eight errors is presented in Table 1.3.

Table 1.3 Taxonomy of systems errors (adapted from Adams & Hester, 2012)

Error	Definition	Issue
Type III (γ)	Solving the wrong problem precisely	Wrong problem
Type IV (δ)	Inappropriate action is taken to resolve a problem as the result of a correct analysis	Wrong action
Type V (ε)	Failure to act when the results of analysis indicate action is required	Inaction
Type VIII (η)	Incorrectly implementing the correctly decided action	Incorrect implementation
Type I (α)	Rejecting the null hypothesis when the null hypothesis is true	False positive
Type II (β)	Failing to reject the null hypothesis when the null hypothesis is false	False negative
Type VI (θ)	Inferring causation when only correlation exists	Unsubstantiated inference
Type VII (ζ)	An error that results from a combination of the other six error types, often resulting in a more complex problem than initially encountered	System of errors

Recalling the TAO approach, we can see when individuals may be prone to these errors. *Thinking* is prone to the Type III error, *acting* to the Type IV, V, or VIII error, and *observation* to the Type I, II, or VI errors. In order to correctly address a problem, all of these errors must be avoided as follows:

1. The Type III error must be overcome; that is, the correct problem must be formulated. Thinking carefully about a situation allows us to ensure we have formulated the correct problem prior to action and observation. Avoidance of the Type III error (by *thinking systemically*) is the focus of Part II of this book.
2. Once we have thought systemically about our problem, we must now act (or not). This offers the opportunity for four possible outcomes:
    (a) We act incorrectly, when action is warranted (committing a Type IV error).
    (b) We fail to act, when action is warranted (committing a Type V error).
    (c) We incorrectly implement the correct action (committing a Type VIII error).
    (d) We act correctly, when action is warranted (committing no error).
    Thus, we must choose the appropriate course of action for a particular problem, given that choosing not to act is also a feasible choice. This can only be achieved if we first think systemically about our problem, ensuring our ensuing actions appropriately address the problem we are dealing with. The avoidance of these errors (by *acting systemically*) is the focus of Part III of this book
3. Finally, we must observe the effects of our actions (or lack thereof). This must include consideration of avoiding the Type I and Type II errors by conducting appropriate statistical analyses and making appropriate conclusions based on these analyses. Further, we must avoid the Type VI error by ensuring our conclusions are supported by evidence and not by conjecture. The avoidance of errors in observation (by *observing systemically*) is the focus of Part IV of this book.

To demonstrate the potential interaction of these errors with the TAO approach, Table 1.4 illustrates the TAO approach applied to reasoning about a disease.

The timeline in Table 1.4 can continue, ad infinitum. That is, you may continue to think, act, and observe with respect to your headache problem. This series of steps is shown graphically in Fig. 1.2 in a manner adapted from Boal and Meckler (2010) and Adams and Hester (2012, 2013) but focused on the probabilities associated with particular paths available to an individual tasked with addressing a complex problem. It is worth noting that Type VIII errors are represented by the different error combinations presented in Fig. 1.2 (i.e., a Type III error followed by a Type I error). Note that $P(\alpha)$, $P(\beta)$, $P(\gamma)$, $P(\delta)$, $P(\varepsilon)$, $P(\theta)$, $P(\zeta)$, and $P(\eta)$ represent the probability of a Type I–VIII error, respectively.

Note that the shaded boxes represent the only scenario in which no errors are committed. It is easy to see, qualitatively, how prone we are to errors based purely on the number of opportunities for us to commit one (or more) errors. Combining

Table 1.4 Example TAO timeline and potential errors

TAO stage	Situation description	Potential error(s)
Think	Recurring headaches cause you to try to figure out their source. Lacking an obvious environmental trigger, you decide to make an appointment to see your primary care provider	Type III
Act	You make an appointment with your doctor based on your thinking	Types IV, V, VIII
Observe	Your doctor observes you, asks you questions, and collects information	Types I, II, VI
Think	Based on the information provided and their own perspectives, the doctor reasons about your condition	Type III
Act	The doctor, with your consent, agrees to schedule you for an MRI	Types IV, V, VIII
Observe	Your insurance company collects the request from your doctor and considers it in concert with your medical history. Given your lack of prior concerns and lack of current evidence, the insurance company denies your claim	Types I, II, VI
Think	Given the reduced options available, your doctor thinks about your situation. Your doctor suggests you go home and start an activity log to keep track of your food, sleep, and activity habits to identify any underlying patterns	Type III
Act	You maintain your activity log for two weeks	Types IV, V, VIII
Observe	You return to the doctor and the doctor observes your activity log, making recommendations based on the results (to include a second attempt at securing insurance approval for an MRI)	Types I, II, VI
And so on…	You can continue to think, act, and observe. Even though the problem may seem resolved (i.e., your headaches go away), there is likely to be an implicit recognition of the danger of their recurrence. Thus, you may devote brain power to the awareness of their presence, no matter how distant they are in memory. The problem, as you see it, may evolve from "How can I make these headaches go away?" to "How can I ensure these headaches do not return?"	Types I–VIII

these error probabilities together, we can devise an equation for the calculation of the probability of a correctly addressed problem. This can be computed as shown in Eq. 1.1.

$$P(\text{correctly addressed problem}) = 1 - [[1 - P(\gamma)][1 - (P(\delta) + P(\varepsilon) + P(\eta))][1 - (P(\alpha) + P(\beta) + P(\theta))]] \tag{1.1}$$

Correctly addressing a problem requires that we think about, act on, and observe the situation appropriately; thus, we do not commit any Type I, II, II, IV, V, VI, or VIII errors (and, by extension, Type VII). While we can calculate $P(\alpha)$ and $P(\beta)$ in a very straightforward manner using statistical techniques, the remaining quantities

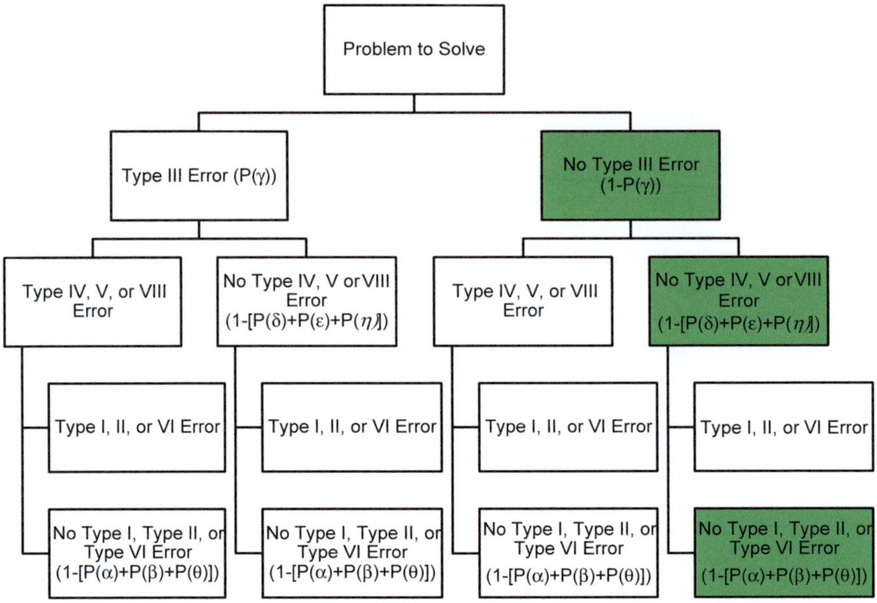

Fig. 1.2 Tree depiction of systems errors

are more difficult, if not impossible, to discern. The essential points are to understand that errors are serial; thus, our approach to understanding is only as strong as its weakest link, be it in our thinking, acting, or observation. Committing any error drastically reduces the likelihood that we have correctly addressed our problem. Thus, we must be diligent in addressing each of these errors.

1.3 Summary

Complex problems demand approaches that can account for their inherent complexity rather than ignore it and hope it goes away. That is the underlying premise of this book. To that end, this chapter introduced the TAO approach to increasing our understanding of a problem. We then discussed a taxonomy for errors that we are prone to when seeking increased understanding.

After reading this chapter, the reader should:

1. Understand the TAO approach; and
2. Have an appreciation for errors and how to avoid them.

References

Ackoff, R. L. (1979). The future of operational research is past. *Journal of Operational Research Society, 30*(2), 93–104.

Ackoff, R. L. (1994a). It's a mistake! *Systems Practice, 7*(1), 3–7.

Ackoff, R. L. (1994b). Systems thinking and thinking systems. *System Dynamics Review, 10*(2/3), 175–188.

Adams, K. M., & Hester, P. T. (2012). Errors in systems approaches. *International Journal of System of Systems Engineering, 3*(3/4), 233–242.

Adams, K. M., & Hester, P. T. (2013). Accounting for errors when using systems approaches. *Procedia Computer Science, 20,* 318–324.

Adams, K. M., & Mun, J. H. (2005). Towards a system of systems methodologies, once again. In *Proceedings of the 26th National ASEM Conference: Organizational Transformation: Opportunities and Challenges* (pp. 502–510). Rolla, MO: American Society for Engineering Management.

Betz, M. A., & Gabriel, K. R. (1978). Type IV errors and analysis of simple effects. *Journal of Educational Statistics, 3*(2), 121–143.

Boal, K., & Meckler, M. (2010). Decision errors of the 4th, 5th and 6th kind. In P. C. Nutt & D. C. Wilson (Eds.), *Handbook of decision making* (pp. 327–348). West Sussex: Wiley.

Cannon, W. (1929). Organization for physiological homeostasis. *Physiological Reviews, 9*(3), 399–431.

Games, P. A. (1973). Type IV errors revisited. *Psychological Bulletin, 80*(4), 304–307. doi:10.1037/h0034832.

Holland, P. W. (1986). Statistics and causal inference. *Journal of the American Statistical Association, 81*(396), 945–960.

Jackson, M. C. (1984). Beyond a system of systems methodologies. *Journal of the Operational Research Society, 41,* 657–668.

Jackson, M. C. (1991). *Systems methodology for the management sciences.* New York: Plenum Press.

Jackson, M. C. (2000). *Systems approaches to management.* New York: Kluwer Academic.

Jackson, M. C. (2003). *Systems thinking: Creative holism for managers.* Chichester, UK: Wiley.

Jackson, M. C., & Keys, P. (1984). Towards a system of systems methodologies. *Journal of the Operational Research Society, 35,* 473–486.

Kaiser, H. F. (1960). Directional statistical decisions. *Psychological Review, 67*(3), 160–167.

Kaufman, N. J., Dudley-Marling, C., & Serlin, R. L. (1986). An examination of statistical interactions in the special education literature. *Journal of Special Education, 20*(1), 31–42.

Kimball, A. W. (1957). Errors of the third kind in statistical consulting. *Journal of the American Statistical Association, 52*(278), 133–142.

Leventhal, L., & Huynh, C.-L. (1996). Directional decisions for two-tailed tests: Power, error rates, and sample size. *Psychological Methods, 1*(3), 278–292. doi:10.1037/1082-989x.1.3.278.

Levin, J. R., & Marascuilo, L. A. (1972). Type IV errors and interactions. *Psychological Bulletin, 78*(5), 368–374.

Levin, J. R., & Marascuilo, L. A. (1973). Type IV errors and games. *Psychological Bulletin, 80*(4), 308–309. doi:10.1037/h0034833.

Marascuilo, L. A., & Levin, J. R. (1970). Appropriate post hoc comparisons for interaction and nested hypotheses in analysis of variance designs: The elimination of Type IV errors. *American Educational Research, 7*(3), 397–421.

Marascuilo, L. A., & Levin, J. R. (1976). The simultaneous investigation of interaction and nested hypotheses in two-factor analysis of variance designs. *American Educational Research Journal, 13*(1), 61–65.

Meyer, D. L. (1991). Misinterpretation of interaction effects: A reply to Rosnow and Rosenthal. *Psychological Bulletin, 110*(3), 571–573. doi:10.1037/0033-2909.110.3.571.

Mitroff, I. (1998). *Smart thinking for crazy times: The art of solving the right problems*. San Francisco, CA: Berrett-Koehler Publishers.

Mitroff, I., & Featheringham, T. R. (1974). On systematic problem solving and the error of the third kind. *Behavioral Science, 19*(6), 383–393.

Mosteller, F. (1948). A K-sample slippage test for an extreme population. *The Annals of Mathematical Statistics, 19*(1), 58–65.

Neyman, J., & Pearson, E. S. (1928a). On the use and interpretation of certain test criteria for purposes of statistical inference: Part I. *Biometrika, 20A*(1/2), 175–240.

Neyman, J., & Pearson, E. S. (1928b). On the use and interpretation of certain test criteria for purposes of statistical inference: Part II. *Biometrika, 20A*(3/4), 263–294.

Neyman, J., & Pearson, E. S. (1933). The testing of statistical hypotheses in relation to probabilities a priori. *Mathematical Proceedings of the Cambridge Philosophical Society, 29* (4), 492–510.

OECD. (2007). *Revised field of science and technology (FOS) classification in the Frascati Manual*. Paris: Organization for Economic Cooperation and Development.

Onwuegbuzie, A. J., & Daniel, L. G. (2003). Typology of analytical and interpretational errors in quantitative and qualitative educational research. *Current Issues in Education, 6*(2).

Reason, J. (1990). *Human error*. Cambridge, UK: Cambridge University Press.

Rosnow, R. L., & Rosenthal, R. (1989). Definition and interpretation of interaction effects. *Psychological Bulletin, 105*(1), 143–146. doi:10.1037/0033-2909.105.1.143.

Rosnow, R. L., & Rosenthal, R. (1991). If you're looking at the cell means, you're not looking at only the interaction (unless all main effects are zero). *Psychological Bulletin, 110*(3), 574–576. doi:10.1037/0033-2909.110.3.574.

Shappell, S., & Wiegmann, D. (2000). *The human factors analysis and classification system—HFACS*. Washington, DC: Office of Aviation Medicine.

Tracz, S. M., Nelson, L. L., Newman, I., & Beltran, A. (2005). The misuse of ANCOVA: The academic and political implications of Type VI errors in studies of achievement and socioeconomic status. *Multiple Linear Regression Viewpoints, 31*(1), 16–21.

Umesh, U. N., Peterson, R. A., McCann-Nelson, M., & Vaidyanathan, R. (1996). Type IV error in marketing research: The investigation of ANOVA interactions. *Journal of the Academy of Marketing Science, 24*(1), 17–26.

Waddington, C. H. (1957). *The strategy of genes: A discussion of some aspects of theoretical biology*. London: George Allen & Unwin.

Chapter 2
Problems and Messes

Abstract As problems have grown more complex, the methods we use to address them must evolve as well. Machine age problems, consisting of simple systems, have traditionally been addressed using a primarily technical perspective. Despite their increased complexity, in systems age problems, a predominantly technical perspective continues to be used at the expense of other complementary perspectives. This myopic approach has often been unsuccessful in solving these problems. The development of multiple perspectives requires those faced with addressing complex problems to include additional perspectives in order to achieve increased understanding. This includes the integration of hard and soft perspectives to ensure that, in addition to a technical perspective, the equally important organizational, political, and human perspectives have been included. The application of multiple perspectives offers a more inclusive framework through which complex problems may be addressed. The integration of technical, organizational, political, and human perspectives widens the aperture through which a problem is viewed, which then increases the likelihood of correctly addressing these complex problems. Embracing these complementary perspectives, guidance is given on how to begin to structure our mess into a number of discrete problems for analysis.

2.1 A Brief Introduction to Complexity

This section will provide a brief discussion on understanding complexity and on the emergence of the systems age and how problems in the systems age are unique from those in the machine age.

2.1.1 Understanding Complexity

Returning to the preface, it is clear that there are many definitions of complexity. While it may seem like an exercise in semantics to differentiate between, for

P.T. Hester and K.M. Adams, *Systemic Decision Making*, Topics in Safety, Risk, Reliability and Quality 33, DOI 10.1007/978-3-319-54672-8_2

example, *complex* and *simple* systems, it is anything but. From a scientific perspective, we wish to understand complexity in a manner that allows us to appropriately deal with it (if indeed there is an it to deal with). We can first begin by looking at the word *complex* itself. The word has origins as a form of the Latin *complector*, meaning *woven together*. From a linguistic perspective, the opposite of the word complex is *simple*, but the linkage between simple and complex is not so straightforward. "A complex system cannot be reduced to a simple one if it was not simple (or perhaps merely complicated) to start off with" (Cilliers, 1998, p. 9). This evokes parallels to the idea of a system as a whole greater than the sum of its parts. Further, "Complex systems are said to be poised at such a position, between order and chaos" (Rickles, Hawe, & Shiell, 2007, p. 935).

> Because complexity results from the interaction between the components of a system, complexity is manifested at the level of the system itself. There is neither something at a level below (a source), nor at a level above (a meta-description), capable of capturing the essence of complexity. (Cilliers, 1998, pp. 2–3)

"A complex system is a system formed out of many components whose behavior is emergent, that is, the behavior of the system cannot be simply inferred from the behavior of its components" (Bar-yam, 1997, p. 10). A complex system

> "must be able to store information concerning the environment for future use; and it must be able to adapt its structure when necessary….Any model of a truly complex system will have to possess these capabilities. In other words, the processes of representation and self-organisation must be simulated by the model." (Cilliers, 1998, p. 10)

Regarding representation, "the structure of the system cannot consist of a random collection of elements; they must have some *meaning*" (Cilliers, 1998, p. 11). Self-organization can be described as "a process whereby a system can develop a complex structure from fairly unstructured beginnings" (Cilliers, 1998, p. 12).

What does analysis of all of these definitions lead us to conclude? Are we any closer to our elusive goal of precisely defining complexity? Seemingly the answer is no. Gershenson (2007) sheds light on the difficult in doing so: "The problem of a strict definition of complexity lies in the fact that there is no way of drawing a line between simple and complex systems independently of a context" (p. 13).

Hester (2016) summarizes the discussion by defining a complex (vice a simple) system as "a purposeful collection of elements whose initial conditions give rise to emergent behavior not present at other levels of system abstraction." The precise distinction drawn is that of the presence of emergence not found in simple systems (but present in complex systems). We utilize this definition for the remainder of this text. So, the next question is, is the rise in problem complexity the result of a natural evolution of problem structure or a fundamental shift in the way in which we have approached problems? That is, are problems actually getting more complex or are we simply tackling more complex problems (given, for example, our advances in computing capabilities as compared to previous generations)? For an answer to these questions, we turn to the notion of the machine and systems ages.

Table 2.1 Ackoff's machine age and systems age characteristics

	Machine age	Systems age
Description	Simple system	Complex system
Boundary	Closed	Open
Elements	Passive parts	Purposeful parts
Observable	Fully	Partially
Method of understanding	Analysis and reductionism	Synthesis and holism

2.1.2 The Machine Age and the Systems Age

Systems and management pioneer Russell Ackoff [1919–2004] (1974b) used the terms *machine age* and *systems age* to refer to eras that he contended were concerned with two fundamentally different types of problems. The machine age was concerned with simple systems (and problems), and the systems age is concerned with complex systems (and problems). Table 2.1 contrasts the most basic characteristics of the machine and systems ages.

Ackoff (1979a) recognized that the technical perspective of the machine age was inadequate for coping with what he termed the *messy* situations present in the systems age, where human activity systems were predominant. He coined the concept of a *mess* and *messes* in 1974 and continued to use the term in 1979 when he solidified the idea in two papers where he was arguing that traditional operations (or in UK terms, operational) research was passé and that a more holistic treatment of systems problems was required (Ackoff, 1974a, 1979a, b). He foresaw that a wide variety of disciplines would be necessary to solve systems problems. Ackoff's (1979a) definition of a mess and messes is worthy of review:

> Because messes are systems of problems, the sum of the optimal solutions to each component problem taken separately is not an optimal solution to the mess. The behavior of the mess depends more on how the solutions to its parts interact than on how they interact independently of each other. But the unit in [operations research] OR is a problem, not a mess. Managers do not solve problems, they manage messes. (p. 100)

The bottom line is that complex problems in the real world must include a definition of human activity in the development of the contextual framework for the problem. For Ackoff (1979a), context was the essential element that modern systems age problem solvers would need to include in each problem formulation if these problems were to be understood and later addressed. He argued that the utility of traditional operations research had been diminished because these techniques, rooted in a machine age paradigm, were unable to account for the complexity caused by humans that were present in almost all systems age problems. Burrell and Morgan (1979) support Ackoff's contention, stating:

> Mechanical models of social systems, therefore, tend to be characterized by a number of theoretical considerations and are thus of very limited value as methods of analysis in situations where the environment of the subject is of any real significance. (p. 61)

In short, the methods and techniques of traditional operations research are "… mathematically sophisticated but contextually naïve and value free" (Hughes & Hughes, 2000, p. 10). Ackoff's work established the need for a clear understanding of specific or relevant context as fundamental to understanding and analyzing systems age problems.

Additional support for Ackoff's notions was provided by Nobel laureate Herb Simon [1916–2001] who addressed what he labeled an *ill-structured problem*. Simon (1973) states that "an ill-structured problem is usually defined as a problem whose structure lacks definition in some respect" (p. 181). A systems age problem is ill-structured when circumstances and conditions surrounding the problem are potentially in dispute, not readily accessible, or lack sufficient consensus for initial problem formulation and bounding. There may be multiple and possibly divergent perspectives or worldviews, rapidly shifting and emergent conditions that render stable solution methods innocuous, and difficulty in framing the problem domain such that the path forward can be engaged with sufficient alignment of perspectives to remain viable. Rittel and Webber (1973) termed this type of problem a *wicked problem*, where:

> The information needed to understand the problem depends upon one's idea for solving it. That is to say: in order to describe a wicked-problem in sufficient detail, one has to develop an exhaustive inventory of all conceivable solutions ahead of time. The reason is that every question asking for additional information depends upon the understanding of the problem—and its resolution—at that time. Problem understanding and problem resolution are concomitant to each other. Therefore, in order to anticipate all questions (in order to anticipate all information required for resolution ahead of time), knowledge of all conceivable solutions is required. (p. 161)

A wicked problem may be contrasted with a *tame* problem, described as "one which can be specified, in a form agreed by the relevant parties, ahead of the analysis, and which does not change during the analysis" (Rosenhead & Mingers, 2001, p. 5). The immediate result of a wicked problem is the questionable ability of traditional approaches based upon a single technical perspective to be successful. Still another articulation of this class of problems comes from Schon, who coined the term *swamp* to describe this class of problems:

> …there is a high, hard ground where practitioners can make effective use of research-based theory and technique, and there is a swampy lowland where situations are confusing "messes" incapable of technical solution. The difficulty is that the problems of the high ground, however great their technical interest, are often relatively unimportant to clients or to the larger society, while in the swamp are the problems of greatest human concern. Shall the practitioner stay on the high, hard ground where he can practice rigorously, as he understands rigor, but where he is constrained to deal with problems of relatively little social importance? Or shall he descend to the swamp where he can engage the most important and challenging problems if he is willing to forsake technical rigor? (Schon, 1983, p. 42)

Ravetz (1971) introduced the idea of the practical problem to contrast with a technical problem. Technical problems have a clearly defined function from the inception of the analysis, which can be solved by experts, whereas *practical problems* have a vague statement of the purpose to be achieved and their output is consensus regarding problem definition, leading to a recommendation for appropriate solution means.

Finally, Ozbekhan (1970) discusses the notion of a *problematique* as a "meta-problem (or meta-system of problems)" (p. 13), in contrast to a standard, well-bounded problem. This new class of problems such as poverty, urban blight, and criminal activity cannot be viewed as problems that exist in isolation. Thus, the problematique arises as a series of interconnected problems for all but the most trivial of problems. Once again, consideration of context is paramount.

The fact that a new class of problems has emerged is clear. The question of how to deal with these messes (beyond simply not applying traditional operations research techniques) is not so clear.

2.2 Dealing with Systems Age Messes

All of the differing articulations of complex problems presented in the previous section describe situations where there are divergent stakeholders, emergent conditions, and nonoptimal solutions to ill-defined problems. Given these difficult conditions, the question becomes, how do we deal with these situations? From our point of view, it seems reasonable to assume that the manner in which a systems age mess is perceived by its stakeholders is a major determinant of the degree of these factors that each of the stakeholders is able to clearly identify as part of the problem analysis.

2.2.1 Scientific Approaches to Complex Problems

Thomas Kuhn defines *paradigm* to be "universally recognized scientific achievements that for a time provide model problems and solutions to a community of practitioners" (Kuhn, 2012, p. xlii). Each scientific community has its own paradigm, including its own ontology, epistemology, axiology, rhetoric, and methodology, that it uses to address a problem (Adams & Hester, 2016). The combination of these factors results in a unique scientific approach, as shown in Fig. 2.1.

Further, relativistic perceptions of complexity add to the difficulty in understanding complex problems. Just like beauty, complexity is in the eye of the beholder. What may be complex to one individual may be simple to another. Take education, for example. A lifelong school administrator may find setting the budget for a given middle school a trivial task, whereas a teacher at the very same school may struggle to keep students out of trouble given a scarcity of after school activities, a direct result of the budget process. A more difficult question is certainly balancing the budget of said school with all others in the district, or perhaps the state or nation. Such a broadening of scope would certainly entail game theory, sociology, economics, and a host of other considerations certainly presenting

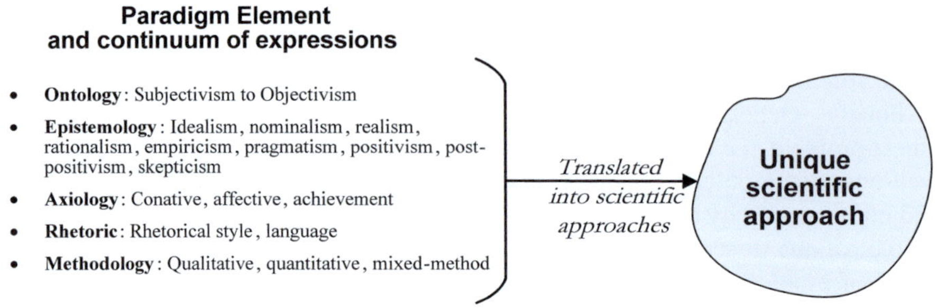

Fig. 2.1 The relationship between paradigm element and scientific approach (Adams & Hester, 2016)

complexities to most, if not all, individuals. So, while the administrator sees a simple budgeting exercise, the educator may see a much more complex problem rife with socioeconomic factors. How can this duality exist? Mitchell (2009) summarizes this phenomenon as "…there is not yet a single science of complexity but rather several different sciences of complexity with different notions of what complexity means" (p. 95).

2.2.2 Perspectives in Complex Problems

Thus, in order to improve our understanding about a complex problem, we must consider numerous perspectives. If we view a problem as simple, it may indeed be simple, or we may not be considering it holistically enough. Because there is not a single true reality or correct perspective of any systems age mess, the systems *principle of complementarity* (Bohr, 1928) must be applied. The principle simply states:

> Two different perspectives or models about a system will reveal truths regarding the system that are neither entirely independent nor entirely compatible.

If we think of a perspective as the state of one's ideas or the known facts, then we can represent the worldview of the observer as a function of the number (i) of perspectives (P_i) utilized to represent the problem under study. Equation 2.1 (Adams & Meyers, 2011) is a mathematical representation of contextual understanding for a limited number of perspectives (n). It is worth noting that this equation is intended to be illustrative, rather than prescriptive. Recalling the earlier discussion of a mess and its properties, our understanding is certainly not a linear summation of constituent perspectives, but rather a complicated relationship that indicates, at least in the abstract, that more perspectives lead to an improved understanding of a complex problem.

$$\text{Contextual Understanding} = \sum_{i=1}^{n} P_i \tag{2.1}$$

Perfect understanding requires complete knowledge of the infinite number of perspectives, a fact that problem solvers struggle to control when bounding messy, ill-structured, or wicked problems. Equation 2.2 (Adams & Meyers, 2011) is a mathematical representation of perfect understanding.

$$\text{Perfect Understanding} = \sum_{i=1}^{\infty} P_i \tag{2.2}$$

A depiction of these concepts is shown in Fig. 2.2. This figure shows that as both time (t) and the number of perspectives increases, our understanding increases dramatically. Perfect understanding is depicted as a plane that we attempt to attain but cannot reach no matter how much time passes or how many perspectives we consider.

Because, by definition, our scope of perspectives is limited, we can never have perfect understanding of a complex problem, and thus, we must strive to increase the value of our contextual understanding. The question naturally arises, then, as to how many perspectives are sufficient. There are two answers: (1) the academic perspective and (2) the practical perspective. The academic answer, as quantified by Eq. 2.2, is that there are never enough perspectives for a problem. While this is true, we must strive for perspective saturation (Glaser & Strauss, 1967). That is, we should continue to gather perspectives until we no longer obtain new or insightful information. The practical answer, however, says the act of gathering perspectives is typically undertaken until we run out of resources (e.g., time and money), which is often well in advance of having collected a sufficient number of perspectives. Practical constraints limit the number of perspectives that we are able to consider and many of us erroneously only consider a singular perspective, our own, when

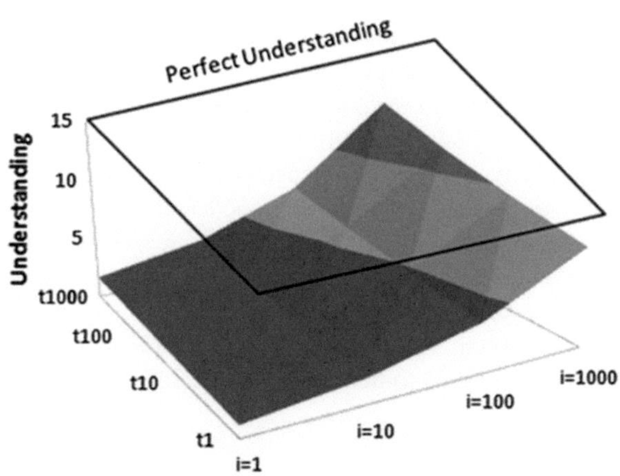

Fig. 2.2 Depiction of increased understanding as a function of time (t) and perspectives (i)

addressing a problem; however, it is clear that it is useful to obtain numerous perspectives as appropriate and available.

It is exceedingly important, then, that we choose the perspectives that we incorporate carefully. We must seek those that have the ability to add to our understanding rather than those viewpoints that confirm our own (a phenomenon known as confirmation bias that we will revisit in Chap. 15). Further, the more disparate our perspectives, the more potentially enlightening the information we obtain. In this way, we can treat this effort in the same manner we would treat a hypothesis test. We wish to collect information that has the potential to disconfirm our hypothesis. If the hypothesis that we have formulated stands up to scientific scrutiny, in this case multiple perspectives, then we have greater confidence in its validity. If not, then perhaps our initial assumptions were incorrect. At the very least, conflicting perspectives may demand additional investigation.

Our ideas about the inclusion of multiple perspectives are echoed by two outstanding systems thinkers, Ian Mitroff and Harold Linstone. Mitroff is a long-time advocate for systemic thinking (Mitroff, Alpaslan, & Green, 2004; Mitroff & Kilmann, 1977) and was the first to formally characterize the Type III error (Mitroff, 1998; Mitroff & Betz, 1972; Mitroff & Featheringham, 1974; Mitroff & Silvers, 2010), as discussed in Chap. 1. Linstone has been a strong proponent of the use of multiple perspectives in problem investigation (Linstone, 1985, 1989; Linstone et al., 1981). In their book *The Unbounded Mind* (Mitroff & Linstone, 1993), they make this important point:

> "everything interacts with everything," that all branches of inquiry depend fundamentally on one another, and that the widest possible array of disciplines, professions, and branches of knowledge—capturing distinctly different paradigms of thought—must be consciously brought to bear on the problem. (p. 91)

2.3 Holistic Understanding

Holistic understanding of systems age messes requires problem solvers to formally account for elements contained in both hard and soft approaches to complex problems. A hard system perspective includes notions such as objectivity, unitary viewpoints, and quantitative assessment; while a soft systems perspective evokes subjectivity, pluralistic perspectives, and qualitative assessments. The attributes of the hard and soft systems approaches are depicted in Table 2.2.

The contrast between the views represented by the soft and hard systems approaches leads to significantly different perspectives of the problems encountered by the problem solver or problem solving team. The soft perspective considers organizational, managerial, policy, political, and human factors, while the hard perspective tends to deal with only technical elements, those that can be reduced to

Table 2.2 Attributes of hard and soft systems approaches (Adams & Meyers, 2011, p. 167)

Attributes	Hard systems view	Soft systems view
Worldview	A real world exists external to the analyst	Perspectives of reality are dynamic and shifting
Data	Factual, truthful, and unambiguous data can be gathered, observed, collected, and objectively analyzed	Data are subjective in collection and interpretation—analysis strives for transparency
System	The system in focus is unaffected by either the analysis or the analyst	The system in focus is affected by both the analysis and the analyst
Analysis results	The results of analysis are replicable	Results of analysis are *credible* and capable of compelling *reconstruction*
Value	The analysis can be conducted free of value judgments	The analysis and interpretation of analysis is value-laden
Boundaries	The system in focus can be bounded and the analysis can be controlled—this is both possible and desirable	Bounding of the system in focus is problematic, control of the analysis is questionable—emergence is dominant

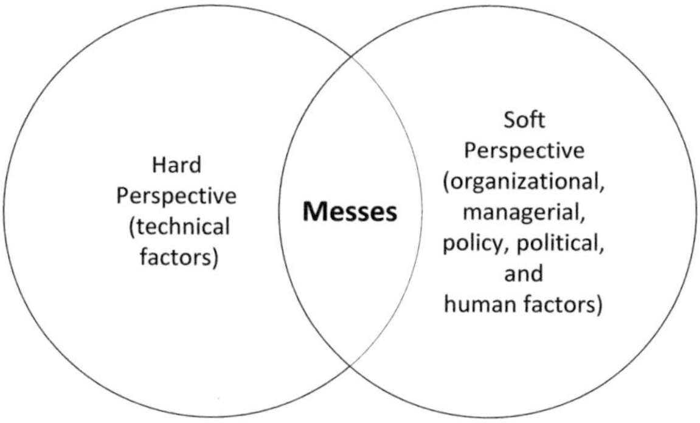

Fig. 2.3 Messes as the intersection between hard and soft perspectives

objective measures. The hard perspective is more appropriate as a stand-alone approach for dealing with machine age problems concerned primarily with technical solutions, whereas the soft perspective is more concerned with social systems, ones that are primarily devoid of technical considerations. Figure 2.3 shows how both approaches contribute to the development of understanding for systems age messes. Messes occur at the intersection of these two perspectives and thus, require both soft and hard perspectives to be considered in order to achieve an appropriate level of understanding.

The most fundamental, and therefore first, step in achieving a holistic understanding of a mess is to first formulate articulate its constituent problems in a manner that is conducive to further exploration.

2.4 What's the Problem?

It is one of the most fundamental questions we are routinely faced with and yet one of the most vexing—*what's the problem*? In order to begin a discussion of *problems*, we first define what we intend when we use the term. Smith (1988) defines three criteria for a problem: (1) a gap between current and desired state, (2) there is some difficulty in bridging that gap, and (3) someone must wish to bridge the gap. While it seems straightforward, in practice, it is anything but. "In any complex, real world, situation, there are an unlimited number of concerns which could be identified as problems, but there are none which absolutely must be so identified" (Smith, 1988, p. 1491).

Duncker (1945) offers a complementary definition of the term:

> A problem arises when a living creature has a goal but does not know how this goal is to be reached. Whenever one cannot go from the given situation to the desired situation simply by action, then there has to be recourse to thinking. (By action we here understand the performance of obvious operations.) Such thinking has the task of devising some action which may mediate between the existing and the desired situations. (p. 1)

Similar to Smith (1988), Sage (1992) succinctly defines a problem as "an undesirable situation or unresolved matter that is significant to some individual or group and that the individual or group is desirous of resolving" (p. 232). In defining a problem in this manner, Sage focuses on a problem as something which is undesirable. Sage (1992) goes on to define four basic characteristics of problems. The first three of his criteria are the same as Smith (1988), but he adds the following criterion:

> The situation is regarded as resolvable by an individual or group, either directly or indirectly. Solving a problem would constitute a direct resolution. Ameliorating or dissolving a problem, by making it go away, is an indirect resolution of a problem. (p. 232)

So, a problem is difficult to resolve, but resolution is perceived as achievable.

Pidd (2009) offers another take on terminology, distinguishing between puzzles, problems, and messes, noting the following:

Puzzles are a

> …set of circumstances where there is no ambiguity whatsoever once some thought has been given to what is happening or needs to be done. The issues that need to be faced are entirely clear, the range of options is completely known, and there exists a single correct solution to the puzzle. (Pidd, 2009, pp. 43–44)

Examples of puzzles include jigsaw puzzles and crossword puzzles. Arriving at a solution is straightforward as it requires us to apply known methods to arrive at a singular correct answer. This class of situations is of no interest in the context of this book and will not be discussed further.

Problems have "no single answer that is definitely known to be correct…it depends on how you, or someone else, decides to construe it" (Pidd, 2009, p. 44). This again evokes notions of context. Thus, there may be agreement about the issue to be addressed, but there may be numerous, equally valid, solutions to the problem.

An added complication is that the solution approach used to address a problem may yield unique results from another, equally appropriate approach. In this case, the problem is well structured, but "considerable ingenuity and expertise may be needed to find an acceptable, let alone optimal solution" (Pidd, 2004, p. 7).

A problem has one or more *owners*, or those that recognize its existence as a problem. The idea of ownership is reflected in Smith's definition of a problem: "Problems are conceptual entities defined as a way of allocating attention to unsatisfactory aspects of reality that one hopes to improve through action" (1988, p. 1492). Ownership involves a desire to see a problem resolved and a willingness to allocate resources (i.e., time, money, and intellect) to do so. It is worth noting that the owner of a problem is not necessarily the decision maker. An individual or group can be the owner of a problem yet not have the resources to resolve it. The decision maker is the individual who has the authority to allocate resources in an effort to resolve a given problem. One example where the problem owner is not the decision maker is that in which a teenager desires to go out and see a movie with his friends on a Saturday night. That is to say, he wishes to move from his current state of boredom to an ideal state of enjoyment with friends. However, if he lacks the authority to go out and he is denied his request for doing so due to being grounded, then his decision has been made for him. Thus, he is not the decision maker. He does not control the resources (e.g., the car or the money to buy a ticket). Another example might involve commuters who drive along a failing highway daily (thus they *own* the problem) and yet whose elected officials reside in an altogether different geographical region (i.e., the state capital) and thus, have no problem ownership, yet they control the financial means to remedy the situation through state budget allocation (thus, they are the decision makers).

Messes are systems of problems "with multiple stakeholders who may hold quite different views on what is feasible and desirable" (Pidd, 2009, p. 44). There may also be debate as to what the definition of the issue is. Thus, both the problem formulation and methods to address it are potentially in conflict. As opposed to problem definition, mess articulation cannot easily be represented in a succinct form. "Indefinite goals seem to be an important factor in producing the weakness of structure in many ill-structured problems" (Greeno, 1976, p. 480). The delineation between a mess and a problem boils down to this definition. The goal (i.e., objective) of a mess is unable to be stated succinctly as it does not exist. If it can be stated as a singular objective, then it is merely a problem, albeit perhaps a complex one. Each constituent problem, however, should be capable of being captured as a concise statement of an objective such as *Find the best route to work* or *Select a job*. A mess is better articulated not linguistically, as in the case of a problem, but graphically. While we can certainly identify a mess linguistically, (i.e., this traffic is a *mess*), this simple statement fails to capture the intricacies that only a graphical depiction can. We will return to this notion later in Chap. 5 as we discuss complex systems modeling.

Others view the notion of a problem more pragmatically. Newell, Shaw, and Simon (1959), studying problem solving and formulation, define a problem as existing "whenever a problem solver desires some outcome or state of affairs that he does not immediately know how to attain" (p. 1). This perspective motivated their work in developing a General Problem Solver, their attempt to generate a universal problem-solving computer algorithm. This work introduced the notion of means-ends analysis, whereby a goal is established (this can be thought of as Smith, Duncker, and Sage's notions of a goal or desired state) for a situation. This desired state is contrasted with a current state. Your problem represents your difference, or delta, between the two. If your current state is equal to your desired state, then you do not have a problem. Newell et al. (1959) discuss a simple example which explains means-ends analysis:

> I want to take my son to nursery school. What's the difference between I have and what I want? One of distance. What changes distance? My automobile. My automobile won't work. What's needed to make it work? A new battery. What has new batteries? An auto repair shop. I want the repair shop to put in a new battery; but the shop doesn't know I need one. What is the difficulty? One of communication. What allows communication? A telephone…And so on. (pp. 8–9)

The universe of acceptable decisions available to you to move from your current state to desired state is your *problem space*. This problem space may include several intermediate steps which each move your current state some amount closer to your desired end state. Identification of the delta between our current and desired states is a useful and practical means for us to articulate our problem. Readers interested in more information on means-ends analysis, problem-solving computer algorithms, and early developments in artificial intelligence are referred to Newell and Simon (1972).

Before proceeding, it is imperative that a consensus on terminology be reached. Thus, with this discussion in mind, we adopt the following definitions for the remainder of this text:

> A *problem* is an undesirable situation without a clear resolution that an individual or group wishes to see resolved. The same problem can be owned by more than one person (which may lead to discordance regarding the resolution of said problem).

Adopting the definition of a complex system found in Hester (2016):

> A *mess* is a purposeful collection of problems whose initial conditions give rise to emergent behavior not present at other levels of problem abstraction. A mess is comprised of two or more uniquely owned problems that interact in some capacity.

A point of clarification is necessary before proceeding. Two problems with the same owner do not represent a mess. Any two problems owned by the same individual or group can be redefined as a multiobjective problem. Thus, there must be multiple problem owners (whose problems interact) to constitute a mess.

We now turn to the issue of problem structuring in an effort to understand how to formulate problems for investigation and potential resolution.

2.5 Problem Structuring

Research has shown the importance of focusing on avoiding the Type III error before attempting to address a problem (Adams & Hester, 2012, 2013; Hester & Adams, 2014). This is not a novel idea to practitioners. "The risk of solving the 'wrong problem' is generally acknowledged and discussed by practitioners" (Woolley & Pidd, 1981, p. 197). Yet, we often fail to correctly identify a problem before attempting to address it. Why?

> Three principal reasons why persons fail to identify accurately problems and their causes are: (1) the problem solver doesn't actually perceive the problem - he is blind to it; (2) the wrong problem or the wrong causes of it (or both) are identified; and (3) the problem identification phase is skipped over and ignored-efforts are immediately made to solve 'the problem'. (Watson, 1976, p. 88)

The most fundamental, and therefore first, step in achieving a holistic understanding of a mess is to first articulate its constituent problems in a manner that is conducive to further exploration. This process is known as problem structuring (or alternatively, as problem formulation, identification, or definition). Problem structuring methods (PSMs) are a class of methods that help a series of divergent stakeholders to understand the complex problem they face *before* attempting to resolve it, thereby hoping to avoid a Type III error. Woolley and Pidd (1981) highlight the importance of structuring in dealing with messes, describing it as "the process of arriving at a sufficient understanding of the components of a particular problem to proceed to some sort of useful operational research work" (p. 198). It helps us to reveal the *real problem*, as opposed to the *perceived* or *reported problem*. This is important as problem articulation is a subjective undertaking. Just as there many models that can be used to describe a given situation, there are many problems that can be identified to describe a given situation (Ackoff, 1974a).

Problem structuring is fundamentally a systems thinking-derived concept. Pidd (1988) elaborates:

> The aim is to take the richness of the presenting mess, and from this to extract research tasks which can be regarded as reasonable units [see Fig. 2.4]. This does not imply that these research tasks, and the issues which they address, are disjoint. To imply that would be totally to ignore all the insights of systems thinking. Rather, these tasks are linked together by assumptions which form a structure which permits further detailed research and which prevents the research tasks from being isolated from one another. (p. 116)

Rosenhead (2006) discusses the situations for which PSMs are advantageous as those having multiple actors, differing perspectives, partially conflicting interests, significant intangibles, and perplexing uncertainties. Even knowing these basic characteristics does not make problem structuring any easier. It is not a straightforward endeavor, for many of the reasons we have talked about so far, e.g., any time we have multiple divergent perspectives, the complexity of our situation increases substantially. Vennix (1996) agrees, stating of messy problems:

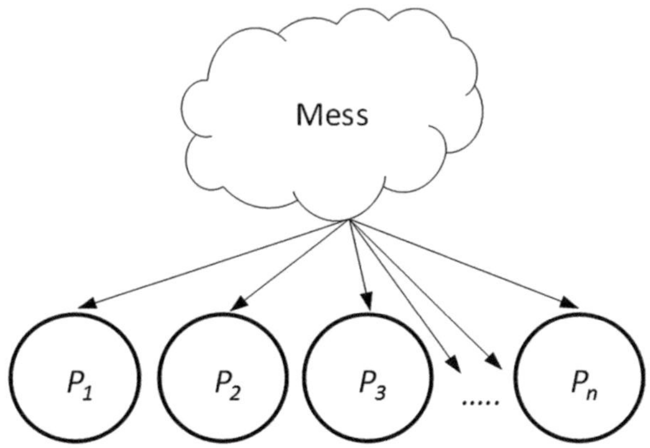

Fig. 2.4 Problem structuring illustration (adapted from Pidd, 1988)

> One of the most pervasive characteristics of messy problems is that people hold entirely different views on (a) whether there is a problem, and if they agree there is, and (b) what the problem is. In that sense messy problems are quite intangible and as a result various authors have suggested that there are no objective problems, only situations defined as problems by people. (p. 13)

As such, problem identification is not trivial. Further, the question of problem identification can have different levels of importance depending on the situation that we are facing—discerning that the stomach pains we are experiencing are really appendicitis likely is more important than choosing what we will have for dinner, and yet both situations may be perceived to meet our earlier definition of a problem as an undesirable situation without a clear resolution that an individual or group wishes to see resolved. Indeed, problems are omnipresent and, often times, overwhelming.

To assist individuals in dealing with their problems (or more appropriately, their messes), we suggest modern approaches to reductionist problem solving are insufficient, not because they suggest we decompose a problem, but because, after analysis of this singular problem, they often ignore the reintegration of this problem into the context of which it is a part. Just like no man is an island, no problem exists in isolation. Our appendicitis problem must also consider insurance, transportation to the doctor, family history, alcohol and drug use, and diet, while our dinner choice must consider our finances, social obligations, fellow diners, availability of cuisine, allergies, and time constraints.

In order to identify and formulate our problem (and surrounding mess), one must appreciate the underlying purpose of its associated system. It is in our best interest to ensure that our stated problem truly reflects the concerns of relevant stakeholders. This is sometimes easier said than done as we do not always have complete latitude over this exercise, however. In fact, our problem may be predefined by some authority (such as a customer) or the organization in which we work. Using our earlier terminology, the decision maker (i.e., customer) may not be the owner of the problem (i.e., the user who wishes to see the problem resolved but does not have the financial means to do so). Hammond, Keeney, and Raiffa (2002) agree, urging decision makers to consider the trigger, the initiating force, behind their problems. They caution, "Most triggers come from others…or from circumstances beyond your control…Because they're imposed on you from the outside, you may not like the resulting decision problems" (pp. 18–19). In this case, at a minimum, we should work with other stakeholders to refine the problem in a manner conducive to gaining further understanding. If we can influence our problem formulation, we need to consider what triggered the problem so that we can ensure we have identified the root problem.

Hammond et al. (2002) echo the importance of problem formulation: "The way you state your problem frames your decision. It determines the alternatives you consider and the way you evaluate them. Posing the right problem drives everything else" (p. 15).

In all, problem formulation is neither trivial nor to be taken lightly. "Defining the problem is sometimes the most difficult part of the process, particularly if one is in a rush to 'get going'" (Blanchard, 2004, p. 48); recall the notion of humans' bias for action discussed in the Preface. Hammond et al. (2002) warn of the pitfalls of taking problem formulation lightly:

> Too often, people give short shrift to problem definition…In their impatience to get on with things, they plunge into the other elements of decision making without correctly formulating the problem first. Though they may feel like they're making progress in solving their problem, to us they seem like travelers barreling along a highway, satisfied to be going 60 miles an hour - without realizing they're going the wrong way. (p. 26)

One final point on problem formulation. We should be careful to specify a problem that is unique enough to be relevant to our concerns, yet not so specific that it predefines a solution. This is important because a true problem may have predispositions toward a solution, but if we already have a solution, then we do not have a problem; rather, we have a puzzle and its resolution is merely a matter of correct implementation.

Only once we have formulated our problems and are satisfied they are representative of the concerns we wish to explore, can we begin to change our way of thinking about, acting on, and observing the problems in question. At this point, we are ready to make systemic decisions. This is reflected in the modification of Chap. 1's TAO Process Figure as shown in the systemic decision making process in Fig. 2.5.

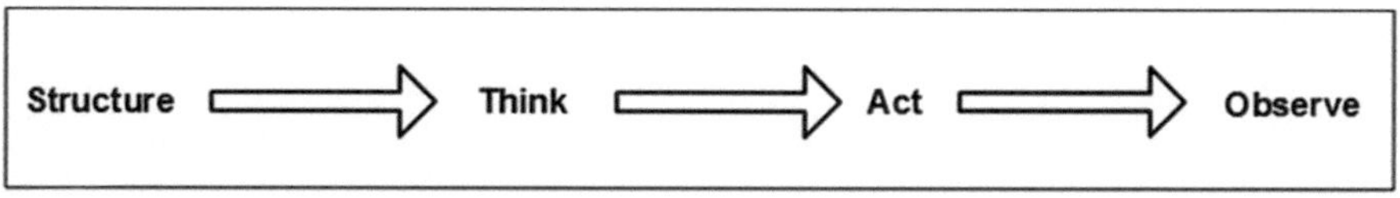

Fig. 2.5 Systemic decision making steps

2.6 Summary

The prevalence of complex problems is increasing, no matter how we measure it. Simply adopting a single technical perspective has been unsuccessful in addressing these ill-structured, wicked, or messy problems. The application of multiple perspectives and the inclusion of soft techniques offers a more inclusive framework through which complex problems may be viewed.

The integration of technical, organizational, political, and human perspectives during problem structuring and resolution widens the aperture and provides an increased probability of correctly addressing systems age problems. Finally, it is worth noting that the range of variability of individual perspectives, objectives, and perceived interests may be so divergent that sufficient alignment necessary to move forward may be unattainable. Many traditional approaches assume a unitary perspective where there is assumed agreement on the problem. We have found that most systems age problem domains have deeply rooted or philosophical divergence which add to the difficulty in developing a mutually agreeable problem formulation. Divergence may involve such issues as allocation of scarce resources, power distribution, control, personal preferences or interests, and other areas that may exist at a tacit level. Assuming alignment in systems age problems may be problematic.

In order to move forward, we must decompose the messes we wish to further understand into tractable problems about which we may reason, making sure to pay deliberate attention to their structuring, and then reconstruct them in order to obtain systemic understanding of our mess. Simply decomposing them, as many methods do, is insufficient, as it fails to holistically consider the context in which each problem operates.

After reading this chapter, the reader should:

1. Understand the difference between systems age problems and machine age messes;
2. Appreciate the importance of considering multiple perspectives in a system's effort;
3. Understand the characteristics of hard and soft perspectives; and
4. Be able to identify a mess and structure its constituent problems.

References

Ackoff, R. L. (1974a). *Redesigning the future: A systems approach to societal problems*. New York: Wiley.

Ackoff, R. L. (1974b). The systems revolution. *Long Range Planning, 7*, 2–20.

Ackoff, R. L. (1979a). The future of operational research is past. *Journal of Operational Research Society, 30*(2), 93–104.

Ackoff, R. L. (1979b). Resurrecting the future of operational research. *Journal of the Operational Research Society, 30*(3), 189–199.

Adams, K. M., & Hester, P. T. (2012). Errors in systems approaches. *International Journal of System of Systems Engineering, 3*(3/4), 233–242.

Adams, K. M., & Hester, P. T. (2013). Accounting for errors when using systems approaches. *Procedia Computer Science, 20,* 318–324.

Adams, K. M., & Hester, P. T. (2016). *Complex problems: The philosophical underpinnings for a transdisciplinary approach.* Paper presented at the 2016 Industrial and Systems Engineering Research Conference, Norcross, GA.

Adams, K. M., & Meyers, T. J. (2011). Perspective 1 of the SoSE methodology: Framing the system under study. *International Journal of System of Systems Engineering, 2*(2/3), 163–192. doi:10.1504/IJSSE.2011.040552.

Bar-yam, Y. (1997). *Dynamics of complex systems.* Reading, MA: Addison-Wesley.

Blanchard, B. S. (2004). *Systems engineering management* (3rd ed.). Hoboken, NJ: Wiley.

Bohr, N. (1928). The quantum postulate and the recent development of atomic theory. *Nature, 121* (3050), 580–590. doi:10.1038/121580a0.

Burrell, G., & Morgan, G. (1979). *Sociological paradigms and organizational analysis: Elements of the sociology of corporate life.* London: Heinemann.

Cilliers, P. (1998). *Complexity and postmodernism: Understand complex systems.* New York: Routledge.

Duncker, K. (1945). On problem solving. *Psychological Monographs, 58*(5), 1–113.

Gershenson, C. (2007). *Design and control of self-organizing systems.* Mexico City, Mexico: CopIt ArXives.

Glaser, B. G., & Strauss, A. L. (1967). *The discovery of grounded theory, strategies for qualitative research.* Chicago: Aldine Publishers.

Greeno, J. G. (1976). Indefinite goals in well-structured problems. *Psychological Review, 83*(6), 479–491.

Hammond, J. S., Keeney, R. L., & Raiffa, H. (2002). *Smart choices: A practical guide to making better life decisions.* New York: Broadway Books.

Hester, P. T. (2016). *Unraveling complexity.* Paper presented at the 2016 Industrial and Systems Engineering Research Conference, Norcross, GA.

Hester, P. T., & Adams, K. M. (2014). *Systemic thinking: Fundamentals for understanding problems and messes.* Cham, Switzerland: Springer International.

Hughes, A. C., & Hughes, T. P. (Eds.). (2000). *Systems, experts, and computers: The systems approach in management and engineering, World War II and after.* Cambridge, MA: MIT Press.

Kuhn, T. S. (2012). *The structure of scientific revolutions: 50th anniversary edition* (4th ed.). Chicago: University of Chicago Press.

Linstone, H. A. (1985). Multiple perspectives: Overcoming the weaknesses of MS/OR. *INTERFACES, 15*(4), 77–85.

Linstone, H. A. (1989). Multiple perspectives: Concept, application, and user guidelines. *Systemic Practice and Action Research, 2*(3), 307–331.

Linstone, H. A., Meltsner, A. J., Adelson, M., Mysior, A., Umbdenstock, L., Clary, B., … Shuman, J. (1981). The multiple perspective concept: With applications to technology assessment and other decision areas. *Technological Forecasting and Social Change, 20*(4), 275–325. doi:10.1016/0040-1625(81)90062-7.

Mitchell, M. (2009). *Complexity: A guided tour.* New York: Oxford University Press.

Mitroff, I. I. (1998). *Smart thinking for crazy times: The art of solving the right problems.* San Francisco: Berrett-Koehler Publishers.

Mitroff, I. I., Alpaslan, M. C., & Green, S. E. (2004). Crises as ill-structured messes. *International Studies Review, 6*(1), 165–194. doi:10.1111/j.1521-9488.2004.393_3.x.

Mitroff, I. I., & Betz, F. (1972). Dialectical decision theory: A meta-theory of decision-making. *Management Science, 19*(1), 11–24.

Mitroff, I. I., & Featheringham, T. R. (1974). On systematic problem solving and the error of the third kind. *Behavioral Science, 19*(6), 383–393.

Mitroff, I. I., & Kilmann, R. H. (1977). Systemic knowledge: Toward an integrated theory of science. *Theory and Society, 4*(1), 103–129. doi:10.1007/BF00209746.

Mitroff, I. I., & Linstone, H. A. (1993). *The unbounded mind: Breaking the chains of traditional business thinking*. New York: Oxford University Press.

Mitroff, I. I., & Silvers, A. (2010). *Dirty rotten strategies: How we trick ourselves and others into solving the wrong problems precisely*. Stanford, CA: Stanford University Press.

Newell, A., Shaw, J. C., & Simon, H. A. (1959). *Report on a general problem-solving program (Technical Report No. P-1584)*. Santa Monica, CA: RAND Corporation.

Newell, A., & Simon, H. A. (1972). *Human problem solving*. Englewood Cliffs, NJ: Prentice-Hall.

Ozbekhan, H. (1970). *The predicament of mankind: Quest for structured responses to growing world-wide complexities and uncertainties (A proposal)*. Rome, Italy: Club of Rome.

Pidd, M. (1988). From problem-structuring to implementation. *The Journal of the Operational Research Society, 39*(2), 115–121.

Pidd, M. (2004). Complementarity in systems modelling. In M. Pidd (Ed.), *Systems modelling: Theory and practice*. West Sussex, UK: Wiley.

Pidd, M. (2009). *Tools for thinking: Modelling in management science* (3rd ed.). West Sussex, UK: Wiley.

Ravetz, J. R. (1971). *Scientific knowledge and its social problems*. Oxford, UK: Oxford University Press.

Rickles, D., Hawe, P., & Shiell, A. (2007). A simple guide to chaos and complexity. *Journal of Epidemiology and Community Health, 61,* 933–937.

Rittel, H. W. J., & Webber, M. M. (1973). Dilemmas in a general theory of planning. *Policy Sciences, 4*(1), 155–169.

Rosenhead, J. V. (2006). Past, present and future of problem structuring methods. *Journal of the Operational Research Society, 57*, 759–765.

Rosenhead, J. V., & Mingers, J. (2001). *Rational analysis for a problematic world revisited* (2nd ed.). Chichester, England: Wiley.

Sage, A. P. (1992). *Systems engineering*. New York: Wiley.

Schon, D. A. (1983). *The reflective practitioner: How professionals think in action*. New York, NY: Basic Books.

Simon, H. A. (1973). The structure of ill structured problems. *Artificial Intelligence, 4*(3/4), 181–201.

Smith, G. (1988). Towards a heuristic theory of problem structuring. *Management Science, 34*(12), 1489–1506.

Vennix, J. (1996). *Group model building: Facilitating team learning using system dynamics*. Chicester, UK: Wiley.

Watson, C. E. (1976). The problems of problem solving. *Business Horizons, 19*(4), 88–94.

Woolley, R. N., & Pidd, M. (1981). Problem structuring—A literature review. *The Journal of the Operational Research Society, 32*(3), 197–206.

Chapter 3
Systemic Thinking

Abstract As machine age problems have given way to systems age messes, the underlying complexity associated with understanding these situations has increased exponentially. Accordingly, the methods we use to address these situations must evolve as well. Unfortunately, however, many antiquated methods for dealing with situations remain prominent. The underlying paradigm for solving problems with many prominent approaches such as systems engineering and operations research can be characterized as *systematic thinking*. While quite appropriate for machine age problems, it lacks the theoretical rigor and systemic perspective necessary to deal with systems age messes. Thus, a new paradigm of *systemic thinking*, conceptually founded in systems theory, is necessary. This chapter provides a brief historical background on the development of systems approaches, contrasts systems approaches, and their underlying paradigm with systemic thinking, and introduces practical guidelines for the deployment of a systemic thinking approach that will provide the foundation for the remainder of this book.

3.1 A Brief Background of Systems Approaches

Problem solvers have been approaching complex problems using a predominantly technical perspective since the advent of large-scale systems in the fledgling radio, television, and telephone industries in the USA during the 1930s. This was a result of the recognized need for an approach to deal with problems encountered during the development of modern telecommunications services. The Radio Corporation of America (RCA) and its subsidiary, the National Broadcasting Company (NBC), were interested in the expansion of their television broadcast domain. At the same time, the Bell Telephone Company was interested in the expansion of their long-distance telephone network. Both companies initiated technical studies aimed at increasing their markets through the use of new broadband technologies that were beginning to emerge in the early 1940s.

P.T. Hester and K.M. Adams, *Systemic Decision Making*, Topics in Safety, Risk, Reliability and Quality 33, DOI 10.1007/978-3-319-54672-8_3

Most of the exploratory studies and experimentation in the commercial sector were interrupted by the Second World War. During the Second World War, the American military used large numbers of scientists and engineers to help solve complex logistical and strategic bombing problems related to the war effort. Many of these efforts made significant contributions to the philosophy and techniques of the field known as operations research (OR). Heyer (2004) provides some historical context on the development of this field:

> OR rose to prominence during World War II largely due to the British military. In the days leading up to World War II, British military management assembled a group of scientists to apply a scientific approach to military operations to determine the most advantageous ways to deploy their massive materiel and manpower. Soon after, the United States military began engaging in OR using specialists from fields such as chemistry, mathematics, and engineering to create management techniques for allocating scarce resources and to achieve both military and industrial goals (Carter & Price, 2001). In the 1950s various academic societies were born in both Britain (who today prefer the term operational research) and the United States (who prefer the term management science [MS]) for operations researchers (those who practice OR) to promote, develop and exchange ideas in the field. These professional societies remain active today and the field of OR has grown even larger and more diverse. (Heyer, 2004, p. 1)

Much of this work formed the basis of the two earliest books on engineering for systems (i.e., systems engineering), written by Harry H. Goode [1909–1960] of the University of Michigan and Robert E. Machol [1917–1998] of Purdue University (Goode & Machol, 1957) and Arthur D. Hall [1925–2006] of Bell Telephone Laboratories (Hall, 1962). Goode and Machol list 145 references and make no reference to any other books on the engineering of systems. The closest they come is to reference two texts on Operations Research (McCloskey & Trefethen, 1954; Morse & Kimball, 1951). Hall lists two texts on the engineering of systems (Flagle, Huggins, & Roy, 1960; Goode & Machol, 1957) and two on Operations Research (Churchman, Ackoff, & Arnoff, 1957; Morse & Kimball, 1951). It is interesting to note that the book by Flagle et al. (1960) retained Operations Research in the lead position in the title, despite its focus on the engineering of systems.

While OR methods were quite useful and appropriate for the fairly structured problems leading up to World War II, modern, complex problems have vexed traditional OR. Rosenhead (1996) elaborates on this issue:

> Since the late 1960s, analysts have actively debated claims for the objectivity of OR/MS models and the limitations imposed on OR/MS practice by its concentration on well-defined problems…Consistent with this, standard formulations of OR methodology (for example, formulate, model, test, solve, and implement) take as their foundation the possibility of a single uncontested representation of the problem situation under consideration. (Rosenhead, 1996, p. 118)

Munro and Mingers (2002) discuss the limitations of a traditional OR/MS approach:

> The typical assumptions made by a hard OR/MS method are: that there is a single decision maker (or at least a consensual group) with a clear objective-if there are multiple objectives these are usually reduced to a single metric; that the nature of the problem is agreed, even

> though a good solution may be difficult to find; that the most important factors can be quantified and reliable data collected; that a model, often mathematical or computer-based, can be used to generate solutions, and that this does not need to be transparent to the client (s); that the role of the OR person is one of expert analyst; and that future uncertainties can be modelled using probability theory.
>
> Soft methods can be characterised by generally not making these assumptions. Typically, a range of decision- makers or stakeholders will be recognised, potentially with differing and possibly conflicting objectives and definitions of the problematic nature of the situation; many important factors will not be able to be quantified; transparency and accessibility of the model will be very important, thus often ruling out mathematical models; the OR person's role will often be one of facilitator with a group of participants; and uncertainties will not simply be reduced to probabilities. (p. 369)

Over time, the constraint imposed by forcing problems to be addressed from a purely technical or mathematically derived perspective became limiting. "In the 1970s, 80s, and 90s, it had become obvious that some organizational problems could not be solved by pure logic, employing hard OR. Indeed problems have continued to become more complex and increasingly difficult to model mathematically" (Heyer, 2004, p. 3). Heyer (2004) continues:

> Underlying these difficulties is the fact that organisations are made up of people and people simply do not act like machines. Soft OR has sought to readdress this by understanding that people are an integral part of organisations and that these people each bring to the organisation their own worldviews, interests and motivations. Furthermore, soft OR understands the difficulties involved in the predictability of human behaviour. Soft OR techniques invariably employ a researcher whose role it is to ensure the study group contains key stakeholders; to act as a facilitator of the process; to orchestrate discussions; and be seen as open, independent and fair.... In very general terms, therefore, soft OR methods are those that structure a problem, as opposed to hard OR that seeks to solve it. Soft OR uses predominantly qualitative, rational, interpretative and structured techniques to interpret, define, and explore various perspectives of an organisation and the problems under scrutiny. (p. 4)

Thus, fairly structured problems can be handled with traditional (hard) operations research, but more complex problems involving humans with varied viewpoints require a different approach. The complex nature and pluralism of perspectives in these problems lead to a fundamentally different issue that must be addressed before hard operations research techniques can be employed, namely of ensuring there is ample agreement on the problem to be addressed before attempting to address it. Curtis, Dortmans, and Ciuk (2006) discuss this phenomenon: "...before we can do the 'problem right' we have to do the 'right problem'. But how do you know if you are doing the 'right problem'? Indeed it may only become apparent later in the piece that the wrong problem was being addressed" (pp. 1300–1301). In other words, before addressing our problem, we have to identify our problem (recall our discussion of problem structuring from Chap. 2). This is all in order to avoid the ever-present Type III error (Mitroff, 1998; Mitroff & Featheringham, 1974; Mosteller, 1948). This simple and yet profound distinction illustrates the dichotomy between soft and hard OR: that of doing the problem right (hard OR) versus doing the right problem (soft OR). This notion is illustrated in Fig. 3.1. The focus of these

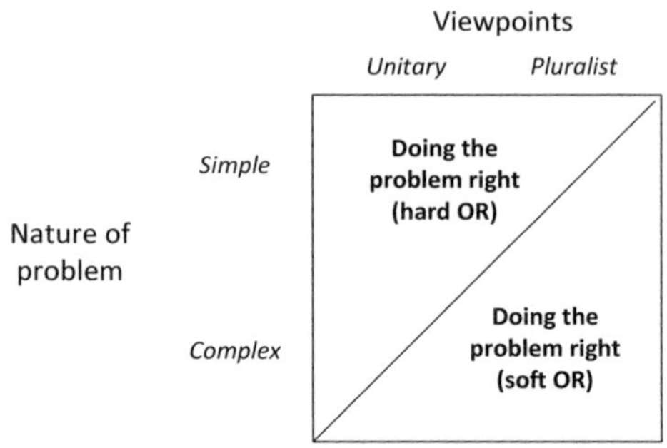

Fig. 3.1 Intersection of OR disciplines and problem characteristics [adapted from Curtis et al. (2006) and Jackson and Keys (1984)]

increasingly complex and pluralistic problems has continued to shift from doing the problem right to an emphasis on doing the right problem. A successful method for dealing with complex problems must be able to handle both doing the problem right as well as doing the right problem.

Many systems methods are in use to address complex problems beyond the aforementioned systems engineering and operations research. Jackson (2003) portrays systems methods using a typology that has four types: (1) goal seeking and viability; (2) exploring purposes; (3) ensuring fairness; and (4) promoting diversity, which are presented in Table 3.1.

While this list is useful, it no doubt generates the question of how to make sense of these methods. That is, how do we decide when faced with a complex problem, which method or methods are appropriate? Jackson and Keys (1984) and Jackson (1990) proposed a relationship between a problem's inherent complexity and its participants (i.e., stakeholders) and the type of methodology that could be used to address the problem. The Jackson-Keys grid of problem contexts has two axes with the following characteristics (Jackson, 1991, p. 31; 2000, p. 351; 2003, p. 18):

1. Problem complexity—mechanical or simple and systemic or complex
2. Participants—unitary (stakeholders agree on a common set of goals), pluralistic (stakeholders cannot agree on a common set of goals, but they can compromise), or coercive (stakeholders cannot agree on a common set of goals and decisions are made through power or force).

The 2×3 grid's utility is that a number of problem solution approaches or methodologies exist within the intersection of these six characteristics. The Jackson-Keys grid and the related systems-based approaches are presented in Fig. 3.2.

Adams and Mun (2005) extended this taxonomy to include the generally accepted four problem types (see, e.g., Kurtz & Snowden, 2003) and we have modified this taxonomy to consistently apply complexity terms. This classification is shown in Fig. 3.3.

Many of the methods presented in these taxonomies are not appropriate for systems age messes due to: a focus on systematic approaches to gaining understanding (e.g., Systems Engineering), a purely hard OR approach (e.g., OR, systems dynamics), or a focus on problem structuring at the expense of assessment (e.g.,

Table 3.1 Systems-based methods based upon Jackson's framework

Approach	Systems method	Primary proponent(s) of the method
Type A: Goal seeking and viability	Operations research	Hillier and Lieberman (2009)
	Systems analysis	Gibson, Scherer, and Gibson (2007)
	Systems engineering	Blanchard and Fabrycky (2011), Sage (1992)
	System dynamics	Forrester (1961, 1969, 1971), Maani and Cavana (2000)
	Soft systems thinking	Senge (1990)
	Viable system model	Beer (1979), Beer (1981, 1985)
	Complexity theory	Kauffman (1995), Stacey (1992)
Type B: Exploring purposes	Social systems design	Churchman (1968,1979)
	Strategic assumption and surfacing technique (SAST)	Mason (1969), Mason and Mitroff (1981), Mitroff, Barabba, and Kilmann (1977), Mitroff and Emshoff (1979), Mitroff, Emshoff, and Kilmann (1979)
	Interactive planning	Ackoff (1974, 1981)
	Soft systems methodology	Checkland (1993), Checkland and Scholes (1999)
Type C: Ensuring fairness	Critical systems heuristics	Ulrich (1983), Ulrich (1988)
	Team syntegrity	Beer (1995)
Type D: Promoting diversity	Participatory appraisal of needs and the development of action (PANDA)	Taket and White (1993), White and Taket (1997, 2000)
	Total systems intervention	Flood and Jackson (1991)

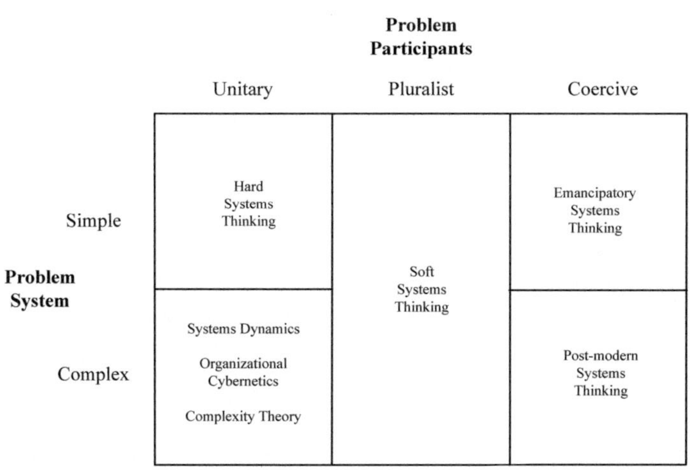

Fig. 3.2 Jackson-Keys grid of systems approaches and problem context (Jackson, 2003, p. 24)

Systems \ Participants	Unitary	Pluralist	Coercive
Chaotic	?	?	?
Complex	Organizations-as-Systems	?	?
Complicated	Systems Engineering (SE)	Soft Systems Methodology and Interactive Planning	System of Systems Engineering (SoSE) Total Systems Intervention (TSI)
Simple	Operations Research (OR) and Systems Analysis (SA)	Social Systems Design and SAST	Emancipatory Systems Thinking

Fig. 3.3 Classification of systems approaches

SSM). We need an approach that appreciates the need for both structuring and assessment (i.e., integrates hard and soft OR), appreciates the complexity of the underlying problem and its subjectivity, and is theoretically based for defensibility and repeatability.

Thus, a new paradigm of *systemic thinking*, conceptually founded in systems theory, is necessary. This new paradigm must be discipline-agnostic and theoretically derived, two foundations upon which our perspective of systemic thinking is founded. The goal is to demonstrate utility in helping individuals to increase their understanding about problems and messes of any size, complexity, or discipline.

3.2 What Is Systemic Thinking?[1]

Systemic thinking, as a term, has gained traction in recent literature (e.g., Boardman & Sauser, 2013; Hester & Adams, 2013; Midgley, 2012; Mingers, 2010), but it is our belief that the term has been used without specificity or universality. Our goal in this book is to articulate our unique perspective on systemic thinking which differentiates it from those systems approaches previously identified, and to demonstrate its utility in helping individuals to increase their understanding about

[1] Much of the text presented in Sect. 3.3 appeared previously in Hester and Adams (2013). Although we have retained the copyright to this text, the authors wish to acknowledge this publication.

Table 3.2 Characteristics of systematic versus systemic thinking

Element	Systematic thinking	Systemic thinking
1. Age or era	Machine	Systems
2. Unit of analysis	Problem	Mess (system of problems)
3. Mathematical formulation	Optimization	Fuzzy math
4. Goal	Problem solution	Resolution or increased understanding
5. Underlying philosophy	Reductionism	Constructivism and reductionism
6. Epistemology	Analysis	Synthesis and analysis
7. Ontology	Objective	Subjective
8. Discipline scope	Multi and interdisciplinary	Inter and transdisciplinary
9. Participants	Unitary	Pluralistic or coercive

problems and messes of any size, complexity, or discipline. The characteristics differentiating systematic thinking and systemic thinking, as we see them, are outlined in Table 3.2, with a discussion of each of the nine elements shown above.

3.2.1 Age or Era

The first distinguishing characteristic separating systematic and systemic thinking concerns the age or era each is designed to address. The machine age was concerned with simple systems and the systems age is concerned with complex systems, or more appropriately for purposes of systemic thinking, messes. Chapter 2 provided a detailed discussion of this distinction. Ackoff (1979) speaks of the inability of *machine age* paradigms to appropriately handle *systems age* messes. The relevant takeaway is that, when we are faced with a mess, we will be unable to appropriately address it with methods designed for solving machine age problems. While these methods, such as operations research and systems engineering, certainly have their place, this place is not in addressing systems age messes, which require methods and an accompanying theoretical basis, that appreciate their complex nature.

3.2.2 Unit of Analysis

Systematic thinking focuses on a singular problem. Due to its broader scope, systemic thinking has a larger, more abstract unit of analysis, that of a mess (Ackoff, 1979). A mess represents a system of problems. Thus, many problems are contained in a mess, but their analysis is not merely summative. As a result,

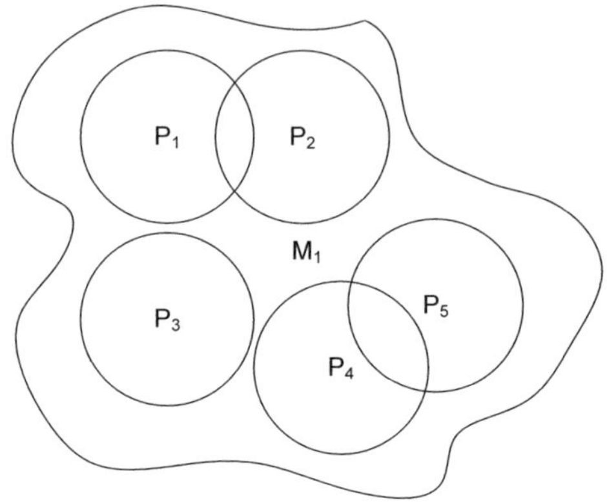

Fig. 3.4 Depiction of mess and constituent problems

analysis of a mess is significantly more complicated than a singular problem. This relationship is depicted in Fig. 3.4.

In Fig. 3.4 there are five problems, P_1, P_2, P_3, P_4, and P_5 and a mess, M_1, consisting of these five problems and their problem context. Succinctly, $M_1 = \mathrm{f}(P_1, P_2 \ldots P_5) \neq \sum P_i$. It is in the interaction of these constituent problems and their associated context where the mess truly arises:

> Problems are elements abstracted from messes; therefore, problems are to messes what atoms are to planets….the behavior of the mess depends more on how the solutions to its components problems interact than on how they act independently of each other. (Ackoff, 1977, pp. 4–5)

Viewing this mess as a whole truly requires a systemic perspective.

3.2.3 Mathematical Formulation

When analyzing a complex situation, it is imperative to think about global criteria associated with the desired end state of the analysis. That is, as an individual wishing to address a complex problem, am I searching for a globally optimal, "best (maximum or minimum) value of the objective function" (Taha, 2011, p. 3), a singular *solution* to a problem, or am I merely seeking a satisfactory resolution to my problem? The answer, as always, depends.

Given the relatively constrained focused of a singular problem and its objective (s), it is easy to conceive that the stopping criteria for a problem analysis using a systematic thinking paradigm is optimization. The end goal of this machine age problem is to develop a best answer to the problem at hand. Thus, we speak of the *best* design for a structural component of a larger system, or the *best* portfolio selection from among a number of choices. Systemic thinking, however, requires a more delicate balancing act to be observed. Given that any systemic thinking effort will involve two or more constituent problems, and the solution to each problem assessed independently represents a unique global solution to the mess, we must consider the *principle of suboptimization* (Hitch, 1953) in our analysis of these messes. Maximizing overall mess performance (i.e., optimizing the mess) requires that its constituent problem solutions be constrained, thus violating the notion of suboptimization. Ackoff (1977) echoes the difficulty in achieving an optimal solution to a mess:

> There is an important systems principle, familiar to all of you, that applies to messes and problems: that the sum of the optimal solutions to each component problem considered separately is not an optimal solution to the mess…. It is silly to look for an optimal solution to a mess. It is just as silly to look for an optimal plan. Rather we should be trying to design and create a process that will enable the system involved to make as rapid progress as possible towards its ideals, and to do so in a way which brings immediate satisfaction and which inspires the system to continuous pursuit of its ideals. (pp. 4–5)

Thus, if each system (i.e., problem) chooses to pursue (and thus, optimize) its own interests, then the mess will necessarily operate at less than maximum performance. Balancing the interests of constituent problems is one of the most difficult aspects of systemic thinking. A mechanism for doing so is known as *satisficing*. Satisficing is a term coined by Simon (1955, 1956) to describe how individuals make rational choices between available options and within a constrained environment. Simon argued that decision makers are rarely able to obtain and evaluate all the information which could be relevant to the making of a decision. Instead, they work with limited and simplified information to reach acceptable compromises (you *satisfice*, a portmanteau of satisfy and suffice) rather than to obtain a globally optimal strategy where a particular objective is wholly maximized. This relaxation from optimal-seeking problem solution approaches represents a departure from traditional OR solution techniques, one appropriate for mess analysis. Instead, a more appropriate approach is more qualitative and inclusive of multiple perspectives, one that invokes a mathematical underpinning such as fuzzy cognitive mapping (discussed at length in Chap. 5).

3.2.4 Goal

Given systematic thinking's focus on the problem as a unit of analysis and optimization as its desired end state, it is clear that the goal of a systematic thinking endeavor is to determine a problem solution. As such, a problem solution effort aims to determine the globally best answer to the particular problem of interest and recognizes that there is a preferred solution for the endeavor in question. Systemic thinking endeavors, however, are not so straightforward. Given their focus on satisficing and messes, it is clear that a singular view of *best* is not only not achievable, but also not necessary. Instead, the goal of a systemic thinking endeavor is to resolve or increase our understanding of a mess (recall the notion of perfect understanding discussed in the previous chapter; the assumption that we'll have complete understanding of our mess is both arrogant and foolhardy). Resolution may occur in pockets of our mess (i.e., singular problems), while increased understanding does not presuppose that our situation will reach a conclusive state. Rather, we may end up trapped in a do-loop until conditions within our situation's environment change. Thus, the question we must ask is, how are we going to improve our situation (either by resolve it or increasing our understanding of it)? This exploration may lead to a set of solutions, each of which may apply to the constituent problems of a mess, or it may lead simply to a greater understanding of the mess being faced. This increased knowledge may manifest itself in a recognition that we cannot do anything to improve or alter the current state. More importantly, perhaps, is the understanding that we may not want to intervene, for fear that we'll upset the *dynamic equilibrium* (D'Alembert, 1743) of the underlying system. The field of cybernetics and the systems principles of *homeostasis* (Cannon, 1929) and *homeorhesis* (Waddington, 1957) inform systems practitioners that systems have the ability to self-regulate to maintain a stable condition. Often times, intervention will cause negative feedback rather than improvement. A full understanding of this concept helps us to avoid the Type IV error (Boal & Meckler, 2010) that we introduced in Chap. 1, where the correct analysis leads to an inappropriate action taken to resolve a problem. So, in achieving increased understanding we may learn that inaction is the best action. Hester (2012) puts the notion of increased understanding in context by introducing the principle of *finite causality*, stating:

> …the outcome of the operation of any system is neither infinitely good nor infinitely bad. As more information is gained, the expected bounds surrounding this range of potential outcomes narrows, but never…meets at a point; in other words, it never reaches an optimal solution. Rather, the best we can hope to achieve is a set of potential outcomes that are boundedly rational and, by definition, neither infinitely good nor infinitely bad. (p. 274)

So, we should not despair at the lack of a singular optimal solution, but rather continue to work toward increased understanding in an effort to reduce the bounds on our solution.

3.2.5 Underlying Philosophy

Philosophy is based in a worldview which ultimately drives the understanding of a mess. Aerts et al. (1994) define worldview as "….a system of co-ordinates or a frame of reference in which everything presented to us by our diverse experiences can be placed" (p. 9).

Ackoff (1999) discusses the concept of a worldview as:

> Every culture has a shared pattern of thinking. It is the cement that holds a culture together, gives it unity. A culture's characteristic way of thinking is imbedded in its concept of the nature of reality, its world view. A change of world view not only brings about profound cultural changes, but also is responsible for what historians call a "change of age." An age is a period of time in which the prevailing world view has remained relatively unchanged. (p. 4)

This consistency in worldview is what Checkland (1993) refers to as *weltanschauung*, the image or model of the world that provides meaning. Each of these definitions hints at the idea of a worldview as a shared perspective or frame of reference for understanding the world. Ackoff's (1979) talk of a transition in ages implies a shift in philosophical worldview. The philosophical worldview has changed from reductionism in the machine age to constructivism in the systems age.

Reductionism, first introduced to Western civilization by René Descartes [1596–1650] in his *Discourse on Method* and later expanded by Isaac Newton [1643–1727] in his *Principia Mathematica* focuses on reducing a system to its barest elements in order to provide for an understanding of a system. Focusing on biological complexity, Mazzocchi (2008) discusses several limitations of applying a purely reductionist perspective to understanding complex phenomena:

- …the reductionist approach is not able to analyse and properly account for the emergent properties that characterize complex systems… (p. 11)
- …reductionism favours the removal of an object of study from its normal context. Experimental results obtained under given particular conditions or from a particular model—such as a mouse, in vitro cell cultures or computer models—are often extrapolated to more complex situations and higher organisms such as humans. But this extrapolation is at best debatable and at worst misleading or even hazardous. (p. 12)
- …reductionism is also closely associated with determinism—the concept that every phenomenon in nature is completely determined by preexisting causes, occurs because of necessity, and that each particular cause produces a unique effect and vice versa. This, naturally, also sustains the idea of predictability…. Nonetheless, complex…systems cannot be fully understood on a purely deterministic basis. (p. 12)
- …to better understand complex…systems and their adaptive behaviour, we need to consider the phenomenon of self-organization.... (p. 12)

Mazzocchi (2008) continues:

> An epistemological rethink is needed to instigate a paradigm shift from the Newtonian model that has dominated science, to an appraisal of complexity that includes both holism and reductionism, and which relaxes determinism in favour of recognizing unpredictability as intrinsic to complex systems. (p. 13)

It is clear that much is to be gained from adapting a worldview focused on holism, or constructivism. This perspective focuses on assembling system components into a purposeful whole in order to provide for an understanding of the entire system. However, this isn't the only way to gain understanding. Within the construct of systemic thinking, we must first use reductionism to deconstruct our mess into discernible elements, understand these individual elements, and then use constructivism to rebuild them in an effort to gain a holistic understanding of our mess. This unique worldview, focused on the use of *both* reductionism and constructivism, underlies systemic thinking and helps to provide for its epistemological basis, discussed in the following section.

3.2.6 Epistemology

Epistemology refers to the theory of knowledge and thus, addresses how knowledge is gained about a particular situation. It is informed by a particular worldview and thus, given their divergent worldviews, the epistemology underlying systematic and systemic thinking is quite divergent as well. Ackoff (1979) succinctly describes the steps in analysis as:

> … (1) taking apart the thing to be understood, (2) trying to understand the behavior of the parts taken separately, and (3) trying to assemble this understanding into an understanding of the whole… (p. 8)

Analysis relies on observation, experimentation, and measurement for its knowledge gathering. It is largely quantitative in its attempts to explain and understand the world.

On the other end of the epistemological spectrum is synthesis. Synthesis involves identification of a system to be studied. It then explores the environment in which the system resides, in order to understand its behaviors and purpose. Thus, rather than decomposing the system, synthesis aggregates a system into larger and larger systems in order to infer meaning. Synthesis relies on understanding, complementarity of perspectives (Bohr, 1928), and social construction for its meaning. Its emphasis on understanding (vice solution) and complementary, subjective evaluation of meaning should be comforting to individuals who focus on messes.

Neither epistemology alone is sufficient. We must invoke both synthesis and analysis, as appropriate, in order to increase our understanding of our mess and its constituent problems.

3.2.7 Ontology

Ontology refers to the nature of reality. Systematic thinking focuses on an objective reality in which there is a universally correct answer. Reality is concrete and this reality can be verified in a scientifically defensible manner using a positivist approach (Morgan & Smircich, 1980). Systemic thinking, however, rejects this notion in favor of subjectivity. Reality is seen as personal, and objective, with a focus on insight and revelation (i.e., understanding) (Morgan & Smircich, 1980). As we invoke complementarity (Bohr, 1928) as a necessary requirement for understanding complex problems, objectivity becomes increasingly difficult, if not impossible (and less important). Messes have pluralist, if not, coercive, stakeholders, and our focus shifts from an objective reality to a focus on an increased shared understanding.

3.2.8 Disciplinary Scope

Although the terms are often erroneously used interchangeably, multidisciplinary, interdisciplinary, and transdisciplinary each have a unique meaning (see, e.g., Lawrence, 2004; Whitfield & Reid, 2004; Young, 1998). A succinct summary of the three terms is provided by Choi and Pak (2006):

> We conclude that the three terms are used by many authors to refer to the involvement of multiple disciplines to varying degrees on the same continuum. Multidisciplinary, being the most basic level of involvement, refers to different (hence "multi") disciplines that are working on a problem in parallel or sequentially, and without challenging their disciplinary boundaries. Interdisciplinary brings about the reciprocal interaction between (hence "inter") disciplines, necessitating a blurring of disciplinary boundaries, in order to generate new common methodologies, perspectives, knowledge, or even new disciplines. Transdisciplinary involves scientists from different disciplines as well as nonscientists and other stakeholders and, through role release and role expansion, transcends (hence "trans") the disciplinary boundaries to look at the dynamics of whole systems in a holistic way. (p. 359)

A graphical depiction of multidisciplinary, interdisciplinary, and transdisciplinary is shown in Fig. 3.5. Note that D_1 and D_2 in the figures refer to Discipline 1 and Discipline 2, respectively.

An interdisciplinary, or perhaps transdisciplinary, scope is required for systemic thinking, whereas a multidisciplinary, or perhaps interdisciplinary, is sufficient for systematic thinking problems. This is further demonstrated by the holistic perspective demanded by systemic thinking. A multidisciplinary perspective represents too narrow a focus for understanding the bigger picture encouraged by a systemic lens. We must reason beyond traditional disciplinary boundaries to address the messes that are the focus of systemic thinking.

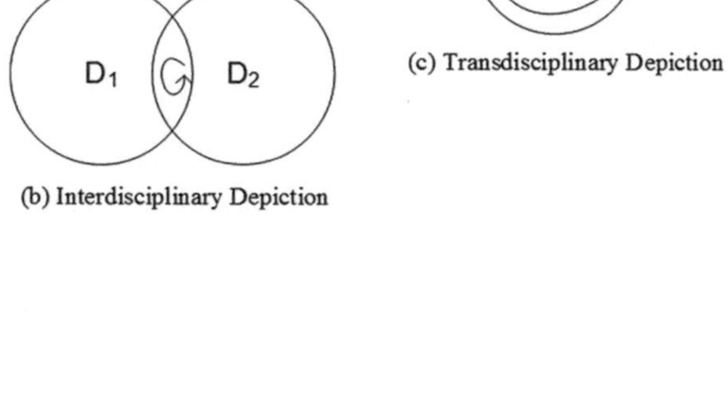

Fig. 3.5 Multidisciplinary, interdisciplinary, and transdisciplinary depictions

3.2.9 Participants

The number of participants within a systemic thinking endeavor can be classified as Jackson (1990, p. 658) does as follows:

- *Unitary* if they all agree on a common set of goals for the system and make their decisions in accordance with these goals.
- *Pluralistic* if they cannot all agree on a common set of goals and they make decisions which support differing objectives, but an accommodation or compromise can be reached upon which all agree.
- *Coercive* if decisions are achieved by the exercise of power and domination of one or more groups over others. In the case where coercive behavior is demonstrated it is impossible for any compromise solution to bring about a genuine accommodation among the parties.

The inherent complexity of a mess means that a unitary perspective is not achievable. If it is able to be achieved, then the issue of concern is a problem and not a mess (as all stakeholders can be treated as one problem owner). Thus, the objective nature of a systematic thinking approach necessitates a unitary perspective, whereas the subjectivity inherent in the many interacting problems (each with a unique owner) present in a complex mess mean that at best a pluralistic representation can be achieved, and, at worst, a coercive representation is necessary.

3.3 A Multimethodology for Systemic Decision Making

Addressing each of the thinking, acting, and observation steps in systemic decision making requires a unique method, and together they require a combination of methods known as a multimethodology, an approach that involves "linking or combining methods or techniques together" (Munro & Mingers, 2002, p. 369). All of the methods discussed earlier in the chapter, such as operations research and systems engineering, may be invoked but none alone is sufficient to fully embrace a systemic thinking paradigm.

Our multimethodology begins with problem structuring as outlined in the previous chapter. While problem structuring is necessary, it is only a first step and it is not typically a static construct. As more information is obtained regarding a problem, its "perceived problem structure will alternatively increase and decrease" (Millet & Gogan, 2006, p. 435). Decision making for complex problems involves four processes, as described by Millet and Gogan (2006, p. 435): (1) groping, (2) structuring, (3) adjusting, and (4) unstructuring.

Groping involves incremental steps taken to give structure to a highly unstructured problem. *Structuring* involves a major shift from an unstructured state to a structured state. Changes may occur such as a significant reduction in the number of available alternatives or the potential solution processes. *Adjusting* involves incrementally adding or removing constraints while also maintaining problem structure. *Unstructuring* involves a major shift from a structured to an unstructured state. This is often done deliberately as the result of newly obtained knowledge that fundamentally changes the problem formulation, in an effort to wipe the slate clean and start anew.

The use of each of these processes may occur in a very nonlinear fashion and each may be required a number of times during a decision making process. Indeed, "all four processes can contribute to the quality of solutions, and that decision making may cycle through a complex sequence of groping, structuring, adjusting, and unstructuring before resolution is achieved. The dialectical perspective can help managers prepare for the inevitable churn of convergence and divergence" (Millet & Gogan, 2006, p. 435).

Thus, although we may conceive of the stages of systemic decision making as beginning with problem structuring, then, per Fig. 2.5, proceeding linearly to thinking, then acting, and finally, observing, the process is rarely that ordered. These two conflicting paths, idealized vs. real systemic decision making, are shown in Fig. 3.6. Instead of following a linear Structuring-Thinking-Acting-Observing path, the illustrative example shown in the figure follows the following path:

Structuring-Thinking-Structuring-Thinking-Acting-Thinking-Acting-Observing-Thinking-Structuring-Thinking-Structuring-Thinking-Acting-Observing-Structuring-Thinking-Acting-Observing-Acting-Observing-Acting-Observing

Thus, what is perceived conceptually to be a four-step process takes 23 steps in this illustrative example. The aim of this text is to provide a semi-structured approach for progressing between four distinct stages (structuring, thinking, acting, and observing) so that deviations from a linear approach are made purposefully and not haphazardly and without purpose. Thus, we aim not to minimize the number of

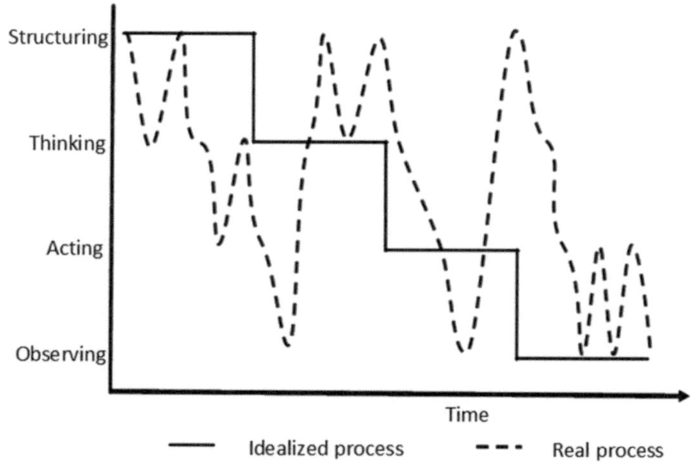

Fig. 3.6 Real versus idealized systemic decision making

steps necessary but rather to inform the reader and ensure that effort is placed effectively in making systemic decisions.

A more accurate reflection of Fig. 2.5 must include the systems principle of *feedback*, a concept which we will return to many times throughout this book. An updated version of this figure complete with feedback is provided in Fig. 3.7. Two major feedback loops emerge. Output of the observe stage can result in alternative actions being undertaken, as well as a rethinking of the problem. More attention will be paid to this notion later in the observe section of the text. Additionally, thinking about, acting on, or observing our mess may cause us to fundamentally reconsider its structure. This concept was addressed in Chap. 2.

Thus, we begin any exercise by asking the most fundamental initial question, namely *What problems are we trying to solve*? Each mess will contain *many* problems, and we must think systemically about each in order to reason about our mess (there are n problems shown in the mess depiction in Fig. 3.8, with all problems beyond P_3 being grayed out, suggesting either they weren't identified or purposefully chosen to be ignored for the purposes of the analysis). Systemic decision making is a series of structuring and unstructuring activities, or a duality of synthesis and analysis. Each of the selected problems (P_1–P_3 in the case of Fig. 3.8) is then analyzing using the methods detailed in Chaps. 6–11.

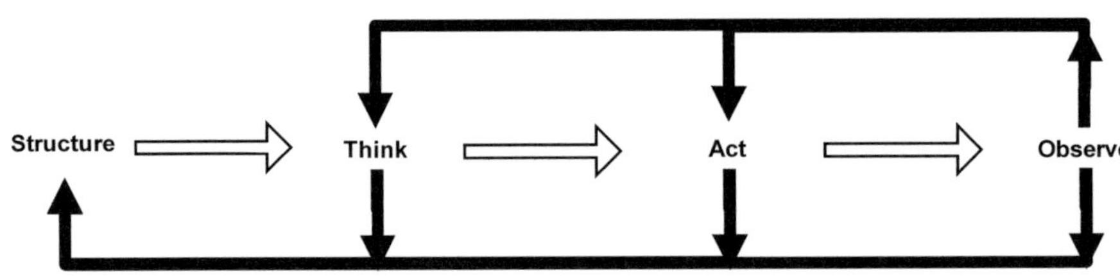

Fig. 3.7 Systemic decision making process with feedback

Fig. 3.8 Mess and problem depiction

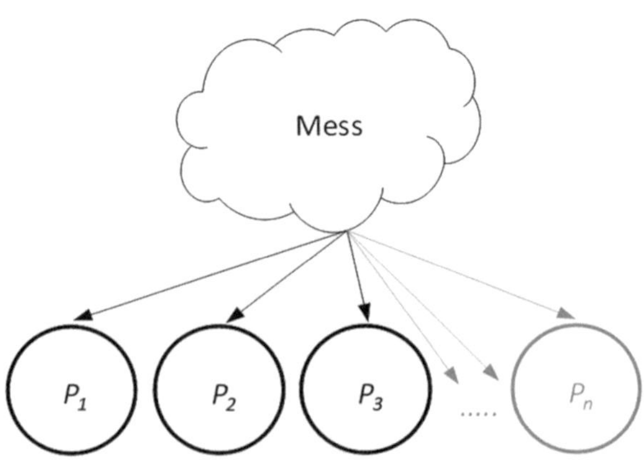

Given that systemic decision making is exploratory in its approach, there is no one, linear progression of steps (highlighting the importance of feedback). However, in the absence of any predisposition for acting otherwise, the authors suggest starting with the *Who* step (Chap. 6) and proceeding through the chapters in order. This will allow the reader the best opportunity for understanding the authors' approach to systemic decision making. It is important to note, however, that any step can lead to any other (as depicted in Fig. 3.7). This pattern is likely to be mess-dependent, however, and attempting to always follow the same path may prove problematic. While we suggest in the absence of other guidance to begin the thinking stage with stakeholder analysis and management (*Who?*), there may be reason to do otherwise. For example, stakeholders of your mess may be predetermined, with their roles clearly defined. Thus, it may behoove us to explore the *What* or the *Why* first. There is no wrong answer.

These perspectives are then reintegrated as detailed in Chap. 12, in order to provide for understanding at the mess level. This increased understanding acts as an input to the act and observe stages of the TAO approach (the focus of Parts III and IV of the text, respectively).

The flexibility of this approach owes itself to its foundation on the theoretical framework of systems theory. Systems theory provides the foundational underpinning for systemic thinking. This generalized theoretical underpinning provides rigor for the use of this approach by way of systemic thinking. This theory and its historical origins are discussed in detail in the following chapter.

3.4 Summary

Systems age messes are much grander and more complex than their machine age problem predecessors. Thus, accompanying methods to understand them must also account for this additional complexity. Practice shows that this is not the case and many methods and their underlying paradigms of systematic thinking are still quite prevalent in today's world. This chapter introduced a methodology for systemic thinking and contrasted it with traditional systematic thinking. The aim of the remainder of this book is to present the multimethodology underlying systemic decision making such that the reader, upon completion, will understand how to put the approach into practice in a manner which will garner increased understanding (and potential resolution) for systems age messes.

After reading this chapter, the reader should:

1. Understand the evolution of systems approaches;
2. Be able to articulate the distinction between systematic and systemic thinking; and
3. Articulate the systemic decision making process.

References

Ackoff, R. L. (1974). *Redesigning the future: Systems approach to societal problems*. New York: Wiley.

Ackoff, R. L. (1977). Optimization + Objectivity = Opt Out. *European Journal of Operational Research, 1*(1), 1–7.

Ackoff, R. L. (1979). The future of operational research is past. *Journal of Operational Research Society, 30*(2), 93–104.

Ackoff, R. L. (1981). *Creating the corporate future: Plan or be planned for*. New York: Wiley.

Ackoff, R. L. (1999). *Re-creating the corporation: A design of organizations for the 21st century*. New York: Oxford University Press.

Adams, K. M., & Mun, J. H. (2005). The application of systems thinking and systems theory to systems engineering. In *Proceedings of the 26th National ASEM Conference: Organizational Transformation: Opportunities and Challenges* (pp. 493–500). Virginia Beach, VA: American Society for Engineering Management.

Aerts, D., Apostel, L., De Moor, B., Hellemans, S., Maex, E., Van Belle, H., & Van Der Veken, J. 1994. *World views: From fragmentation to integration*. Brussels: VUB Press.

Beer, S. (1979). *The heart of the enterprise*. New York: Wiley.

Beer, S. (1981). *Brain of the firm* (2nd ed.). New York: Wiley.

Beer, S. (1985). *Diagnosing the system for organizations*. Chichester: Wiley.

Beer, S. (1995). *Beyond dispute: The invention of team syntegrity*. New York: Wiley.

Blanchard, B. S., & Fabrycky, W. J. (2011). *Systems engineering and analysis* (5th ed.). Upper Saddle River, NJ: Prentice Hall.

Boal, K., & Meckler, M. (2010). Decision errors of the 4th, 5th and 6th kind. In P. C. Nutt & D. C. Wilson (Eds.), *Handbook of decision making* (pp. 327–348). West Sussex: Wiley.

Boardman, J., & Sauser, B. (2013). *Systemic thinking: Building maps for worlds of systems*. New York: Wiley.

Bohr, N. (1928). The quantum postulate and the recent development of atomic theory. *Nature, 121* (3050), 580–590.

Cannon, W. (1929). Organization for physiological homeostasis. *Physiological Reviews, 9*(3), 399–431.

Carter, M. W., & Price, C. C. (2001). *Operations research: A practical introduction*. Boca Raton, FL: CRC Press.

Checkland, P. B. (1993). *Systems thinking, systems practice*. New York: Wiley.

Checkland, P. B., & Scholes, J. (1999). *Soft systems methodology in action*. Chichester, UK: Wiley.

Choi, B. C. K., & Pak, A. W. P. (2006). Multidisciplinarity, interdisciplinarity and transdisciplinarity in health research, services, education and policy: 1. Definitions, objectives, and evidence of effectiveness. *Clinical and Investigative Medicine, 29*(6), 351–364.

Churchman. C. W.1968. The systems approach (Reviwed and Updated ed.). New York: Dell Publishing.

Churchman, C. W. (1979). *The Design of Inquiring systems: Basic concepts of systems and organization*. New York: Basic Books.

Churchman, C. W., Ackoff, R. L., & Arnoff, E. L. (1957). *Introduction to operations research*. New York: Wiley.

Curtis, N. J., Dortmans, P. J., & Ciuk, J. (2006). 'Doing the right problem' versus 'doing the problem right': Problem structuring within a land force environment. *Journal of the Operational Research Society, 57*(11), 1300–1312.

D'Alembert, J. (1743). *Traité de dynamique*. Paris: David l'Ainé.

Flagle, C., Huggins, W., & Roy, R. (1960). *Operations research and systems engineering*. Baltimore: The Johns Hopkins Press.

Flood, R. L., & Jackson, M. C. (1991). *Creative problem solving: Total systems intervention*. New York: Wiley.

Forrester, J. W. (1961). *Industrial dynamics*. Cambridge, MA: MIT Press.
Forrester, J. W. (1969). *Urban dynamics*. Cambridge, MA: MIT Press.
Forrester, J. W. (1971). *World dynamics*. Cambridge, MA: MIT Press.
Gibson, J. E., Scherer, W. T., & Gibson, W. E. (2007). *How to do systems analysis*. New York: John Wiley & Sons.
Goode, H., & Machol, R. (1957). *Systems engineering: An introduction to the design of large-scale systems*. New York: McGraw-Hill Book Company.
Hall, A. D. (1962). *A methodology for systems engineering*. Princeton: D. Van Nostrand Company Inc.
Hester, P. T. (2012). Why optimization of a system of systems is both unattainable and unnecessary. *International Journal of System of Systems Engineering, 3*(3/4), 268–276.
Hester, P. T., & Adams, K. M. (2013). Thinking systemically about complex systems. *Procedia Computer Science, 20,* 312–317.
Heyer, R. (2004). *Understanding soft operations research: The methods, their application and its future in the defence setting (DSTO-GD-0411)*. Edinburgh, Australia: DSTO Information Sciences Laboratory.
Hillier, F. S., & Lieberman, G. J. (2009). *Introduction to operations research* (9th ed.). New York: McGraw-Hill.
Hitch, C. J. (1953). Sub-optimization in operations problems. *Journal of the Operations Research Society of America, 1*(3), 87–99.
Jackson, M. C. (1990). Beyond a system of systems methodologies. *Journal of the Operational Research Society, 41*(8), 657–668.
Jackson, M. C. (1991). *Systems methodology for the management sciences*. New York: Plenum Press.
Jackson, M. C. (2000). *Systems approaches to management*. New York: Kluwer Academic.
Jackson, M. C. (2003). *Systems thinking: Creative holism for managers*. Chichester: Wiley.
Jackson, M. C., & Keys, P. (1984). Towards a system of systems methodologies. *Journal of the Operational Research Society, 35,* 473–486.
Kauffman, S. A. (1995). *At home in the universe: The search for laws of self-organization and complexity*. New York: Oxford University Press.
Kurtz, C. F., & Snowden, D. J. (2003). The new dynamics of strategy: Sense-making in a complex-complicated world. *IBM Systems Journal, 42*(3), 462–483.
Lawrence, R. J. (2004). Housing and health: From interdisciplinary principles to transdisciplinary research and practice. *Futures, 36,* 487–502.
Maani, K. E., & Cavana, R. Y. (2000). *Systems thinking, systems dynamics: Understanding change and complexity*. Auckland, NZ: Prentice-Hall.
Mason, R. O. (1969). A dialectical approach to strategic planning. *Management Science, 15*(8), B403–B414.
Mason, R. O., & Mitroff, I. I. (1981). *Challenging strategic planning assumptions: Theory, cases and techniques*. New York: Wiley.
Mazzocchi, F. (2008). Complexity in biology: Exceeding the limits of reductionism and determinism using complexity theory. *EMBO Reports, 9*(1), 10–14.
McCloskey, J., & Trefethen, F. (1954). *Operations research for management*. Baltimore: Johns Hopkins Press.
Midgley, G. (2012). *Systemic intervention: Philosophy, methodology, and practice*. New York: Springer.
Millet, I., & Gogan, J. L. (2006). A dialectical framework for problem structuring and information technology. *The Journal of the Operational Research Society, 57*(4), 434–442.
Mingers, J. (2010). *Realising systems thinking: Knowledge and action in management science*. New York: Springer.
Mitroff, I. I. (1998). *Smart thinking for crazy times: The art of solving the right problems*. San Francisco, CA: Berrett-Koehler Publishers.
Mitroff, I. I., Barabba, C. P., & Kilmann, R. H. (1977). The application of behavioral and philosophical techniques to strategic planning: A case study of a large federal agency. *Management Science, 24*(1), 44–58.

Mitroff, I. I., & Emshoff, J. R. (1979). On Strategic assumption-making: A dialectical approach to policy and planning. *The Academy of Management Review, 4*(1), 1–12.
Mitroff, I. I., Emshoff, J. R., & Kilmann, R. H. (1979). Assumptional analysis: A methodology for strategic problem solving. *Management Science, 25*(6), 583–593.
Mitroff, I. I., & Featheringham, T. R. (1974). On systematic problem solving and the error of the third kind. *Behavioral Science, 19*(6), 383–393.
Morgan, G., & Smircich, L. (1980). The case for qualitative research. *The Academy of Management Review, 5*(4), 491–500.
Morse, P., & Kimball, G. (Eds.). (1951). *Methods of operations research*. New York: Wiley.
Mosteller, F. (1948). A K-sample slippage test for an extreme population. *The Annals of Mathematical Statistics, 19*(1), 58–65.
Munro, I., & Mingers, J. (2002). The use of multimethodology in practice–results of a survey of practitioners. *The Journal of the Operational Research Society, 53*(4), 369–378.
Rosenhead, J. V. (1996). Whats the problem? An introduction to problem structuring methods. *Interfaces, 26*(6), 117–131.
Sage, A. P. (1992). *Systems engineering*. New York: Wiley.
Senge, P. M. (1990). *The fifth discipline: The art & practice of the learning organization*. New York: Doubleday.
Simon, H. A. (1955). A behavioral model of rational choice. *Quarterly Journal of Economics, 69* (1), 99–118.
Simon, H. A. (1956). Rational choice and the structure of the environment. *Psychological Review, 63*(2), 129–138.
Stacey, R. D. (1992). *Managing the unknowable: Strategic boundaries between order and chaos in organizations*. San Francisco: Jossey-Bass.
Taha, H. A. (2011). *Operations research: An introduction* (9th ed.). Upper Saddle River, NJ: Prentice Hall.
Taket, A. R., & White, L. A. (1993). After OR: An agenda for postmodernism and poststructuralism in OR. *Journal of the Operational Research Society, 44*(9), 867–891.
Ulrich, W. (1983). *Critical heuristics of social planning: A new approach to practical philosophy*. New York: Wiley.
Ulrich, W. (1988). Systems thinking, systems practice, and practical philosophy: A program of research. *Systems Practice, 1*(2), 137–163. doi:10.1007/BF01059855.
Waddington, C. H. (1957). *The strategy of genes: A discussion of some aspects of theoretical biology*. London: George Allen & Unwin.
White, L. A., & Taket, A. R. (1997). Beyond appraisal: Participatory appraisal of needs and the development of action (PANDA). *Omega, 25*(5), 523–534.
White, L. A., & Taket, A. R. (2000). Exploring the use of narrative analysis as an operational research method: A case study in voluntary sector evaluation. *Journal of the Operational Research Society, 51*(6), 700–711.
Whitfield, K., & Reid, C. (2004). Assumptions, ambiguities, and possibilities in interdisciplinary population health research. *Canadian Journal of Public Health, 95,* 434–436.
Young, C. A. (1998). Building a care and research team. *Journal of the Neurological Sciences, 160* (Suppl 1), S137–S140.

Chapter 4
Systems Theory

Abstract In the last chapter, you were introduced to a systemic thinking methodology, one that is very different from the traditional, systematic methods for dealing with problems. The construct for systemic thinking is *holistic*, using both reductionism (i.e., to deconstruct problems and messes into understandable elements) and constructivism (i.e., to rebuild problems and messes to understand the whole). This unique systemic perspective, focused on the use of both reductionism and constructivism, and which underlies all aspects of systemic thinking, is built upon a science-based foundation labeled ***systems theory***.

4.1 Overview

Systems theory is a very scientific sounding term that has been used inconsistently in a variety of disciplines. Further, few of the disciplines that mention systems theory provide any type of formal definition for the term. Consequently, the term, systems theory, is often subject to confusion when used between disciplines. To alleviate much of the confusion, and formalize the term, our notion of systems theory has both a syntactic definition and a supporting construct (Adams, Hester, Bradley, Meyers, & Keating, 2014). The supporting construct includes seven axioms. In addition, there are large number of science-based propositions (i.e., systems principles), which give each axiom the power necessary to explain phenomena observed in real-world systems endeavors. The authors believe that the seven axioms, along with their corresponding science-based propositions, can be applied to all systems and can serve as the foundation for systemic thinking. We purport that each of the principles and axioms of systems theory serves to improve understanding when dealing with systems and their attendant problems and messes.

However, before exposing you to our formal definition and construct for systems theory, we will attempt to provide you with an historical perspective and classification of earlier, major, historical streams of thought that have addressed systems theory.

P.T. Hester and K.M. Adams, *Systemic Decision Making*, Topics in Safety, Risk, Reliability and Quality 33, DOI 10.1007/978-3-319-54672-8_4

4.2 Historical Roots of Systems Theory

Systems theory has been mentioned in a wide variety of disciplines that range from clinical psychology (Bowen, 1966; Plas, 1986) to chaos (Lorenz, 2005). The mention of systems theory, without a firm definition or construct, in such a wide variety of disciplines, has contributed to both inconsistent interpretation and misunderstanding in its application. Our classification of the historical development of systems theory, in six streams of thought (and their principal contributors), is presented in Table 4.1.

Each of the streams of thought from Table 4.1 is briefly discussed in the following sections.

4.2.1 General Systems Theory

The proponents of what is classified as general systems theory (GST) were Ludwig von Bertalanffy, Kenneth Boulding, Anatol Rapport, and Ralph Gerard. In 1954, they founded the Society for General Systems Research (SGSR). The purpose of the society was outlined in its original bylaws as follows (Hammond, 2002):

1. To investigate the isomorphy of concepts, laws, and models from various fields, and to help in useful transfers from one field to another;
2. To encourage development of adequate theoretical models in fields which lack them;
3. To minimize the duplication of theoretical effort in different fields; and
4. To promote the unity of science through improving communications among specialists. (pp. 435, 436)

Table 4.1 Historical classifications for systems theory (Adams, Hester, & Bradley, 2013, p. 4102)

Stream of thought	Major contributor(s) with selected references
1. General systems theory	Bertalanffy (1949, 1950, 1968), Boulding (1956)
2. Living systems theory	Miller (1978)
3. Mathematical systems theory	Mesarovic (1967), Wymore (1967, 1993), Klir (1968)
4. Cybernetics	Rosenblueth, Wiener & Bigelow (1943), Wiener (1965), Ashby (1947, 1952, 1956), Forrester (1961, 1969, 1971)
5. Social systems theory	Parsons (1970, 1979, 1991), Buckley (1967, 1998), Luhmann (1995, 2012)
6. Philosophical systems theory	Laszlo (1972, 1973, 1996), Bunge (1979, 1997, 1999, 2004)

For a more detailed discussion of these streams of thought and their relationship with systems theory, the reader is encouraged to review Adams (2012), Adams et al. (2013)

The SGSR bylaws were modified to include the practical application of systems concepts and models in planning and decision making processes (Hammond, 2002). However, founders and members of the SGSR had significant differences and the stated goals and objectives for the SGSR and GST diverged to the point where their unified theory for general systems became muddled and of reduced utility as a theory for systems practitioners.

> GST received a cool reception from the established sciences. It was criticized for dealing in metaphors, for being philosophical speculation, and for being incapable of falsification. As a result, the claims of GST were not taken seriously in the courts of academia and public opinion. (Bela Banathy in the foreward to Bausch, 2001, pp. vii, viii)

Due to these circumstances, and the need to reflect its broadening scope, in 1988 the SGSR was renamed the International Society for Systems Science. Today, general systems theory is spoken of in the past tense and it serves as a guide for the improved understanding of systems.

4.2.2 Living Systems Theory

Living systems theory describes living systems, how they are organized, how they work, how they evolve, and how they die. James Grier Miller [1916–2002], the originator of living systems theory (Miller, 1978), defines living systems as being open systems (i.e., they interact richly with their environment) that exhibit self-organization and have the special characteristic of life, thereby including both biological and social systems. A primary construct of living systems theory is the hierarchy and organization for systems which includes a hierarchy of eight levels and 20 processes which are integrated into a table of 160 cells. This 160-cell matrix can be used as a guide to classify all living systems.

4.2.3 Mathematical Systems Theory

The proponents of mathematical systems theory use the rigor of mathematics to construct models that explain systems. Early work in using axiomatic methods to describe systems was conducted by Mesarovic (1964), Klir (1968), Friedman (2005, 2006), Friedman and Leondes (1969a, b, c), and Wymore (1967, 1993). However, it is the utilization of set theory, first advocated by Mesarovic and fully expanded by Wymore and Fertig and Zapata (1978) that have gained traction in recent years and are worthy of additional mention.

Albert Wayne Wymore [1927–2011], a mathematician, was an early advocate for systems engineering and served as the first chair of the systems engineering program at the University of Arizona. Wymore's work focused upon the use of set theory to characterize and classify systems. The characterization capitalized on the

Table 4.2 Uses of set theory to describe systems (chronological order)

References	Brief description
1. Wolff (2000)	Proposes the adoption of *conceptual system theory* to solve fundamental problems concerning the notion of 'state' and 'time' in systems
2. Shell (2001)	Proposes a theory of systems design, using formal constructs and set theory notation
3. Vajna, Clement, Jordan, and Bercsey (2005)	Proposes *autogenetic design theory* (ADT) to describe systems design from the evolutionary view as a continuous optimization of a basic solution by observing starting conditions, boundary conditions, and constraints
4. Thompson (2006)	Proposes *axiomatic general systems behavioral theory* (A-BST) to describe the logical definition of general system and as a critical definition in devising mathematical models for predicting results in behavioral systems
5. Ford, Colombi, Jacques, and Graham (2009)	Proposes, through theory and application, that a quantitative, business process-constrained system classification, based upon the set theory, may be adopted for use in a variety of systems engineering uses
6. Sherwin (2010)	Proposes the desirability of a shift toward a holistic approach over reductionist approaches in the understanding of complex phenomena encountered in science and engineering by using an argument based on set theory to analyze three examples that illustrate the shortcomings of the reductionist approach

fact that systems are constructed from unique elements which interact with one another and the environment through inputs and outputs and have discrete states. Wymore's work has provided the foundation for additional work in describing systems through mathematics, and Table 4.2 provides six references to recent work where set theory was used to describe systems.

4.2.4 Cybernetics

The original proponent of cybernetics, Norbert Wiener [1894–1964] used the concepts of regulation and command as his central thought (Wiener, 1965). Both of these concepts, more commonly characterized as communications and control, rely on feedback within a system for the transmission of operational properties related to the systems' performance. Feedback is the mechanism that controls, guides, or steers the system to ensure performance of its goals. In fact, the term cybernetics comes from the Greek word *kybernetes*, for pilot or steersman.

W. Ross Ashby [1903–1972], a physician, expanded upon Wiener's work and used the human body as a model for understanding systems (Ashby, 1947, 1956, 1962).

Finally, Jay Forrester of MIT developed a technique (system dynamics) for modeling complex systems which operationalizes the concepts of cybernetics

(Forrester, 1961, 1969, 1971). The feedback principle is the foundation for system dynamics which uses causal loop diagrams that contain information feedback and circular causality to model the dynamic interplay in the real-world system under consideration.

The concepts espoused by the proponents of cybernetics serve as the foundation for the interdisciplinary field of controls and control theory—the design and analysis of feedback systems (Åström & Kumar, 2014; Åström & Murray, 2008). The emphasis on feedback is focused upon ensuring that desired outcomes in man-made processes are maintained through the use of mathematical transfer functions (i.e., differential equations), which provide adjustments to maintain systems within desired parameters. Cybernetics also extends to the systems dynamics methodology (Sterman, 2000, 2001), which has gained popularity due to its ability to quantitatively address complex issues and problems through modeling and highly visual dynamic simulations.

4.2.5 Social Systems Theory

Social systems theory uses relationships between human beings to form the structural elements for social systems. Talcott Parsons [1902–1979] stated that it was the actions of the human actors that constituted the system (Parsons, 1970, 1979, 1991). This contrasts sharply with the ideas of Niklas Luhmann [1927–1988] who considered communication processes as the elements which constituted the social system (Luhmann, 1995, 2012). The work done in social systems theory provides a systems-based foundation for the analysis of human–organizational systems and is too vast to properly treat in a short description. Bausch (2001, 2002) provides an excellent survey of this field.

4.2.6 Philosophical Systems Theory

Not surprisingly, the proponents of philosophical systems theory chose to approach systems from a higher level. Ervin Laszlo "proposes a systems language that enables the understanding between scientific disciplines now separated by specialized concepts and terms" (Adams et al., 2013, p. 4107). Laszlo (1972, 1973, 1996) was fundamentally interested in ensuring that systems practitioners are not thwarted in their efforts to communicate, which is most often caused by the trappings and limitations of the unique language and concepts attributable to a specific discipline. The ability to think about systems at the philosophical level, using language, concepts, ideas, and terms that are uniformly accepted and understood, increases the chance that each perspective may contribute, in a meaningful way, to an improved understanding of the complex system under study.

Mario Bunge's approach (Bunge, 1979, 1997, 1999, 2004) focuses on what he terms *systemism*, where mechanism is a process of a system and may not be separated from the system. Bunge states:

> Mechanism is to system as motion is to body, combination (or dissociation) to chemical compound, and thinking to brain. [In the systemic view], agency is both constrained and motivated by structure, and in turn the latter is maintained or altered by individual action. In other words, social mechanisms reside neither in persons nor in their environment - they are part of the processes that unfold in or among social systems…All mechanisms are system-specific: there is no such thing as a universal or substrate-neutral mechanism. (Bunge, 1999, p. 58)

Bunge's utilization of mechanism (a process of a system) as a means for explaining a system is unique, expansive, and philosophical in nature.

4.2.7 Historical Roots of Systems Theory Summary

The six streams of thought that have dominated systems theory do not provide a generally accepted canon of general theory that applies to all systems. However, each identifies some notions and elements that apply to all systems. The next section of this chapter will provide a more focused definition and supporting construct for systems theory.

4.3 Systems Theory

As we have already mentioned, the term *systems theory*, although used frequently in the systems literature, is a weakly defined term. As such, it is open to much misinterpretation and sharp attacks. In order to cogently present a theory for systems, any theory must contain both a syntactic definition (i.e., words) and a supporting construct.

We have proposed this syntactic definition for systems theory:

> a unified group of specific propositions which are brought together to aid in understanding systems, thereby invoking improved explanatory power and interpretation with major implications for systems practitioners. (Adams et al., 2014, p. 114)

The construct for systems theory is seven interconnected axioms (Adams et al., 2014, pp. 116–119). The seven axioms are as follows:

1. *Centrality Axiom* Central to all systems are two pairs of propositions: emergence and hierarchy, and communication and control. The centrality axiom's propositions describe the system by focusing on (1) a system's hierarchy and its demarcation of levels based on emergence and (2) systems control which

requires feedback of operational properties through communication of information.
2. *Contextual Axiom* System meaning is informed by the circumstances and factors that surround the system. The contextual axiom's propositions are those which bound the system by providing guidance that enables an investigator to understand the set of external circumstances or factors that enable or constrain a particular system.
3. *Goal Axiom* Systems achieve specific goals through purposeful behavior using pathways and means. The goal axiom's propositions address the pathways and means for implementing systems that are capable of achieving a specific purpose.
4. *Operational Axiom* Systems must be addressed in situ, where the system is exhibiting purposeful behavior. The operational axiom's propositions provide guidance to those that must address the system in situ, where the system is functioning to produce behavior and performance.
5. *Viability* Axiom Systems have key parameters that must be controlled to ensure continued existence. The viability axiom addresses how to design a system so that changes in the operational environment may be detected and affected to ensure continued existence.
6. *Design Axiom* Systems design is a purposeful imbalance of resources and relationships. Resources and relationships are never in balance because there are never sufficient resources to satisfy all of the relationships in a systems design. The design axiom provides guidance on how a system is planned, instantiated, and evolved in a purposive manner.
7. *The Information Axiom* Systems create, possess, transfer, and modify information. The information axiom provides understanding of how information affects systems.

Each axiom of the theory contains a number of propositions that support the axiom. These scientific propositions originated in a wide variety of the 42 scientific fields and as integrated in the axioms of systems theory serve as *principles* that may be applied to real-world systems. Note that there are 6 major fields of science depicted in the major sectors in Fig. 4.1: (1) natural sciences, (2) engineering and technology, (3) medical and health sciences, (4) agricultural sciences, (5) social sciences, and (6) humanities. Each of the six fields has a number of individual fields, which are represented by the 42 minor sectors (OECD, 2007).

Figure 4.1 is unique in that it also includes a series of inner rings which indicate the type and level of *knowledge contribution* that is being made. The knowledge contributions are hierarchical and structured as shown in Table 4.3.

The structure of knowledge is important. As knowledge contributions move from the philosophical level to the level of technique, they become less generalizable and easier to use. Conversely, as knowledge contributions move from the level of a technique toward the philosophical level, they lose specificity, are harder to use, and increase in generalizability. This concept is depicted in Fig. 4.2.

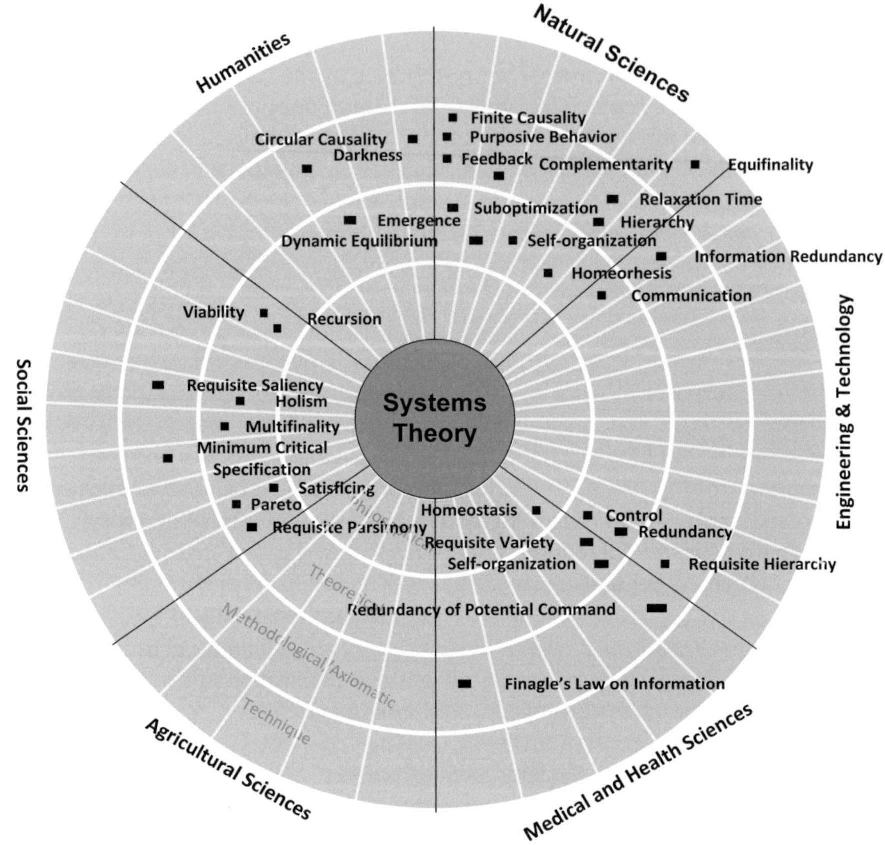

Fig. 4.1 Systems theory and the major fields of science. Updated version of Adams et al. (2014, p. 120)

Table 4.3 Structure for knowledge contributions (adapted from Adams et al., 2014, p. 113)

Level	Basic description
Philosophical	The emerging system of beliefs providing grounding for theoretical development
Theoretical	Research focused on explaining phenomena related to scientific underpinnings and development of explanatory models and testable conceptual frameworks
Methodological	Investigation into the emerging propositions, concepts, and laws that define the field and provide high-level guidance for design and analysis
Technique	Specific models, technologies, standards, and tools for implementation

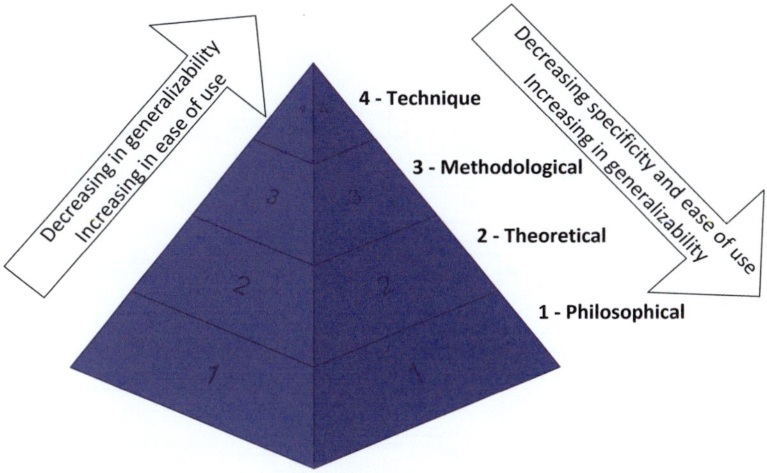

Fig. 4.2 Hierarchical nature of knowledge contributions

To summarize, our notion of systems theory is a unified group of axioms and supporting propositions (depicted in Fig. 4.1), linked with the aim of achieving understanding of systems. Systems theory can help systems practitioners to invoke improved explanatory power and predictive ability by using the seven axioms and their supporting propositions (from the 42 fields of science) as the foundation for *systemic thinking* related to the formulation, analysis, and solution of systems problems. It is in this manner that systems theory provides the truly transdisciplinary foundation for systemic thinking as described in Chap. 3.

The seven axioms and the 33 supporting propositions for systems theory will be discussed briefly in the sections that follow.

4.4 Centrality Axiom

The centrality axiom states:

> Central to all systems are two pairs of propositions; emergence and hierarchy, and communication and control. (Adams et al., 2014, p. 116)

The centrality axiom has four principles: (1) emergence; (2) hierarchy; (3) communications; and (4) control.

4.4.1 Emergence

Emergence is expressed simply by the statement that the whole is more than the sum of the parts. More formally:

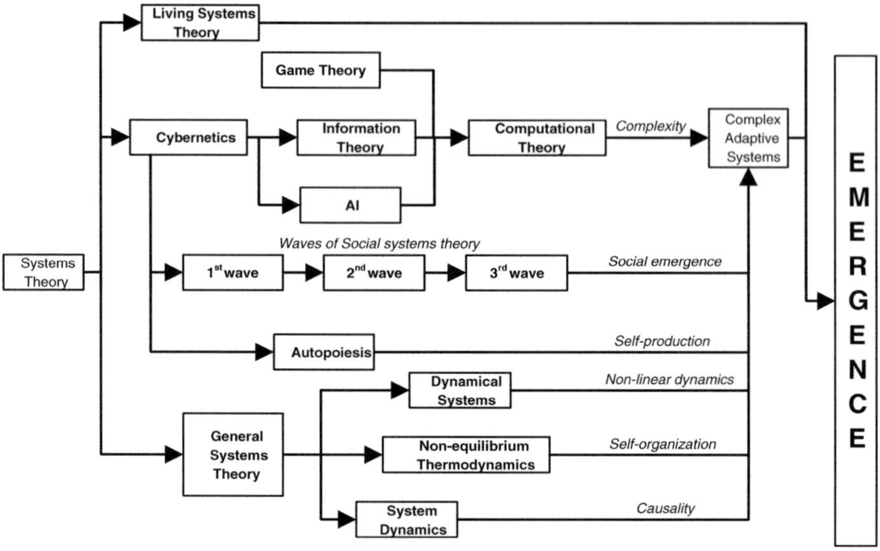

Fig. 4.3 Roots of emergence

> Emergence is the principle that whole entities exhibit properties which are meaningful only when attributed to the whole, not its parts – e.g. the smell of ammonia. Every model of human activity system exhibits properties as a whole entity which derive from it component activities and their structure, but cannot be reduced to them. (Checkland, 1999, p. 314)

Emergence is a concept that has a wide reach and roots in a number of mathematical and scientific disciplines, as shown in Fig. 4.3.

For the practitioner, emergence has an immediate benefit in both the design and analysis of systems. During design endeavors, the presence of emergence serves to demark the change in system hierarchy, where new properties have arisen due to the interaction of systems elements. In analysis, emergence is a property that must be accounted for when addressing state changes and system's operations.

4.4.2 Hierarchy

"Hierarchy is the principle according to which entities meaningfully treated as wholes are built up of smaller entities which are themselves wholes…and so on. In a hierarchy, *emergent properties* denote the levels" (Checkland, 1999, p. 314). The hierarchy principle is used in all aspects of systems design and analysis. Systems in design start from a high-level concept and are then developed by allocating functions to subsystems and components and so on. During analysis, a system is broken into smaller parts, understood, and then reassembled. In a systems hierarchy, emergent properties denote the transition from one level to another. More formally:

Fig. 4.4 Three-level system hierarchy

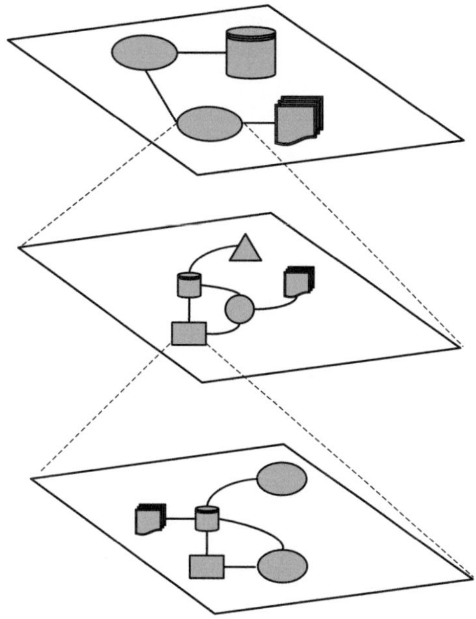

> … there exists a hierarchy of levels of organization, each more complex than the one below, a level being characterized by emergent properties which do not exist at the lower level. (Checkland, 1999, p. 78)

A simple three-level systems hierarchy is depicted in Fig. 4.4.

For the practitioner, hierarchy also has immediate benefits in both the design and analysis of systems. During design endeavors, hierarchy may be used to limit complexity by partitioning the system into understandable subsystems and lower-level components. This is most often accomplished through functional decomposition and requires application of the *law of requisite parsimony* (discussed in Sect. 4.9.1). In analysis, the presence of a new hierarchical level should indicate the presence of emergent properties that do not exist at the lower levels of the hierarchy.

4.4.3 Communications

Communications and control are the pair set that enable transmission of operational properties related to a systems' performance. Without the ability to communicate, essential operating properties, as the elements of control, could not be transmitted to the system, potentially affecting its viability. Communications (and control) are essential:

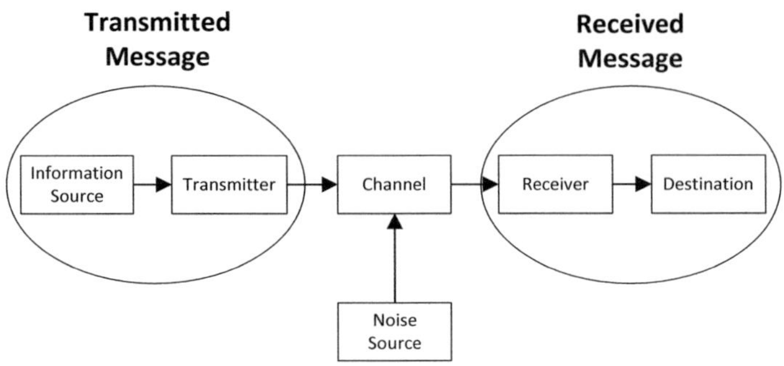

Fig. 4.5 Communications system block diagram

> … a hierarchy of systems which are open must entail processes of communication and control if the systems are to survive the knocks administered by the systems' environment. (Checkland, 1999, p. 83)

Communication is, in a very broad sense, "all of the procedures by which one mind may affect another" (Shannon & Weaver, 1998/1949, p. 3). A simple communications system can be symbolically represented using a schematic or block diagram. The block diagram in Fig. 4.5, adapted for enhanced clarity from the original figure in Shannon (1948, p. 381), contains all of the elements required for successful communication.

The block diagram elements in Fig. 4.5 are as follows:

1. *Information Source*: The information source produces and then selects the *transmitted message* from a number of possible messages. The messages may be text, spoken words, pictures, music, images, etc.
2. *Transmitter*: The transmitter converts or encodes the message into a signal.
3. *Channel*: This is the medium over which the signal travels. The medium may be an electric current on a wire, sound pressure in air or water, etc.
4. *Noise Source*: "In the process of being transmitted, it is unfortunately characteristic that certain things are added to the signal which were not intended by the original information source. These unwanted additions may be distortions of sound (in telephony, for example) or static (in radio), or distortions in shape or sending of picture (television), or errors in transmission (telegraphy or facsimile), etc. All of these changes in the transmitted signal are called *noise*" (Shannon & Weaver, 1998/1949, pp. 7, 8).
5. *Receiver*: The receiver changes or decodes the transmitted signal back into a message.
6. *Destination*: This is the message supplied to the *destination* by the receiver.

68 years ago, Shannon's formal depiction and mathematical theory associated with communications networks were revolutionary (Verdú, 1998). Shannon's work is the "Magna Carta of the information age. Shannon's discovery of the

fundamental laws of data compression and transmission marks the birth of Information Theory" (Verdú, 1998, p. 2057).

While Shannon's work on communications theory contains many important elements, perhaps the most important element was his mathematical construct for the measurement of *information* in a communicated message. The information we are referring to is not the traditional *semantic information* or *meaning*, but the information represented by "one's freedom of choice when one selects a message …the amount of information is defined, in the simplest cases, to be measured by the logarithm of the number of available choices" (Shannon & Weaver, 1998/1949, p. 9). Because modern communications systems rely on electrical current as the primary drivers for the communications channels, a measure related to electrical current has been adopted to measure information. Electrical current flow can be modeled as being in one of two possible states: (1) flowing (on) or (2) not flowing (off). The associated current waveforms have amplitudes that can represent the current flow by using a value of 1 for *on* and 0 for *off.* As a result, the base 2 or binary system was selected for use and the unit of information became the binary digit (either a 0 or a 1). Shannon later contracted the term binary digit into the word *bit*. Note that although Shannon (1948) was the first to use the term bit in a published paper, in this paper he gives credit for the contraction of the term *binary digit* into the word *bit* to a suggestion made by his Bell Lab colleague, Dr. John W. Tukey. Based upon this theory, it is the number of digits required to construct a message that represents the information content. This concept is the foundation for modern information theory, discussed further in section on the information axiom.

4.4.4 Control

The second element of the pair set of communications and control is control. In a discussion of the characteristics of modern systems, one of the earliest texts on systems engineering concludes the section on systems characteristics by stating:

> The last, and most important, characteristic of systems is that they are automatic. Although the degree of automaticity may vary over a wide spectrum, there are no systems in which human beings perform all control functions; conversely, there will probably never be systems in which no human beings are involved. (Flagle, Huggins, & Roy, 1960, p. 538)

This statement clearly defines the essential role of control in a system. Control is classically defined as follows:

> A means or device to direct and regulate a process or sequence of events. (Parker, 1994, p. 117)

Control is the method by which we ensure that the internal operations and processes of a system are "… regulated so that it will continue to meet the expectations of its designers and move in the direction of its goals" (van Gigch, 1974, p. 352). An additional definition defines control as "The process by means of which a whole entity retains its identity and/or performance under changing circumstances" (Checkland, 1999, p. 313).

For the practitioner, the *control principle* is the principle that permits the system to adapt and remain viable. As such, its mention should induce the following themes: (1) automated, (2) gives direction to the system by moving the system toward defined goals, (3) maintains system identity, and (4) does all of this in response to changing circumstances in both the system and its environment.

4.5 The Contextual Axiom

The contextual axiom states:

> System meaning is informed by the circumstances and factors that surround the system. The contextual axiom's propositions are those which bound the system by providing guidance that enables an investigator to understand the set of external circumstances or factors that enable or constrain a particular system. (Adams et al., 2014, p. 119)

The contextual axiom has three principles: (1) holism; (2) darkness; and (3) complementarity.

4.5.1 Holism

Holism is the philosophical position which holds that understanding a system is based not solely in terms of the functions of the component parts, but by viewing the system as a whole. It may be thought of as being in direct opposition to the scientific position of reductionism that states that systems can be explained by reduction to their fundamental parts. More formally:

> It is very important to recognize that the whole is not something additional to the parts: it is the parts in a definite structural arrangement with mutual activities that constitute the whole. The structure and the activities differ in character according to the stage of development of the whole; but the whole is just this specific structure of parts with their appropriate activities and functions. (Smuts, 1961 (1926), p. 104)

The holism described by Smuts, while still accurate, has taken on a new prominence with the study of complexity. Scientific organizations such as the *Santa Fe Institute* and *the New England Complex Systems Institute* have focused on holism as an underlying principle for understanding complex systems.

> The past three centuries of science have been predominantly reductionist, attempting to break complex systems into simple parts, and those parts, in turn into simpler parts. The reductionist program has been spectacularly successful, and will continue to be so. But it has often left a vacuum: How do we use the information gleaned about the parts to build up a theory of the whole? The deep difficulty here lies in the fact that the complex whole may exhibit properties that are not readily explained by understanding the parts. The complex

> whole, in a completely nonmystical sense, can often exhibit collective properties, 'emergent' features that are lawful in their own right. (Kauffman, 1995, pp. vii, viii)

Holism is an equal and complementary (Bohr, 1928, 1937) partner with reductionism in the scientific method. By understanding this relationship, systems practitioners gain improved understanding through both analytic and synthetic perspectives.

> Holism has been around for a long time, but because of the apparent success of the traditional scientific method, has had to take second place to reductionism. Holism deserves to be reinstated as an equal and complementary partner to reductionism. It encourages the use of transdisciplinary analogies, it gives attention to both structure and process, it provides a powerful basis for critique, and it enables us to link theory and practice in a learning cycle. As a result, there is evidence that holism can help managers make a success of their practice and address broad, strategic issues as well as narrow, technical ones. (Jackson, 2006, pp. 647, 648)

For the practitioner, holism helps to shape our worldview and is an essential first step in systemic thinking. There are four essential benefits that holism provides to those who adopt it as an element of systemic thinking:

> 1. The ability to develop and make use of transdisciplinary analogies.
> 2. The capacity to recognise the importance of both process and structure in system development and maintenance, and their interdependence.
> 3. It provides a good basis for critique.
> 4. The 'theoretical awareness' to which it gives rise. (Jackson, 2006, pp. 650, 651)

4.5.2 Darkness

System darkness states that "no system can be known completely" (Skyttner, 2001, p. 93). This is based upon the fact that the human observer has limited sensory capabilities and may never be able to truly see all aspects of a system. This does not mean giving up, but does provide some humility to the scientific observer when treating observations as absolutes. For practitioners, it is important in that:

> Each element in the system is ignorant of the behavior of the system as a whole, it responds only to information that is available to it locally. This point is vitally important. If each element 'knew' what was happening to the system as a whole, all of the complexity would have to be present in that element. (Cilliers, 1998, pp. 4, 5)

For the practitioner, the important point to take away is that "all complex systems are by definition open and so it is nigh on impossible to know how the system's environment will affect the system itself—we simply cannot model the world, the Universe and everything" (Richardson, 2004a, p. 77).

4.5.3 *Complementarity*

Complementarity addresses the aspect that no single perspective or view of a system can provide complete knowledge of the system. Niels Bohr [1885–1962], the 1922 Nobel laureate in physics, coined this term during his experiments on particle physics. Bohr stated that if two concepts are complementary, an experiment that clearly illustrates one concept will obscure the other complementary one. For example, an experiment that illustrates the particle properties of light will not show any of the wave properties of light (Adams, 2011, p. 128).

Once again, this does not mean giving up, but requires the observer to gain additional perspectives in order to improve understanding. In the limit, an infinite number of perspectives will reveal perfect understanding. Computer scientist and systems thinker Gerald Weinberg expresses this very well in his idea of a general law of complementarity.

> 'Absolute complementarity', [...] depends on the idea that there is no alternative to accept but a 'fundamental, integral' limitation on observation. If, 'for whatever reason', observers do not make infinitely refined observations, then between any two points of view there will generally be complementarity. Since in almost every case there will be 'some' reason for stopping short of infinitely refined observations, we can remove this condition and create the 'General Law of Complementarity'. (Weinberg, 2001, p. 120)

For the practitioner, it is important to understand that an infinite number of observations or perspectives are not realistic, but inform the practitioner that each additional observation, and perspective, of a system will reveal additional truths in a never-ending quest to overcome system darkness.

4.6 The Goal Axiom

The goal axiom states:

> Systems achieve specific goals through purposeful behavior using pathways and means. The goal axiom's principles address the pathways and means for implementing systems that are capable of achieving a specific purpose. (Adams et al., 2014, p. 119)

The goal axiom has four principles: (1) equifinality; (2) multifinality; (3) purposive behavior; and (4) satisficing.

4.6.1 *Equifinality and Multifinality*

An essential difference between most man-made and living systems can be expressed by the principle of equifinality, a principle that can be summed up by the famous idiom, *all roads lead to Rome*. Most man-made systems are closed systems,

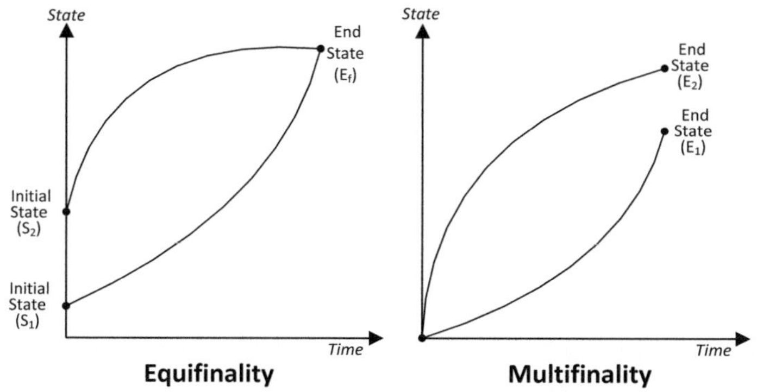

Fig. 4.6 Contrast between the principles of equifinality and multifinality

while living systems are open or vital systems. Open systems are exchanging materials with the environment and can exhibit equifinal behavior. However, a closed system must obey the second law of thermodynamics which states that entropy (the property of matter that measures the degree of randomization or disorder at the microscopic level) can be produced but never destroyed (Reynolds & Perkins, 1977).

Equifinality states:

> If a steady state is reached in an open system, it is independent of the initial conditions, and determined only by the system parameters, i.e. rates of reaction and transport. (Bertalanffy, 1968, p. 142)

This can be sharply contrasted with multifinality where "similar initial conditions may lead to dis-similar end-states" (Buckley, 1967, p. 60) . Figure 4.6 shows that multifinality is a state in which similar initial conditions lead to dissimilar end states, and equifinality is a state in which dissimilar initial conditions lead to a similar end state.

For the practitioner, these two principles provide formality for the notions that any two endeavors may (1) have dissimilar initial states, but can achieve the same end state or (2) have similar initial states, but can achieve dissimilar end states.

4.6.2 *Purposive Behavior*

All man-made systems display purposive behavior. Purposive behavior is defined as follows:

> Purposeful behavior is meant to denote that the act or behavior may be interpreted as directed to the attainment of a goal-i.e., to a final condition in which the behaving object reaches a definite correlation in time or in space with respect to another object or event. (Rosenblueth et al., 1943, p. 18)

In man-made systems, purposive behavior is a function of the systems' mission, goals, and objectives. Churchman and Ackoff (1950) noted a number of similarities in purpose-built objects (i.e., man-made systems). Three of these similarities are important elements of this systems principle.

1. Presence of Choice: "The basis of the concept of purpose is the awareness of voluntary activity" (Rosenblueth et al., 1943, p. 19). Choice is essential to identify purpose.
2. Inclusion of Time: "Purposive behavior can only be studied relative to a period of time" (Churchman & Ackoff, 1950, p. 35).
3. Production Requirement: "The purposive object or behavior is at least a potential producer of some end-result (end, objective, goal)" (Churchman & Ackoff, 1950, p. 35).

In summary, purposive behavior, to which all man-made systems prescribe, requires the system to have choices and to produce some end result over a period of time.

For the practitioner, in order to provide a complete view of the objectives of a system, an understanding of the system's purpose is necessary. Comprehension of purpose, through formal statement of the system's mission, goals, and lower-level supporting objectives, provides the foundation for framing the systems design that is an instantiation of its purpose.

4.6.3 Satisficing

Herbert A. Simon [1916–2001], the 1978 Nobel laureate in economics, questioned the utility of traditional economic and statistical theories of rational behavior and their applicability as the foundation for human decision making. He stated:

> Both from these scanty data and from an examination of the postulates of the economic models it appears probable that, however adaptive the behavior of organisms in learning and choice situations, this adaptiveness falls far short of the ideal of 'maximizing' postulated in economic theory. Evidently, organisms adapt well enough to 'satisfice', they do not, in general, 'optimize'. (Simon, 1956, p. 129)

Simon's observation is keen and utilizes elements of the contextual axiom to propose that humans do not have complete information for decision making and that best results are not optimal but satisficing in nature. Once again, it DOES NOT mean ignoring the optimum by not striving for the most satisfactory in the decisions that support a system's purpose, goal, or objectives. It does mean knowing that there is incomplete information with which to make the optimal decision and that any solution will be, at best, a *satisficing*, or mostly satisfactory, solution. In other words, satisficing can be thought of as the best possible solution given the information, which is always incomplete, that you have at the present time.

For the practitioner, satisficing solutions are the norm in approaching complex systems with systemic methodologies. The idea for, and concepts surrounding, optimization should be reserved for designs and approaches that may be formulated in a manner where a mathematical solution may be obtained. Practitioners will find that this is most often restricted to situations which address smaller system elements through solutions that utilize reductionist approaches (i.e., machine age problems).

4.7 The Operational Axiom

The operational axiom states:

> Systems must be addressed 'in situ', where the system is exhibiting purposeful behavior. The operational principles provide guidance to those that must address the system in situ, where the system is functioning to produce behavior and performance. (Adams et al., 2014, p. 119)

The operational axiom has seven principles: (1) dynamic equilibrium; (2) relaxation time; (3) basins of stability; (4) self-organization; (5) homeostasis and homeorhesis; (6) suboptimization; and (7) redundancy.

4.7.1 Dynamic Equilibrium

Dynamic equilibrium is the principle that states "for a system to be in a state of equilibrium, all subsystems must be in equilibrium. All subsystems being in a state of equilibrium, the system must be in equilibrium" (Adams, 2011, p. 134). As a result of this principle, we know that systems will stay in their initial condition until some sort of interaction is made with them.

For the practitioner, understanding the forces which contribute to a system's states is an essential element of understanding. Failure to properly account for forces present within the system and those to and from its environment can result in incomplete models and faulty solutions during systems endeavors.

4.7.2 Relaxation Time

The relaxation time principle states that "system stability is possible only if the system's equilibrium state is shorter than the mean time between disturbance" (Adams, 2011, p. 134). In the top portion of Fig. 4.7, the system does not achieve equilibrium based on its standard relaxation time (shown in the lower portion of the figure) because it has been perturbed by another disturbance before it can achieve equilibrium. This second disturbance places the system in a more complex series of amplitude shifts and decreased related relaxation times. Figure 4.7 is a depiction of relaxation time.

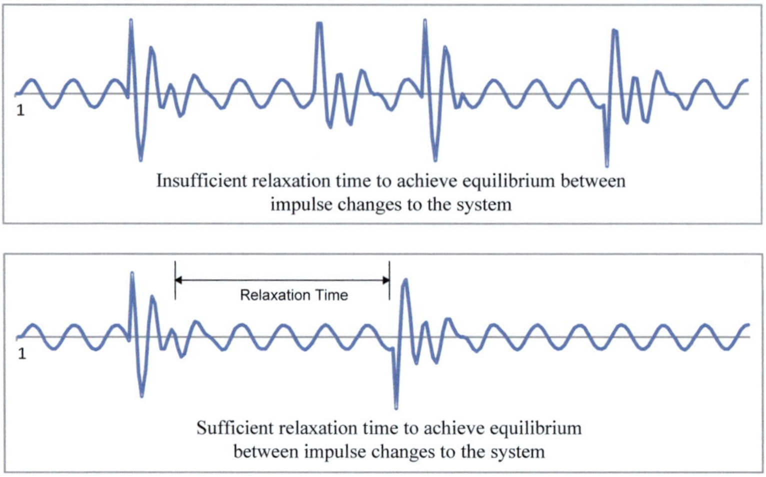

Fig. 4.7 Relaxation time

For the practitioner, like the case of dynamic equilibrium, it is important to understand the forces present within the system and those to and from its environment and the relaxation time associated with each perturbing force. "The number of disturbances over time determines whether the system can possibly maintain internal stability and return to an equilibrium state" (Adams, 2011, p. 134). The principle of relaxation time may be applied to many generalized systems as an aid during the analysis of specific system states.

4.7.3 Basins of Stability

Stuart Kauffman (Kauffman, 1990, 1993), a well-known complex systems researcher at the *Santa Fe Institute*, states that complex systems have three regimes: (1) order, (2) chaos, and (3) phase transition.

Order is where the system is stable (i.e., in equilibrium). This is referred to as a basin of stability. The basin is not a permanent place or state. The complex system may be subject to change (i.e., through self-organization of external impetus) and will shift from order to chaos. The period of time during the shift is labeled the transition phase and signifies that the system is moving to or from order to chaos. A system in order or chaos is fairly easy to identify. However, it is the *thresholds of instability*, the areas between chaos and order that are difficult to recognize.

For the practitioner, this is an important concept to understand when working with complex systems. The system's regime is important because the both the method of analysis and mode of intervention are functions of its position vis-à-vis order and chaos. This notion will be expanded upon in Chap. 10 when we introduce a sensemaking framework which will address the treatment of order and chaos.

4.7.4 Self-organization

Simply stated, the principle of self-organization is "the spontaneous emergence of order out of the local interactions between initially independent components" (Adams, 2011, p. 138).

Self-organization is a well-established principle in the physical sciences (Nicolis & Prigogine, 1977). Self-organization is the characteristic and ability of a system (and its constituent parts) to determine its structure and features. A leading cybernetician, W. Ross Ashby [1903–1972], proposed what he called the principle of self-organization (Ashby, 1947) when he noted that "dynamic systems, independently of their type or composition, always tend to evolve towards a state of equilibrium" (Adams, 2011, p. 136).

For the practitioner, knowledge of this principle provides insight into the functioning of most of the complex systems surrounding the world today. Attempts to manage or control self-organizing systems may run into severe limitations because, by design, self-organizing systems resist external changes. In fact, efforts at control often achieve results very different from the desired effect and may even result in the loss of viability and eventual destruction of the system.

4.7.5 Homeostasis and Homeorhesis

Homeostasis has played an important role in the development of the field of cybernetics. The term was created to describe the reactions in humans which ensure the body remains in steady state (Cannon, 1929, 1967/1932).

The principle of homeostasis is "the property of an open system to regulate its internal environment so as to maintain a stable condition, by means of multiple dynamic equilibrium adjustments controlled by interrelated regulation mechanisms" (Adams & Mun, 2005, p. 497).

Homeostasis may be used to depict how a system may superficially appear to be unchanged over time. If Fig. 4.8 is examined superficially, the number of elements and structure at time (a) and time (b) appears to be the same. However, when the observer more carefully examines the system, they recognize that input, output, and cell elements have changed, representing the actual exchange of materials, information, and energy.

Homeorhesis is a dynamic extension of the idea presented in homeostasis. In the case of homeorhesis, the equilibrium is dynamic, where in homeostasis the equilibrium is static (Willander, Mamontov, & Chiragwandi, 2004). The term *homeorhesis* is attributed to Waddington (1957, 1968) who described the regulation in a living particle as moving along some defined time path, from its initial creation through various life stages that end at senescence.

> The regulation that occurs in such particle is a regulation not necessarily back to a static stable equilibrium, as in homeostasis, but to a more general stable mode, some future

Fig. 4.8 Homeostasis in cell at **a** time t and **b** time $t + s$

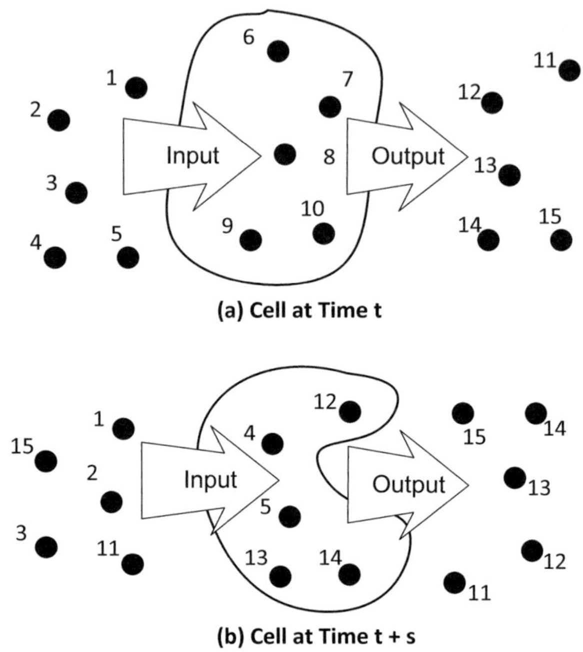

stretch of the time path. The appropriate notion to describe this process is homeorhesis. (Waddington, 1957, p. 32)

Homeorhesis is the self-regulating process through which the living particle, cell, or organism is maintaining its internal stability while adjusting dynamical conditions required for its survival. The stability attained as a result of homeorhesis is dynamic, which makes sense in environments where conditions are continuously changing.

For the practitioner, both homeostasis and homeorhesis have direct application in man-made complex systems. Whenever a dynamic system is using a process or mechanism to transfer energy, material, or information, these principles may be utilized to explain its behavior (Chen & Aihara, 2002; Yates & Iberall, 1982).

4.7.6 Suboptimization

The principle of suboptimization was recognized during analysis and optimization experiences by those conducting operations research in support of a number of localized and global efforts Second World War. Renowned RAND scientist, DoD Comptroller, and University of California President, Charles Hitch [1910–1995] found that efforts at optimization related to the detection and sinking of German U-boats during the localized Battle of the Atlantic involved lower-level criteria than those used to prosecute the larger global war as a whole.

> The optimal (or less ambitiously, good) solutions sought by operations research are almost always "sub-optimizations" in the sense that the explicit criteria used are appropriate to a low (or at least not the highest) level with which the researcher and his client are really concerned. (Hitch, 1952, p. 1; 1953, p. 87)

This elegant principle may be restated such that "if each subsystem, regarded separately, is made to operate with maximum efficiency, the system as a whole will not operate with utmost efficiency" (Adams, 2011, p. 135).

For the practitioner, this principle is important during both the design and development and the operation and maintenance of subsystems and the larger system of systems in which they belong. By applying this principle, the systems practitioner acknowledges that attempts at optimization within each subsystem independently will not in general lead to an overall system optimum. In fact, improvement of a particular subsystem may actually worsen the overall performance of the larger system.

4.7.7 Redundancy

Simply stated, the redundancy principle is the duplication of critical components or functions of a system with the intention of increasing reliability of the system (Pahl, Beitz, Feldhusen, & Grote, 2011). Redundancy is a critical element in a number of nonfunctional systems requirements, specifically robustness and survivability (Adams, 2015).

For the practitioner, the introduction of redundancy in the operational axiom is to ensure that the system has excess resources in order to operate successfully. Recognizing that operational systems exist in the real world, where they are often subject to changing resources, unstable environments, and changing requirements, levels of redundancy are provided to ensure stability in the system. The practitioner must also recognize that redundancy stands in sharp contrast with the principle of minimum critical specification (Cherns, 1976, 1987), which requires the designer to design as little as possible and only specify what is essential. However, all designs require some level of redundancy to ensure continued operation in the face of changing resources and perturbations to the system.

4.8 The Viability Axiom

The viability axiom states:

> Key parameters in a system must be controlled to ensure continued existence. The viability principles address how to design a system so that changes in the operational environment may be detected and affected to ensure continued existence. (Adams et al., 2014, p. 119)

The viability axiom has six principles: (1) viability; (2) requisite variety; (3) requisite hierarchy; (4) feedback; (5) circular causality; and (6) recursion.

4.8.1 Viability Principle

Our definition of viability, from a systems perspective, is *the continued existence of a system—the ability to survive*. Our definition is founded upon two essential words: (1) *continued* and (2) *existence*, or more specifically, *identity*.

- "'Continued' refers to the length of time a system has to exist in order to be worthy of study. How long this must be is a question of relative time scale between system and observer, and thus relates, at least indirectly, to the typical length of time the observer survives" (Weinberg, 2001, p. 238). To exist means to have an identity.
- "Identity is synonymous with viabilityviability, for nothing remains to be identified that is not viable, and a thing that changes its identity passes out of existence" (Weinberg, 2001, p. 239).

The viable system is one that has an identity—it exists at a specific point in time and within a specific environment.

> When we talk about a viable system, we mean that this system is able to survive, be healthy and develop in its particular system environment. In other words, system viability has something to do with both the system and its properties, and with the system environment and its properties. (Bossel, 2001, p. 24)

Due to the importance of this principle, we will spend some additional space discussing a theoretical construct (i.e., orientation theory) for addressing viability and the minimum set of properties systems must possess to ensure their continued existence—their ability to survive.

4.8.1.1 Orientation Theory

Orientation theory has been formulated and proposed by German environmental and systems scientist Bossel (1977, 1987, 1996, 1998). Orientation theory focuses on properties of systems labeled *orientors* that emerge as a result of evolutionary adaptation of systems to their specific environments. This is a powerful new perspective that may be used to model systems in relation to their environments and how the system must possess properties able to react, overcome, or control environmental forces that act upon the system.

Bossel (1992) posits that:

> A valid real-structure (explanatory, process) model in principle (1) not only allows analysis of the behavioral spectrum even for new environmental conditions, (2) but also permits analysis and prediction of system development and structural dynamics as a consequence of the interaction between system and environment. (p. 261)

The context diagram in Fig. 4.9 is used to depict Bossel's basic environmental forces. In order to account for each of these forces, a successful or viable system

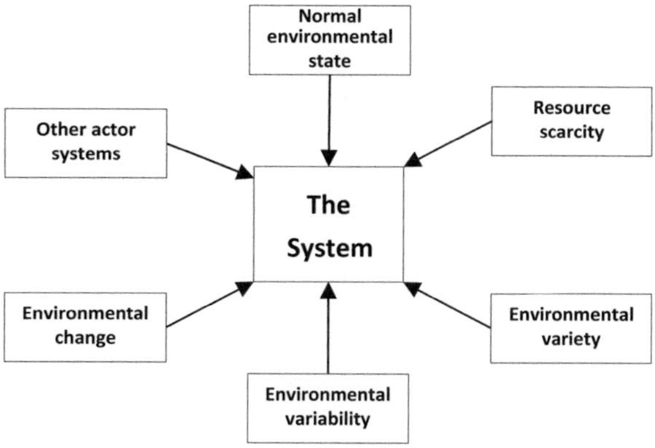

Fig. 4.9 System environmental forces

must, at a minimum, possess discrete processes for coping with the environmental forces.

> Evaluation of the (relative) satisfaction of each of the basic orientors … allows aggregated assessment of system performance and system limitations in a given environment, of coping deficiencies and resulting stress accumulation, of likely behavioral change and structural dynamics. (Bossel, 1992, p. 261)

4.8.1.2 Orientor Hierarchy in Orientation Theory

Orientation theory includes a formal orientor hierarchy to support our goal axiom's principle of purposive behavior—the system's mission, goals, and objectives. Figure 4.10 is a depiction of the orientor hierarchy.

At the highest level of the orientor hierarchy is the *supreme orientor*. Bossel (2007) explains that this is "the overall reason for being and behaving of the system (perhaps: 'viability' or 'sustainability') would be located as the *supreme orientor*" (p. 177). The supreme orientor is decomposed into the dimensions of the *basic orientors* in the next lower level of the orientor hierarchy. "These are the basic operational dimensions which must—in the given overall system and environmental context—enter the decision making process in order to assure satisfaction of the supreme orientor" (Bossel, 2007, p. 177).

4.8.1.3 Basic Orientors in Orientation Theory

The second level in the orientor hierarchy contains the basic orientors. These orientors support the supreme orientor by ensuring that all of the system's environmental forces are accounted for. There are seven environment-determined basic orientors, described in Table 4.4 that directly address the environmental forces in Fig. 4.9.

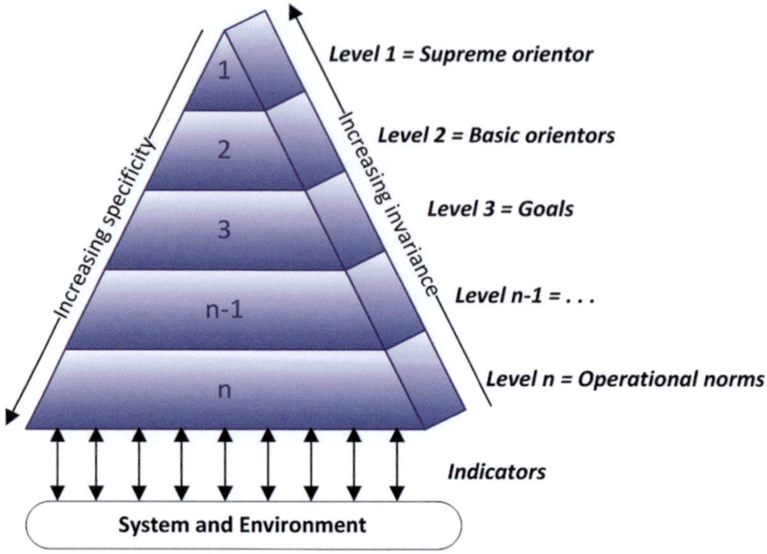

Fig. 4.10 Orientor hierarchy, based upon Fig. 3.6 in (Bossel, 2007, p. 176)

Table 4.4 Environment-determined basic orientors

Basic orientor	Description (Bossel, 2007, p. 185)
1. Existence	The system must be compatible with and able to exist in the normal environmental state. The information, energy, and material inputs necessary to sustain the system must be available
2. Effectiveness	The system should on balance (over the long term) be effective (not necessarily efficient) in its efforts to secure scarce resources (information, matter, energy) from and to exert influence on its environment
3. Freedom of action	The system must have the ability to cope in various ways with the challenges posed by environmental variety
4. Security	The system must be able to protect itself from the detrimental effects of environmental variability, i.e., variable, fluctuating, and unpredictable conditions outside of the normal environmental state
5. Adaptability	The system should be able to learn, adapt, and self-organize in order to generate more appropriate responses to challenges posed by environmental change
6. Coexistence	The system must be able to modify its behavior to account for behavior and interests (orientors) of *other* (actor) *systems* in its environment

Although we not be addressing living systems, in the interest of completeness, there are three additional system-determined basic orientors which are listed in Table 4.5.

Table 4.5 System-determined basic orientors

Basic orientor	Description (Bossel, 2007, pp. 185, 186)
7. Reproduction	Self-producing (autopoietic) systems must be able to fulfill their need to reproduce (either as individuals or as populations)
8. Psychological needs	Sentient beings have certain additional, truly psychological needs that require a minimum of satisfaction, and that cannot be explained by the system/environment interaction alone, such as affection, avoidance of stress or pain, etc.
9. Responsibility	Conscious actors are confronted with having to make choices among options that produce different consequences for themselves and for other affected systems. This requires a normative reference (even if it is only the rule to "flip a coin") amounting to assigning (relative) weights to the "interests" (basic orientors) of affected systems

4.8.1.4 System Types in Orientation Theory

Orientation theory formally addresses eight system types. Each of the eight system types requires some combination of the *basic and system-determined basic orientors.* Table 4.6 describes the eight system types and the basic orientors required for viability.

Orientors numbers in Table 4.6 refer to (1) existence; (2) effectiveness; (3) freedom of action; (4) security; (5) adaptability; (6) coexistence; (7) reproduction; (8) psychological needs; and (9) responsibility.

The types of systems we have and will continue to address in this text are system types 1–6 from Table 4.6. Based upon this condition, the next section will address the application of the six environment-determined basic orientors.

4.8.1.5 Controlling Environmental Forces in Orientation Theory

Armed with the notion of environmental forces and a system's environment-determined basic orientors, we can use a force field diagram to depict the relationship between the restraining environmental forces and the driving orientors. The force field diagram is derived from the work of social psychologist Kurt Lewin [1890–1947]. According to Lewin's theories (Lewin, 1938, 1939, 1943), human behavior is caused by forces such as beliefs, cultural norms, and societal pressure that exist within an individual's life or in society at large. These forces are either driving movement toward a goal (driving forces) or blocking movement toward a goal (restraining forces). Figure 4.11 is a force field diagram that portrays Bossel's environmental forces (i.e., blocking or restraining forces) and the associated driving forces (i.e., basic orientors).

Table 4.6 System qualities and basic orientors (Bossel, 2007, pp. 4, 5)

System type	System description	Orientor number								
		1	2	3	4	5	6	7	8	9
1. Static systems	Inanimate and static	✓								
2. Metabolic system	Require energy, material, or information throughputs for their existence	✓								
3. Self-sustaining systems	Securing necessary resources, protecting itself from adverse influences, and responding selectively ("intelligently") to environmental signals	✓	✓	✓	✓					
4. Self-organizing system	Can change their rigid structure, parameters, or rules, to adapt to changes in and to coevolve with their environment	✓	✓	✓	✓	✓				
5. Nonisolated systems	The existence of other systems in a system's environment will usually force it to modify its behavior in some way	✓	✓	✓	✓	✓	✓			
6. Self-replicating systems	A special form of self-replicating system that can generate systems of their own kind (i.e., autopoietic)	✓	✓	✓	✓	✓	✓	✓		
7. Sentient systems	Systems like animals and human beings that can experience stresses, pain, and emotions that are an important part of their life and development process	✓	✓	✓	✓	✓	✓	✓	✓	
8. Conscious systems	Systems that can reflect about their actions and their impacts (as "actors") and have to make conscious choices among alternatives	✓	✓	✓	✓	✓	✓	✓	✓	✓

4.8.1.6 Systems Viability for the Practitioner

For man-made systems, the six (6) environment-determined basic orientors, (1) existence; (2) effectiveness; (3) freedom of action; (4) security; (5) adaptability; and (6) coexistence may be included as critical design elements during the conceptual design stage when the system's mission, goals, and objectives are developed. An assessment of a systems' orientor satisfaction may be visualized by using a Kiviat diagram (Kolence & Kiviat, 1973), as depicted in Fig. 4.12.

Bossel terms these diagrams *orientor stars*. This method for visualization of viability has been used successfully to portray viability in a number of endeavors (Berardi et al., 2015; Bossel, 1999, 2000).

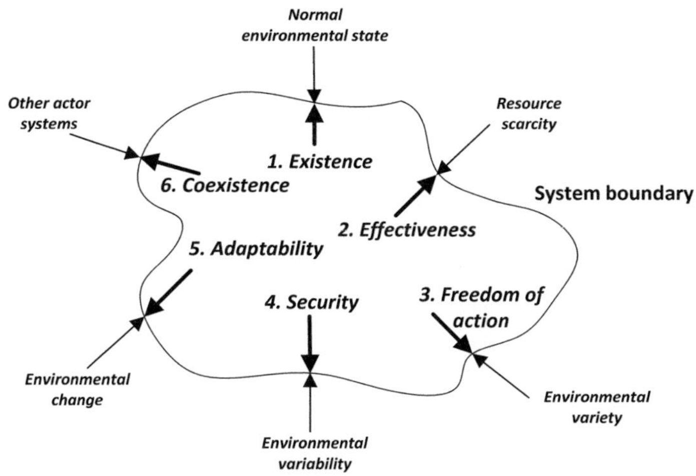

Fig. 4.11 Environmental forces and basic orientors

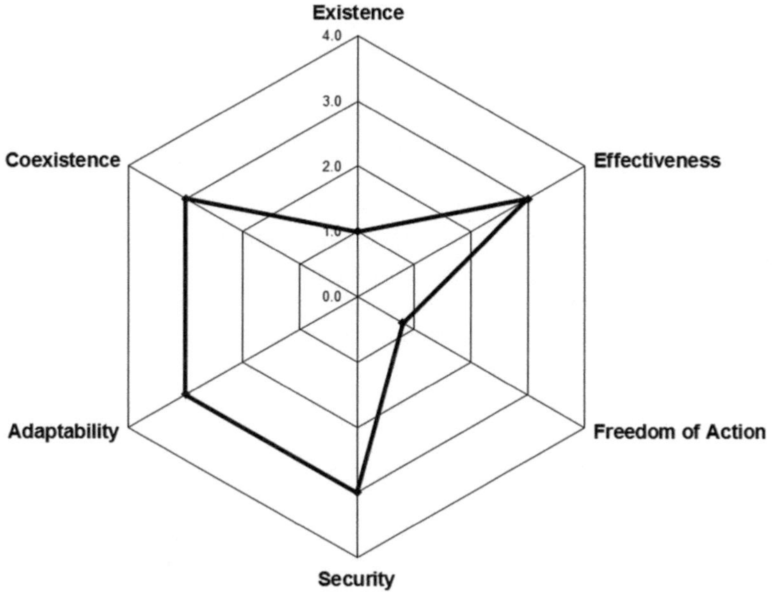

Fig. 4.12 Kiviat diagram or orientor star for visualizing viability

There are a limited number of systems principles that may be applied during a systems design endeavor that ensure system viability. Orientation theory has provided a clear, science-based approach to understanding those forces which directly affect both man-made and living systems. The application of orientation theory and both the physical and behavioral aspects associated with the basic orientors provide assurances that systems design endeavors ensure system survival in changing and potentially hostile environments.

4.8.2 Requisite Variety

Variety is a measure of complexity. Specifically, it is a measure of the number of different system states that may exist. A simple equation for calculating the variety of a system is presented in Eq. 4.1 (Flood & Carson, 1993, p. 26).

$$V = Z^n \tag{4.1}$$

where V = variety or potential number of system states,
Z = number of possible states of each system element,
n = number of system elements.

A simple example shows how the variety measure relates to complexity. Suppose there is a system with six operators working on five different machines where the machines may only have one of two states: on or off. This gives us 30 possible system elements. The formula for variety may be used to calculate the system variety which in this case is 2^{30} or 1,073,741,824. So, for a relatively simple system, the number of states is greater than 1 billion.

Ashby's *law of requisite variety* simply says "variety can destroy variety" (Ashby, 1956, p. 207). There are two methods for controlling variety: (1) properly defining the system boundary and (2) introducing the use of regulators (i.e., variety attenuators). Each method has, as its primary purpose, the reduction of inputs to control the variety and the overall complexity of the system.

For the practitioner, the machine operator example shows that the potential variety rapidly exceeds what is both comprehendible and controllable. Systems practitioners should recognize that variety is a function of the systems inputs and outputs and that in an unbounded or open system, the variety is infinite. Therefore, systems designers must ensure that their designs contain control variety that is greater than or equal to the variety of the element being controlled.

4.8.3 Requisite Hierarchy

In many cases, a regulator of sufficient variety does not exist. In this case, the systems practitioner may apply the *principle of requisite hierarchy*. Requisite hierarchy states that "regulatory ability can be compensated for, up to a certain amount, by a greater hierarchy in organization" (Adams, 2011, p. 142).

For the systems practitioner, this means that in order to supplement the variety in a single regulator, a hierarchy of regulation may need to be constructed.

4.8.3.1 Feedback

Feedback is the central tenet of cybernetics and the foundation for the study of all control mechanisms present in living systems and in man-made systems. Feedback is the basic element that systems use to control their behavior and to compensate for unexpected disturbances.

For the practitioner, the use of feedback, in many forms, is the primary method for maintaining or regulating system parameters. Feedback is an essential element of systems design and in satisfying the objectives and goals of a system.

4.8.4 Circular Causality

The principle of circular causality states:

> An effect becomes a causative factor for future effects, influencing them in a manner particularly subtle, variable, flexible, and of an endless number of possibilities. (Korzybski, 1994, p. 12)

"Circular causality addresses the impact or effects that one system may have on another… The utility of the principle of causality arises from the fact that systems must be treated carefully and that a range of disturbances and events, no matter how seemingly trivial they seem, may directly impact one another" (Adams, 2011, p. 146).

Circular causality refers to a complex of events that reinforce themselves through a series of feedback loops (e.g., causal loops). There are two labels that may be used for these two highly specialized loops:

1. *Virtuous Circles*: "What is a vicious circle for one party, then, is a virtuous circle for another" (Masuch, 1985, pp. 30, 31). A virtuous circle has favorable results and is depicted in Fig. 4.13.
2. *Vicious Circles*: "A deviation amplifying loop (i.e., actions loops) with counterproductive results" (Masuch, 1985, p. 16). A vicious circle has detrimental results and is depicted in Fig. 4.14.

4.8.5 Recursion

The principle of recursion is closely related to the hierarchy principle. "The principle of recursion states that the fundamental laws governing the processes at one

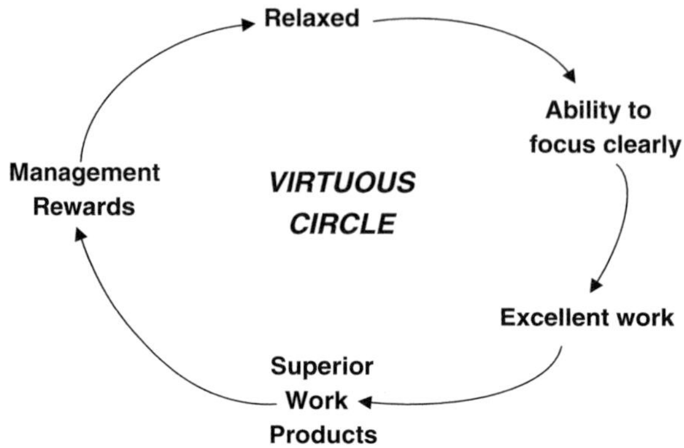

Fig. 4.13 A virtuous circle

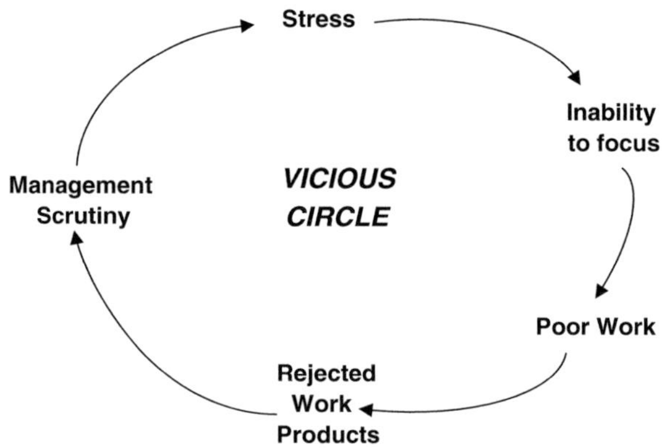

Fig. 4.14 A viscious circle

level are also present at the next higher level. The principle can be expressed by understanding the following" (Adams, 2011, p. 147):

- although level n + 1 is more complex than level n, the fundamental laws present at level n are still present at level n + 1
- when you apply the principle of recursion, you can deduce the fundamental principles of level n + 1 from empirical observations at level n.

For the practitioner, this principle provides help in gaining improved understanding for the presence of properties across the levels of a hierarchy. In software engineering, for example, this principle is termed *inheritance* where "a semantic notion by which the responsibilities (properties and constraints) of a subclass are considered to include the responsibilities of a superclass" (ISO/IEC/IEEE, 2010, p. 175).

4.9 The Design Axiom

The design axiom states:

> System design is a purposeful imbalance of resources and relationships. The design principles provide guidance on how a system is planned, instantiated, and evolved in a purposive manner. (Adams et al., 2014, p. 119)

The design axiom has four principles: (1) requisite parsimony; (2) requisite saliency; (3) minimum critical specification; and (4) power laws.

4.9.1 Requisite Parsimony

The *law of requisite parsimony* is an outcome of a seminal paper by Miller (1956) titled *The Magical Number Seven, Plus or Minus Two: Some Limits on Our Capability for Processing Information.* Miller states that human beings have a limit on the number of items they can process simultaneously and that this number is between five and nine observations. Miller's research showed that the *magical number seven* applies to a number of areas, including (1) span of attention, (2) span of immediate memory, and (3) span of absolute judgment.

For the practitioner, this has practical applications in just about every systems endeavor. As an example, a designer could invoke the *law of requisite parsimony* and ensure that system goals, objectives, concepts, hierarchies, configuration items, design levels, etc., are maintained between five and nine. This is particularly important when judgments are being made. Wherever system hierarchies, levels, partitions, and subdivisions exist, this principle should be invoked.

4.9.2 Requisite Saliency

Early general systems theory proponent Kenneth Boulding [1910–1993] was interested in how designers selected the most important features (i.e., salient features) in a design (1966). After careful research, he proposed the *principle of requisite saliency* which states:

> The situational factors that require consideration in developing a design Target and introducing it in a Design Situation are seldom of equal saliency. Instead there is an underlying logic awaiting discovery in each Design Situation that will reveal the relative saliency of these factors. (Warfield, 1999, p. 34)

Requisite saliency is particularly important during a system's design because techniques such as analysis of alternatives (AoA) and design trade-offs process data into information and information into knowledge and use resulting knowledge to make decisions. Requisite saliency allows the design team to rank the system's design parameters and treat them accordingly.

For the practitioner, all systemic methods and techniques should ensure that analysis, design, and solution processes include a specific provision that reveals relative saliency for all design parameters and associated factors.

4.9.3 Minimum Critical Specification

The principle of minimum critical specification "… has two aspects, negative and positive. The negative simply states that no more should be specified than is absolutely essential; the positive requires that we identify what is essential" (Cherns, 1987, p. 155).

Because engineers invoke the principle of redundancy to ensure both safety and viability, many designs include significant overdesign. By applying this principle, the systems practitioner is bound to ensure designs specify only those elements which are essential.

For the practitioner, there is another highly compelling reason for placing bounds on design activities. This seems to sharply contrast with the principle of redundancy, which requires redundancy for both safety and to absorb shock to the system. However, both principles are important and requisite saliency must be applied. Because of the principle of darkness, where designers never have complete knowledge of a system, many of the benefits planned through specification often become obsolete as the human, social, political, and organizational elements that surround the design become known. Therefore, specifying only what is required, especially in the early design phases, may mitigate the crippling effects caused by the evolving changes in context.

4.9.4 Power Laws

There are a number of observable phenomena that, when their occurrences are plotted, seem to follow some sort of similar pattern. These may be classified as following either one, two, or all three of the following: (1) the Pareto principle; (2) Zipf's law; or (3) power laws.

4.9.4.1 Pareto Principle

The oldest of these three phenomena is attributed to Vilfredo Pareto [1848–1923], an Italian civil engineer and economist, who noticed that there was an inequality in the distribution of incomes in the economies of Italy, England, German states, Paris, and Peru. When he plotted the cumulative distributions of income for these economies on double logarithmic paper (an engineering practice in use at that time), he found that in each case, the result was a straight line with about the same slope (Persky, 1992). In conjunction with this research, his analysis showed that the allocation of wealth among individuals was highly skewed and that a relatively small percentage of the population (20–30%) had amassed far greater wealth (70–80%) than the rest of the population. This has become known as the *Pareto principle* or the *80–20 rule* which says "that in any large complex system 80% of the output will be produced by only 20% of the system. The corollary to this is that 20% of the results absorb 80% of the resources or productive efforts" (Adams, 2011, p. 147). In fact, this is a generalized power law probability distribution.

4.9.4.2 Zipf's Law

The second of these is attributed to George Kingsley Zipf [1902–1950], a Harvard linguist who studied statistical occurrences in language. Zipf's research of word use indicated that the frequency of word usage is inversely proportional to the word's rank in the frequency table. This fact caused him to theorize that the distribution of word use was based solely on the individual's unconscious actions to communicate efficiently, in order to conserve effort. He carried this further and proposed the *principle of least effort* stating:

> In simple terms, the Principle of Least Effort means, for example, that a person in solving his immediate problems will view these against the background of his future problems, as estimated by himself. Moreover, he will strive to solve his problems in such a way as to minimize the total work that he must expend in solving both his immediate problems and his probable future problems. That in turn means that the person will strive to minimize the probable average rate of his work-expenditure (over time). And in so doing he will be minimizing his effort…Least effort, therefore, is a variant of least work. (Zipf, 2012/1949)

This has become known as Zipf's law. In fact, this is another generalized power law probability distribution.

4.9.4.3 Power Laws

This brings us to the general concept of a power law. A power law is characterized by the special relationship between two quantities. Specifically, a relationship where the relative change in a first quantity results in a proportional relative change in the second quantity that varies as a power of the first quantity. This power law relationship is shown in Eq. 4.2 (Bak, 1996, p. 27).

$$N(s) = s^{-\tau} \tag{4.2}$$

The result of this type of special relationship, when plotted on a double logarithmic scale, is a straight line. The generalized logarithmic expression is shown in Eq. 4.3 (Bak, 1996, p. 27).

$$\log N(s) = -\tau \log s \tag{4.3}$$

It turns out that this is very special relationship in the natural world:

> Power laws appear widely in physics, biology, earth and planetary sciences, economics and finance, computer science, demography, and the social sciences. For instance, the distributions of the sizes of cities, earthquakes, solar flares, moon craters, wars and people's personal fortunes all appear to follow power laws. (Newman, 2006, p. 1)

Newman (2006) provides the details for eleven real-world occurrences that follow the power law as well as the associated mathematics.

4.9.4.4 The Practitioner and Power Laws

Practitioners should be alert to the fact that many observations of real-world phenomena may occur according to a pattern other than a normal distribution. In fact, the power law has significance in a number of endeavors and demonstrates that the probability of measuring a particular value of some quantity may vary inversely as a power of that number. This principle may be applied to any number of system problem-solving situations.

4.10 The Information Axiom

The information axiom states:

> Systems create, possess, transfer, and modify information. The information principles provide understanding of how information affects systems. (Adams et al., 2014, p. 119)

The information axiom has five principles: (1) information redundancy; (2) information channel capacity; (3) information entropy; (4) redundancy of potential command; and (5) information inaccessibility.

4.10.1 Information Redundancy

Information redundancy is "the fraction of the structure of the message which is determined not by the free choice of the sender, but rather by the accepted statistical rules governing the use of the symbols in question" number of bits used to transmit a message minus the number of bits of actual information in the message (Shannon & Weaver, 1998/1949, p. 13).

For the practitioner, information redundancy may be viewed from both positive and negative viewpoints. The negative perspective views redundancy as the amount of wasted space used to transmit certain data. The positive perspective may view redundant checksums as a highly desirable method of error detection when communicating over a noisy channel of limited capacity.

4.10.2 Principle of Information Channel Capacity

Claude Shannon [1916–2001], the father of information theory, introduced the concept of a communication channel. In Shannon's model of communications, depicted earlier in Fig. 4.5, a channel was needed to account for the medium and means through which a signal would travel once it left the transmitter. Once he had created the concept of a channel, he addressed the issue of why the received message was not identical to the transmitted message. Shannon reformulated this problem into one of communications reliability, one which he was able to describe using probability theory.

Shannon described the capacity of a channel "in terms of its ability to transmit what is produced out of [the] source of a given information" (Shannon & Weaver, 1998/1949, p. 16). He theorized that channel capacity was a function of (1) transmitter power, (2) bandwidth of the channel, and (3) noise within the channel band. Equation 4.4 shows the Gaussian channel capacity (Shannon, 1948, p. 645).

$$C = W \log_2(1 + \frac{S}{N}) \tag{4.4}$$

where C is the channel capacity in bits per second, W is the channel bandwidth, S is the input power, and N is the noise within the channel band.

This is the *principle of information channel capacity* which states: The maximum capacity of an information channel is a function of its frequency bandwidth W, the average power S used in transmitting, and the noise power N applied to the channel.

For the practitioner, communications channel capacity is an important measure in systems design and analysis endeavors where information capacity is an essential feature in understanding how information flows to and from system elements and the environment. For instance, if the information rate of a transmitting subsystem is greater than the capacity of the selected communications channel, messages from the transmitting subsystem cannot be transmitted over the channel without error.

4.10.3 Principle of Information Entropy

"Full and sole credit is due to Shannon for the introduction of entropy in information theory" (Verdú, 1998, p. 2058). Shannon developed the concept of using entropy in information by stating:

> That information be measured by entropy is, after all, natural when we remember that information, in communication theory, is associated with the amount of freedom of choice we have in constructing a message. (Shannon & Weaver, 1998/1949, p. 13)

From this concept, Shannon was able to relate information entropy meaningfully to the entropy of statistical mechanics developed by Ludwig Boltzmann [1844–1906]. Shannon's equation for information entropy in Eq. 4.5 is very similar to that for statistical entropy.

$$H = -\sum_{i=1}^{n} p_i \log p_i \tag{4.5}$$

where H is entropy and p the probability associated with each of the symbols in each discrete message i.

The *principle of information entropy,* sometimes referred to as *Shannon entropy*, states: That "the entropy of a process is the amount of information in the process" (Gray, 2013, p. xii).

For the practitioner, entropy and channel capacity are important principles in understanding communications flow in complex systems. These principles are particularly important when evaluating the efficacy of communications channel parameters in systems feedback and control. In addition, the principle of information entropy is an essential element of the axiomatic design methodology (Suh, 1998, 2001) which is a systemic method utilized in the design of complex systems.

4.10.4 Redundancy of Potential Command

The studies that produced this principle were associated with the transmission of signals between the brain and the nervous system conducted in the 1950s by Warren McCulloch and his staff at the MIT electronics laboratory. The studies uncovered the importance played by auxiliary information channels during nervous systems transmissions. The researchers found that the auxiliary channel was transmitting, just like the primary channel, so that two signals were being delivered. Neither signal was feedback, but signals based on the primary stimulus. Dual channels transmit redundant information (McCulloch, 1959a).

McCulloch likened this to an actual experience he had during his stint in the US Navy in First World War.

> Every ship of any size or consequence receives information from the others and sweeps the sky for hundreds of miles and water for tens of miles with its own sense organs. In war games and in action, the actual control passes from minute to minute from ship to ship, according to which knot of communication has then the crucial information to commit the fleet to action. This is neither the decentralized command proposed for armies, nor a fixed structure of command of any rigid sort. It is a redundancy of potential command wherein knowledge constitutes authority. (McCulloch, 1959b, p. 226)

The *principle of redundancy of potential command* states that "effective action is achieved by an adequate concatenation of information. In other words, power resides where information resides" (Adams, 2011, p. 151).

For the practitioner, this principle may be utilized during design endeavors to ensure that signals used for feedback are sourced as close to the primary stimulus as possible.

4.10.5 Information Inaccessibility

The information inaccessibility principle is based upon *Finagle's law of information accessibility*, which is less of a scientific law and more of an aphorism, one generally accepted and applied in the public health profession (Badrinath & Yates, 2012; Parrish & McDonnell, 2000). The information inaccessibility principle focuses on data and its processed forms (i.e., information and knowledge—explained in Chap. 9). Finagle's law of information accessibility (Hunt, 1975; Murnaghan, 1974) states:

- The information you have is not what you want.
- The information you want is not what you need.
- The information you need is not what you can obtain.

Further, Opit (1987) adds a fourth law, which states:

- The information you can get costs more than you want to pay.

For the practitioner, this aphorism is a warning to not take data, information, and knowledge for granted when addressing messes and problems and to go the extra mile in ensuring the accuracy, validity, and reliability of data. The information inaccessibility principle should be viewed as an element of improved understanding when dealing with complex systems, their messes, and constituent problems.

4.11 Summary

The significance of this chapter is the extensive integration and transformation of existing information (i.e., scientific propositions from a variety of disciplines) into new knowledge, with specific applicability when addressing problems and messes in complex systems. The improved understanding gained through the use of *systems theory*, its seven axioms, and attendant principles provides systems practitioners with a foundation for systemic thinking.

Systems theory, as described in this chapter, provides the underlying theoretical foundation for understanding systems. Understanding the laws, principles, and concepts that underlie all systems understanding, in conjunction with the thought process developed in systemic thinking, is necessary first step in approaching messes and their constituent problems.

Our concept of systemic thinking is focused on the pragmatic application of the laws, principles, and concepts in the seven axioms of systems theory in order to address complex problems. Application of systems theory will serve to provide the formalism and framework for the inclusion of systems laws, principles, and concepts that can be used in the chapters that follow.

Readers interested in reviewing additional principles of systems theory are encouraged to consult Clemson (1984, pp. 199–257), (Skyttner, 2001, pp. 92–96), (Bechtel & Richardson, 2010; Richardson, 2004a, 2004b, 2005, 2007), Adams (2011), and Whitney, Bradley, Baugh, and Chesterman (2015).

After reading this chapter, the reader should

1. Have a brief notion of the history of the term systems theory;
2. Understand the syntactic definition and axiom construct for systems theory;
3. Be able to articulate how axioms in systems theory are supported by science-based principles;
4. Appreciate that systems principles originate in multiple fields of science; and
5. Be able to articulate how systems principles are invoked by practitioners when addressing messes and constituent problems.

References

Adams, K. M. (2011). Systems principles: Foundation for the SoSE methodology. *International Journal of System of Systems Engineering, 2*(2/3), 120–155. doi:10.1504/IJSSE.2011.040550.

Adams, K. M. (2012). Systems theory: A formal construct for understanding systems. *International Journal of System of Systems Engineering, 3*(3/4), 209–224. doi:10.1504/IJSSE.2012.052684.

Adams, K. M. (2015). *Non-functional requirements in systems analysis and design*. New York: Springer.

Adams, K. M., Hester, P. T., & Bradley, J. M. (2013). A historical perspective of systems theory. In A. Krishnamurthy & W. K. V. Chan (Eds.), *Proceedings of the 2013 Industrial and Systems Engineering Research Conference* (pp. 4102–4109). Norcross, GA: Institute of Industrial Engineers.

Adams, K. M., Hester, P. T., Bradley, J. M., Meyers, T. J., & Keating, C. B. (2014). Systems theory: The foundation for understanding systems. *Systems Engineering, 17*(1), 112–123. doi:10.1002/sys.21255.

Adams, K. M., & Mun, J. H. (2005). The application of systems thinking and systems theory to systems engineering. In *Proceedings of the 26th National ASEM Conference: Organizational Transformation: Opportunities and Challenges* (pp. 493–500). Rolla, MO: American Society for Engineering Management.

Ashby, W. R. (1947). Principles of the self-organizing dynamic system. *Journal of General Psychology, 37*(1), 125–128. doi:10.1080/00221309.1947.9918144.

Ashby, W. R. (1952). *Design for a brain: The origin of adaptive behavior*. London: Chapman & Hall Ltd.

Ashby, W. R. (1956). *An introduction to cybernetics*. London: Chapman & Hall Ltd.

Ashby, W. R. (1962). Principles of the self-organizing system. In H. Von Foerster & G. W. Zopf (Eds.), *Principles of self-organization: Transactions of the University of Illinois symposium* (Vol. 9, pp. 255–278). London: Pergamon Press.

Åström, K. J., & Kumar, P. R. (2014). Control: A perspective. *Automatica, 50*(1), 3–43. doi:10.1016/j.automatica.2013.10.012.

Åström, K. J., & Murray, R. M. (2008). *Feedback systems: An introduction for scientists and engineers*. Princeton, NJ: Princeton University Press.

Badrinath, P., & Yates, J. (2012). The health status of the population. In S. Gillam, J. Yates, & P. Badrinath (Eds.), *Essential public health: Theory and practice* (2nd ed., pp. 77–89). New York: Cambridge University Press.

Bak, P. (1996). *How nature works: The science of self-organized criticality*. New York: Springer Science + Business Media.

Bausch, K. C. (2001). *The emerging consensus in social systems theory*. New York: Springer.

Bausch, K. C. (2002). Roots and branches: A brief, picaresque, personal history of systems theory. *Systems Research and Behavioral Science, 19*(5), 417–428. doi:10.1002/sres.498.

Bechtel, W., & Richardson, R. C. (2010). *Discovering complexity: Decomposition and localization as strategies in scientific research*. Cambridge, MA: MIT Press.

Berardi, A., Mistry, J., Tschirhart, C., Bignante, E., Davis, O., Haynes, L., … & de Ville, G. (2015). Applying the system viability framework for cross-scalar governance of nested social-ecological systems in the Guiana Shield, South America. *Ecology and Society, 20*(3). doi:10.5751/ES-07865-200342.

Bertalanffy, Lv. (1949). General systems theory. *Biologia Generalis, 19*(1), 114–129.

Bertalanffy, Lv. (1950). An outline of general systems theory. *The British Journal for the Philosophy of Science, 1*(2), 134–165.

Bertalanffy, Lv. (1968). *General system theory: Foundations, development, applications* (Revised ed.). New York: George Braziller.

Bohr, N. (1928). The quantum postulate and the recent development of atomic theory. *Nature, 121* (3050), 580–590. doi:10.1038/121580a0.

Bohr, N. (1937). Causality and complementarity. *Philosophy of Science, 4*(3), 289–298.

Bossel, H. (1977). Orientors of nonroutine behavior. In H. Bossel (Ed.), *Concepts and tools of computer-assisted policy analysis* (Vol. 1: Basic concepts, pp. 227–265). Basel: Birkhäuser Verlag.

Bossel, H. (1987). Viability and sustainability: Matching development goals to resource constraints. *Futures, 19*(2), 114–128. doi:10.1016/0016-3287(87)90046-2.

Bossel, H. (1992). Real-structure process description as the basis of understanding ecosystems and their development. *Ecological Modelling, 63*(1–4), 261–276. doi:10.1016/0304-3800(92)90072-M.

Bossel, H. (1996). Deriving indicators of sustainable development. *Environmental Modeling and Assessment, 1*(4), 193–218. doi:10.1007/bf01872150.

Bossel, H. (1998). Ecological orientors: Emergence of basic orientors in evolutionary self-organization. In F. Müller & M. Leupelt (Eds.), *Eco targets, goal functions, and orientors* (pp. 19–33). Berlin, Heidelberg: Springer.

Bossel, H. (1999). *Indicators for sustainable development—Theory, method, applications. A report to the Balaton Group*. Winnipeg, Manitoba, Canada: International Institute for Sustainable Development.

Bossel, H. (2000). Policy assessment and simulation of actor orientation for sustainable development. *Ecological Economics, 35*(3), 337–355. doi:10.1016/S0921-8009(00)00218-4.

Bossel, H. (2001). Assessing viability and sustainability: A systems-based approach for deriving comprehensive indicator sets. *Conservation Ecology, 5*(2), 12.

Bossel, H. (2007). *Systems and models: Complexity, dynamics, evolution, sustainability*. Norderstedt, GE: Books on Demand.

Boulding, K. E. (1956). General systems theory—The skeleton of science. *Management Science, 2* (3), 197–208.

Boulding, K. E. (1966). *The impact of social sciences*. New Brunswick, NJ: Rutgers University Press.
Bowen, M. (1966). The use of family theory in clinical practice. *Comprehensive Psychiatry, 7*(5), 345–374. doi:10.1016/S0010-440X(66)80065-2.
Buckley, W. (1967). *Sociology and modern systems theory*. Englewood Cliffs, NJ: Prentice-Hall.
Buckley, W. (1998). *Society—A complex adaptive system: Essays in social theory*. Amsterdam: Overseas Publishers Association.
Bunge, M. (1979). A systems concept of society: Beyond individualism and holism. *Theory and Decision, 10*(1–4), 13–30. doi:10.1007/BF00126329.
Bunge, M. (1997). Mechanism and explanation. *Philosophy of the Social Sciences, 27*(4), 410–465. doi:10.1177/004839319702700402.
Bunge, M. (1999). *The sociology-philosophy connection*. New Brunswick, NJ: Transaction Publishers.
Bunge, M. (2004). How does it work?: The search for explanatory mechanisms. *Philosophy of the Social Sciences, 34*(2), 182–210. doi:10.1177/0048393103262550.
Cannon, W. B. (1929). Organization for physiological homeostasis. *Physiological Reviews, 9*(3), 399–431.
Cannon, W. B. (1967/1932). *The wisdom of the body* (Revised ed.). New York: W.W. Norton & Company.
Checkland, P. B. (1999). *Systems thinking, systems practice*. New York: Wiley.
Chen, L., & Aihara, K. (2002). A model of periodic oscillation for genetic regulatory systems. *IEEE Transactions on Circuits and Systems I: Fundamental Theory and Applications, 49*(10), 1429–1436. doi:10.1109/TCSI.2002.803354.
Cherns, A. (1976). The principles of sociotechnical design. *Human Relations, 29*(8), 783–792. doi:10.1177/001872677602900806.
Cherns, A. (1987). The principles of sociotechnical design revisited. *Human Relations, 40*(3), 153–161. doi:10.1177/001872678704000303.
Churchman, C. W., & Ackoff, R. L. (1950). Purposive behavior and cybernetics. *Social Forces, 29* (1), 32–39.
Cilliers, P. (1998). *Complexity and postmodernism: Understanding complex systems*. New York: Routledge.
Clemson, B. (1984). *Cybernetics: A new management tool*. Kent, UK: Abacus Press.
Fertig, J. A., & Zapata, R. N. (1978). A mathematical foundation for system synthesis. In G. J. Klir (Ed.), *Applied general systems research: Recent developments and trends* (pp. 209–224). Boston, MA: Springer US.
Flagle, C. D., Huggins, W. H., & Roy, R. H. (1960). *Operations research and systems engineering*. Baltimore: The Johns Hopkins Press.
Flood, R. L., & Carson, E. R. (1993). *Dealing with complexity: An introduction to the theory and application of systems science* (2nd ed.). New York: Plenum Press.
Ford, T. C., Colombi, J. M., Jacques, D. R., & Graham, S. R. (2009). On the application of classification concepts to systems engineering design and evaluation. *Systems Engineering, 12* (2), 141–154. doi:10.1002/sys.20114.
Forrester, J. W. (1961). *Industrial dynamics*. Cambridge, MA: MIT Press.
Forrester, J. W. (1969). *Urban dynamics*. Cambridge, MA: MIT Press.
Forrester, J. W. (1971). *World dynamics*. Cambridge, MA: MIT Press.
Friedman, G. J. (2005). *Constraint theory: Multidimensional mathematical model management*. New York: Springer.
Friedman, G. J. (2006). On the unification of systems engineering. *INSIGHT, 8*(2), 16–17. doi:10.1002/inst.20068216.
Friedman, G. J., & Leondes, C. T. (1969a). Constraint theory, Part I: Fundamentals. *IEEE Transactions on Systems Science and Cybernetics, 5*(1), 48–56. doi:10.1109/TSSC.1969.300244.

Friedman, G. J., & Leondes, C. T. (1969b). Constraint theory, Part II: Model graphs and regular relations. *IEEE Transactions on Systems Science and Cybernetics, 5*(2), 132–140. doi:10.1109/TSSC.1969.300204.

Friedman, G. J., & Leondes, C. T. (1969c). Constraint theory, Part III: Inequality and discrete relations. *IEEE Transactions on Systems Science and Cybernetics, 5*(3), 191–199. doi:10.1109/TSSC.1969.300260.

Gray, R. M. (2013). *Entropy and information theory* (Corrected ed.). New York: Springer-Verlag.

Hammond, D. (2002). Exploring the genealogy of systems thinking. *Systems Research and Behavioral Science, 19*(5), 429–439. doi:10.1002/sres.499.

Hitch, C. J. (1952). *Suboptimization in operations problems*. Retrieved from Santa Monica.

Hitch, C. J. (1953). Sub-optimization in operations problems. *Journal of the Operations Research Society of America, 1*(3), 87–99.

Hunt, P. (1975). Funds position: Keystone in financial planning. *Harvard Business Review, 53*(3), 106–115.

ISO/IEC/IEEE. (2010). *ISO/IEC/IEEE Standard 24765: Systems and software engineering—Vocabulary*. Geneva and New York: International Organization for Standardization and the International Electrotechnical Commission and the Institute of Electrical and Electronics Engineers.

Jackson, M. C. (2006). Creative holism: A critical systems approach to complex problem situations. *Systems Research and Behavioral Science, 23*(5), 647–657. doi:10.1002/sres.799.

Kauffman, S. A. (1990). The sciences of complexity and "origins of order". In *PSA: Proceedings of the Biennial Meeting of the Philosophy of Science Association* (Vol. Volume Two: Symposia and Invited Papers, pp. 299–322). Chicago: The University of Chicago Press.

Kauffman, S. A. (1993). *The origins of order: Self-organization and selection in evolution*. New York: Oxford University Press.

Kauffman, S. A. (1995). *At home in the universe: The search for laws of self-organization and complexity*. New York: Oxford University Press.

Klir, G. J. (1968). *An approach to general systems theory*. Princeton, NJ: Nostrand.

Kolence, K. W., & Kiviat, P. J. (1973). Software unit profiles & Kiviat figures. *ACM SIGMETRICS Performance Evaluation Review, 2*(3), 2–12.

Korzybski, A. (1994). *Science and sanity: An introduction to non-aristotelian systems and general semantics* (5th ed.). Englewood, NJ: Institute of General Semantics.

Laszlo, E. (1972). *Introduction to systems philosophy: Toward a new paradigm of contemporary thought*. New York: Harper Torchbooks.

Laszlo, E. (1973). The rise of general theories in contemporary science. *Journal for General Philosophy of Science, 4*(2), 335–344. doi:10.2307/25170288.

Laszlo, E. (1996). *The systems view of the world: A holistic vision for our time*. Creskill, NJ: Hampton Press.

Lewin, K. (1938). *The conceptual representation and measure of psychological forces*. Durham, NC: Duke University Press.

Lewin, K. (1939). Field theory and experiment in social psychology: Concepts and methods. *American Journal of Sociology, 44*(6), 868–896.

Lewin, K. (1943). Defining the 'field' at a given time. *Psychological Review, 50*(3), 292–310.

Lorenz, E. N. (2005). Designing chaotic models. *Journal of the Atmospheric Sciences, 62*(5), 1574–1587. doi:10.1175/jas3430.1.

Luhmann, N. (1995). *Social systems* (J. Bednarz, Trans. D. Baecker ed.). Stanford, CA: Stanford University Press.

Luhmann, N. (2012). *Theory of society* (R. Barrett, Trans. Vol. 1). Stanford, CA: Stanford University Press.

Masuch, M. (1985). Vicious circles in organizations. *Administrative Science Quarterly, 30*(1), 14–33.

McCulloch, W. S. (1959a). *Embodiments of mind*. Cambridge, MA: MIT Press.

McCulloch, W. S. (1959b). Where fancy is bred. In W. S. McCulloch (Ed.), *Embodiments of mind* (pp. 216–229). Cambridge, MA: MIT Press.

Mesarovic, M. D. (1964). *Views on general systems theory*. New York: Wiley.
Mesarovic, M. D. (1967). General systems theory and its mathematical foundation. In *Record of the IEEE Systems Science and Cybernetics Conference* (Vol. 3, pp. 1–22). New York: Institute of Electrical and Electronics Engineers.
Miller, G. A. (1956). The magical number seven, plus or minus two: Some limits on our capability for processing information. *Psychological Review, 63*(2), 81–97.
Miller, J. G. (1978). *Living systems*. New York: McGraw Hill.
Murnaghan, J. H. (1974). Health-services information systems in the United States today. *New England Journal of Medicine, 290*(11), 603–610.
Newman, M. E. J. (2006). Power laws, Pareto distributions and Zipf's law. *Contemporary Physics, 46*(323), 1–28. doi:10.1080/00107510500052444.
Nicolis, G., & Prigogine, I. (1977). *Self-organization in nonequilibrium systems: From dissipative structures to order through fluctuations*. New York: Wiley.
OECD. (2007). *Revised field of science and technology (FOS) classification in the Frascati Manual*. Retrieved from Paris.
Opit, L. J. (1987). How should information on health care be generated and used? *World Health Forum, 8*(4), 409–438.
Pahl, G., Beitz, W., Feldhusen, J., & Grote, K.-H. (2011). *Engineering design: A systematic approach* (K. Wallace & L. T. M. Blessing, Trans. 3 ed.). Darmstadt: Springer.
Parker, S. (Ed.). (1994). *McGraw-Hill dictionary of engineering*. New York: McGraw-Hill.
Parrish, R. G., & McDonnell, S. M. (2000). Sources of health related information. In S. M. Teutsch & R. E. Churchill (Eds.), *Principles and practice of public health surveillance* (2nd ed., pp. 30–75). New York: Oxford University Press.
Parsons, T. (1970). On building social system theory: A personal history. *Daedalus, 99*(4), 826–881. doi:10.2307/20023975.
Parsons, T. (1979). Concrete systems and "abstracted systems". *Contemporary Sociology, 8*(5), 696–705.
Parsons, T. (1991). *The social system* (New ed.). London: Routledge.
Persky, J. (1992). Retrospectives: Pareto's law. *The Journal of Economic Perspectives, 6*(2), 181–192.
Plas, J. M. (1986). *Systems psychology in the schools*. Elmsford, NY: Pergamon Press.
Reynolds, W. C., & Perkins, H. C. (1977). *Engineering thermodynamics* (2nd ed.). New York: McGraw-Hill.
Richardson, K. A. (2004a). Systems theory and complexity: Part 1. *Emergence: Complexity & Organization, 6*(3), 75–79. doi:10.17357.7ec106df2487e832480906e35d688f00.
Richardson, K. A. (2004b). Systems theory and complexity: Part 2. *Emergence: Complexity & Organization, 6*(4), 77–82.
Richardson, K. A. (2005). Systems theory and complexity: Part 3. *Emergence: Complexity & Organization, 7*(2), 104–114.
Richardson, K. A. (2007). Systems theory and complexity: Part 4. The evolution of systems thinking. *Emergence: Complexity & Organization, 9*(1/2), 166.
Rosenblueth, A., Wiener, N., & Bigelow, J. (1943). Behavior, purpose and teleology. *Philosophy of Science, 10*(1), 18–24.
Schlesinger, L. A., & Heskett, J. L. (1991). Breaking the cycle of failure in services. *Sloan Management Review, 32*(3), 17–28.
Shannon, C. E. (1948). A mathematical theory of communication, Part 2. *Bell System Technical Journal, 27*(4), 623–656.
Shannon, C. E., & Weaver, W. (1998/1949). *The mathematical theory of communication*. Champaign, IL: University of Illinois Press.
Shell, T. (2001). System function implementation and behavioral modeling: A systems theoretic approach. *Systems Engineering, 4*(1), 58–75. doi:10.1002/1520-6858(2001)4:1<58:AID-SYS6>3.0.CO;2-Z.
Sherwin, J. (2010). An approach towards holism in science and engineering. *Journal of Systems Science and Systems Engineering, 19*(3), 285–305. doi:10.1007/s11518-010-5141-y.

Simon, H. A. (1956). Rational choice and the structure of the environment. *Psychological Review, 63*(2), 129–138. doi:10.1037/h0042769.

Skyttner, L. (2001). *General systems theory: Ideas and applications*. Singapore: World Scientific.

Smuts, J. C. (1961 (1926)). *Holism and Evolution*. New York: Viking Press.

Sterman, J. D. (2000). *Business dynamics: Systems thinking and modeling for a complex world*. New York: McGraw-Hill.

Sterman, J. D. (2001). System dynamics modeling: Tools for learning in a complex world. *California Management Review, 43*(4), 8–25. doi:10.2307/41166098.

Suh, N. P. (1998). Axiomatic design theory for systems. *Research in Engineering Design, 10*(4), 189–209. doi:10.1007/s001639870001.

Suh, N. P. (2001). *Axiomatic design: Advances and applications*. New York: Oxford University Press.

Thompson, K. R. (2006). "General system" defined for predictive technologies of A-GSBT (Axiomatic-general systems behavioral theory). *Scientific Inquiry, 7*(1), 1–11.

Vajna, S., Clement, S., Jordan, A., & Bercsey, T. (2005). The autogenetic design theory: An evolutionary view of the design process. *Journal of Engineering Design, 16*(4), 423–440. doi:10.1080/09544820500267781.

van Gigch, J. P. (1974). *Applied general systems theory* (2nd ed.). New York: Harper & Row.

Verdú, S. (1998). Fifty years of Shannon theory. *IEEE Transactions on Information Theory, 44*(6), 2057–2078.

Waddington, C. H. (1957). *The strategy of genes: A discussion of some aspects of theoretical biology*. London: George Allen & Unwin.

Waddington, C. H. (1968). Towards a theoretical biology. *Nature, 218*(5141), 525–527.

Warfield, J. N. (1999). Twenty laws of complexity: Science applicable in organizations. *Systems Research and Behavioral Science, 16*(1), 3–40.

Weinberg, G. M. (2001). *An introduction to general systems thinking* (Silver Anniversary ed.). New York: Dorset House.

Whitney, K., Bradley, J. M., Baugh, D. E., & Chesterman, C. W. (2015). Systems theory as a foundation for governance of complex systems. *International Journal of System of Systems Engineering, 6*(1–2), 15–32. doi:10.1504/IJSSE.2015.068805.

Wiener, N. (1965). *Cybernetics: Or control and communication in the animal and the machine* (2nd ed.). Cambridge, MA: MIT Press.

Willander, M., Mamontov, E., & Chiragwandi, Z. (2004). Modelling living fluids with the subdivision into the components in terms of probability distributions. *Mathematical Models & Methods in Applied Sciences, 14*(10), 1495–1520.

Wolff, K. E. (2000). Towards a conceptual system theory. In B. Sanchez, N. Ada, A. Rashid, T. Arndt, & N. Sanchez (Eds.), *Proceedings of the World Conference on Systemics, Cybernetics, and Informatics* (Vol. II, pp. 124–132). Winter Garden, FL: International Institute of Informatics and Systemics.

Wymore, A. W. (1967). *A mathematical theory of systems engineering—The elements*. New York: Wiley.

Wymore, A. W. (1993). *Model-based systems engineering*. Boca Raton, FL: CRC Press.

Yates, F. E., & Iberall, A. S. (1982). A skeleton of physical ideas for the dynamics of complex systems. *Mathematics and Computers in Simulation, 24*(6), 430–436. doi:http://dx.doi.org/10.1016/0378-4754(82)90639-5.

Zipf, G. K. (2012/1949). *Human behavior and the principle of least effort: An introduction to human ecology*. Eastford, CT: Martino Fine Books.

Chapter 5
Complex Systems Modeling

Abstract Modeling is a necessary mechanism for understanding complex phenomena such as the messes this book is designed to help with. This chapter compares methods available for complex systems modeling. A method is then recommended for use in addressing messes. A framework for the development and use of such a model and an accompanying simulation is then presented. This framework is demonstrated on an example problem, with an eye toward using this approach to first think about, then act on, and finally observe our mess systemically.

5.1 Introduction

We use models to gain understanding about complex phenomena; indeed, modeling is a "purposeful abstraction of reality" (Hester & Tolk, 2010, p. 18). Maria (1997) offers, "a model should be a close approximation to the real system and incorporate most of its salient features. On the other hand, it should not be so complex that it is impossible to understand and experiment with it. A good model is a judicious tradeoff between realism and simplicity" (p. 7). It is a necessary simplification of the real-world system it models.

Figure 5.1 illustrates the methods available for modeling a system. If the option is available and it is feasible (i.e., it is not too dangerous, timely, or costly), we would prefer to experiment with the actual system to improve our understanding. Given the messes this book is intended to address, this is not realistic. These systems are too complex and unwieldy for full-scale experimentation to be undertaken (i.e., imagine experimenting with a nuclear missile attack or catastrophic flood in order to test potential mitigation strategies). For similar scale-driven reasons, a physical model is unobtainable for experimentation purposes. The underlying complexity and divergent perspectives associated with the associated systems make closed-form analytical solutions problematic as well. This leaves us with a simulation in order to gain understanding about our mess.

Mechanics regarding the simulation of a real-world system are not trivial. The decision to create a simulation carries with it the burden of choosing an appropriate

P.T. Hester and K.M. Adams, *Systemic Decision Making*, Topics in Safety, Risk, Reliability and Quality 33, DOI 10.1007/978-3-319-54672-8_5

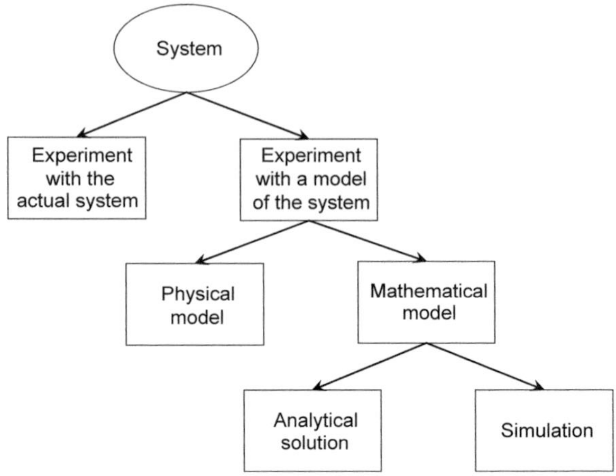

Fig. 5.1 Modeling methods (adapted from Law & Kelton, 2000)

mathematical framework on which to build it. This chapter compares the methods available for complex systems modeling, it outlines the choice of a method considering mess characteristics discussed in previous chapters, and it presents a framework for developing such a model and an accompanying simulation. This framework is then demonstrated on an example problem, with an eye toward using this approach to first think about, then act on, and finally observe our mess systemically.

5.2 The Role of Modeling

While it may be natural to think of first observing the world before we do anything else, the reality is that all of our observations are biased (a topic we will return to in Chap. 15) and we cannot conduct true observation without first thinking about the world (i.e., developing a model). "The first step in the scientific process is not observation but the generation of a hypothesis which may then be tested critically by observations and experiments" (Banerjee, Chitnis, Jadhav, Bhawalkar, & Chaudhury, 2009, p. 127). This stresses the importance of a theory (and accompanying model) before observation. Thus, we must think *before* we observe (as outlined in our model of systemic decision making discussed in Chap. 3). We create a hypothesis. The reason for this lies in the very nature of scientific inquiry; the goal of a scientist is not to prove a hypothesis (or a model) correct, but rather to falsify it. Even in circumstances in which we have not disproven a hypothesis, we do not say it has been proven; rather, we say it has not yet been disproven. This notion is the essence of modeling and one we will return to many times throughout this text.

As the goal of this text is to help the reader make better decisions in a complex environment, it helps us to understand the role of modeling in the systemic decision making process and the overarching purpose of modeling. When we speak of the

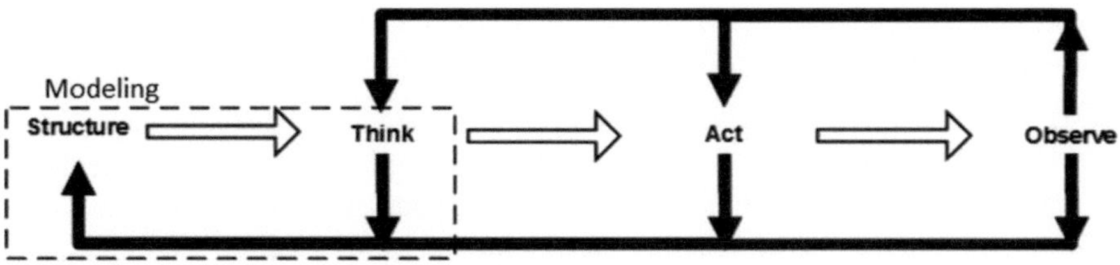

Fig. 5.2 Systemic decision making with modeling

process of modeling as a purposeful abstraction of reality, we are specifically referring to the process that incorporates the structuring and thinking phases of systemic decision making. Action and observation with the real system are beyond the scope of modeling as they extend beyond the conceptual world and into the real world. This notion is captured in Fig. 5.2.

Building a model and experimenting with that model help us to understand the behavior of the real-world phenomena being modeled. We model the structure of the system (i.e., its relationships) and then observe its behavior over time (i.e., we simulate it). Ultimately, we are trying to understand the response of a given system to a proposed stimulus. Complex problems are more difficult to predict and require a more concerted modeling effort due to their emergent behaviors. So, which method is appropriate for modeling a complex problem? This is the question we now turn our attention to.

5.3 Method Comparison

Simulations (a time-evolving realization of a model) first require an underlying model to evaluate. Thus, it is necessary to review available methods and select an appropriate modeling paradigm before developing a simulation. It is desirable to have a method that can help with both problem structuring and problem assessment to span the thinking, acting, and observing phases of mess understanding. We can draw from available methods in both the problem structuring and modeling and simulation literature. Ackermann (2012) and Mingers (2011) identify the most prominent problem structuring methods as soft systems methodology (SSM) (Checkland, 1999; Checkland & Scholes, 1999), strategic options development and analysis (SODA) (Eden & Ackermann, 1998), and strategic choice approach (SCA) (Friend & Hickling, 1987; Friend & Jessop, 1977). The underlying models for each of these methodologies are described by Mingers (2011) as rich pictures (for SSM), cognitive mapping (for SODA), and soft decision analysis (for SCA). Hester and Tolk (2010) identify the most well-known general modeling paradigms as system dynamics (Forrester, 1961; Sterman, 2000), discrete event simulation (Zeigler, Praehofer, & Kim, 2000), and agent-based simulation (Yilmaz & Ören, 2009). These six methods are contrasted with three questions of interest for a comprehensive modeling technique that is applicable to both problem structuring and assessment:

(1) *Does it provide a visual representation of the scenario?* Visualization is necessary in complex scenarios so that stakeholders can view a holistic articulation of a scenario and communicate across disciplinary boundaries. As the saying goes, a picture is worth 1,000 words. Complex scenarios are better understood when accompanied with graphics.
(2) *Does it support simulation?* Simulation is necessary to account for emergent behavior. The inability of humans to predict emergent behavior requires a mechanism such as simulation to explore a myriad of potential scenarios in order to understand what potential outcomes may occur in a given mess.
(3) *Does it support qualitative assessment?* Complex problems have both qualitative and quantitative elements. In many cases, qualitative, or soft, elements dominate problem structuring and assessment. Thus, any method chosen for modeling complex problems must be able to account for qualitative, softer elements of representation and not be strict in its mathematical requirements.

A comparison of the potential methods across these criteria is shown in Table 5.1.

Arguably, the most well-known modeling approach of those listed in Table 5.1 is system dynamics. System dynamics, however, requires significant empirical data, typically unavailable in messes for all but a few of the relevant entities of interest. Thus, system dynamics may be useful, but "since numerical data may be uncertain or hard to come by, and the formulation of a mathematical model may be difficult, costly or even impossible, then efforts to introduce knowledge on these systems should rely on natural language arguments in the absence of formal models" (Carvalho & Tome, 2000, p. 407). The same criticism can be levied on agent-based simulation. Both techniques are useful for visually representing scenarios, as well as for simulating those scenarios, but they are too rigorous in their mathematical requirements to be of use for systems age messes. Discrete event simulation

Table 5.1 Comparison of modeling paradigms/techniques

Paradigm/technique	Visual representation of scenario?	Supports simulation?	Supports qualitative assessment?
Rich picture	Yes	No	Yes
Cognitive mapping	Yes	Yes	Yes*
Soft decision analysis	Yes	No	Yes
System dynamics	Yes	Yes	No
Discrete event modeling	No	Yes	No
Agent-based simulation	Yes	Yes	No

*Denotes fuzzy cognitive mapping

supports a simulation environment, but it does not provide a useful visual representation of a scenario, and it is also too rigorous in its mathematical specification requirements. Rich pictures and soft decision analysis are useful for representing scenarios graphically, as well as incorporating qualitative assessment of stakeholders, but they lack in their ability to support simulation. Simulation is a necessary element of any complex problem assessment, and its absence prevents *what-if* scenarios from being explored. This leaves cognitive mapping as the remaining method that meets all of the specified requirements. Regarding cognitive mapping, Heyer (2004) offers the following:

> Cognitive mapping, a form of influence diagram, is a technique that has been used by a variety of researchers in a variety of settings. Cognitive maps provide a holistic picture of an individual's overall perspective, without the loss of any detail; enabling researchers to move beyond the assumption of internal consistency to the detailed assessment of specific concepts within the map. For OR, this means gaining a better understanding of the clients perception of a problem which is vital for a successful OR study. In cognitive mapping, self-defined constructs represent the causal knowledge of a decision maker in the form of a map of their own subjective world. Cognitive maps can be seen as a model of action-orientated thinking about a situation where arrows signify influences in a line of argument linking cause and effect (Eden, 1992). Cognitive maps can be analysed through interpretative coding (where individual concepts are interpreted); in terms of their content (the meanings they contain); and in terms of the complexity of configuration of the maps (for example, link to node ratio, cluster analyses). (p. 9)

Cognitive mapping's entry in Table 5.1 as it concerns the *qualitative assessment* criteria, is marked with an asterisk, however, as fuzzy cognitive mapping, a special variant of cognitive mapping (FCM), is required to fully support qualitative assessment. Bueno and Salmeron (2009) discuss the distinction between the two:

> Cognitive maps possess, as their main limitation, the impossibility of quantifying relationships among variables. With the purpose of offering a solution to this weakness and enhancing cognitive maps, fuzzy numbers have been conjugated with cognitive maps… FCM substitute the signs (+) and (−) for a fuzzy value between −1 and 1. The zero value indicates the absence of weight. (p. 5222)

Further, in direct comparison with other methods (including decision analysis and system dynamics), Özesmi and Özesmi (2004), in Table 5.2, offer an assessment of FCM.

Table 5.2 FCM compared to other methods

Method	Advantages	Disadvantages	FCM comparison
Multiattribute decision theory	Useful for ranking a finite number of alternatives with conflicting criteria; can aggregate qualitative and quantitative data	Does not allow for feedback loops; alternatives must be prespecified	FCM can suggest alternatives through exploratory analysis
System dynamics	Use differential or difference equations; dynamic models	Require significant empirical data	FCMs are not dynamic models, but they are useful for data-poor situations

"During the past decade, FCMs played a vital role in the applications of diverse scientific areas, such as social and political sciences, engineering, information technology, robotics, expert systems, medicine, education, prediction, environment, and so on" (Papageorgiou & Salmeron, 2013, p. 67). Özesmi and Özesmi (2004, pp. 46–47) discuss the choice of FCM in the context of modeling preferences:

> Why choose FCM over other modeling methods? To answer this question, we must consider the issues of model complexity and the reason for the model. Obviously it is important to have a model that is complex enough for the problem to be solved; however data poor situations limit model complexity. Data is costly and often not available, especially in developing countries, where conservation efforts and management are important but not resolved. The…approach…is not obtained from empirical data but can be used for modeling perception and therefore social ideas of how systems work. This is essential…where the support of many stakeholders is necessary. It is also useful for extension activities to educate stakeholders, if there are any misperceptions.
>
> The main advantage of the multi-step FCM approach is that it is easy to build and gives qualitative results. It does not require expert knowledge in every field but can be constructed based on simple observations by anybody…It does not make quantitative predictions but rather shows what will happen to the system in simulations under given conditions of relationships. The model provides a better summary of relationships between variables instead of articulating how that relationship is in detail.
>
> With FCMs the strengths and signs of relationships can be easily changed and simulations run easily and quickly. Thus, they are ideal tools for theory development, hypothesis formation, and data evaluation. However, FCMs are not substitutes for statistical techniques; they do not provide real-value parameter estimations or inferential statistical tests.

Jetter (2006, p. 511) further discusses the appropriate use of FCMs:

> Adoption of FCMs can furthermore be improved through a better choice of applications: In the past, FCMs have been used for all kinds of problems and in some cases, the reason for choosing FCMs over other modeling techniques (e.g. System Dynamics or Bayesian networks) is all but clear. Future FCM research should focus on problems that FCMs are "good at": they are a powerful means to represent knowledge domains that are characterized by high complexity, by widespread knowledge sources that usually only have partial knowledge, by qualitative information that frequently changes, and by a lack of a commonly accepted "theory" or "truth". They can thus be useful for the analysis of business ecosystems, scenario planning, and the forecasting of market or technology trends and should be increasing applied in these areas.
>
> Like many other models, e.g. System Dynamics models, they can help decision-makers to reflect upon their worldviews and to improve their understanding of the dynamic systems and decision alternative they encounter. Unlike these models, they can handle qualitative concepts with no dimensions and linguistic imprecision and so (relatively) simple to understand that they allow for a strong involvement of the decision-maker in modeling, simulation and interpretation of results.

Amer, Jetter, and Daim (2011) have additional comments about the utility of fuzzy cognitive mapping:

> Cognitive maps are mainly used to analyze and aid the decision-making process by investigating causal links among relevant concepts…The mapping process fosters system thinking and allows experts to better assess their own mental models…The visual nature of

> concept maps facilitates understanding of existing dependencies and contingencies between various concepts. (p. 567)

Additionally, FCMs are scalable, in terms of the maps themselves and the number of participants. "With FCMs you can have as many knowledge sources as wanted with diverse knowledge and different degrees of expertise. These knowledge sources can all be easily combined into one FCM. There is no restriction on the number of experts or on the number of concepts" (Özesmi & Özesmi, 2004, p. 45).

Given its advantages over alternative methods, fuzzy cognitive mapping is advised as a method for modeling and accompanying simulation for messes and their constituent problems. This method is used as part of a larger multimethodology which will incorporate numerous other techniques for populating a cognitive map, discussed in subsequent chapters. We now turn to details regarding the use of fuzzy cognitive maps.

5.4 Fuzzy Cognitive Mapping

Fuzzy cognitive mapping was introduced by Kosko (1986), based on the foundational causal map work of Axelrod (1976) as a way to visually and logically capture the relationships between elements in a problem. Fuzzy cognitive maps (FCMs) are network-based collections of concepts (represented as nodes) and causal relationships (represented as arcs between the concepts). Arcs have weights that indicate both the strength and direction of a causal relationship; thus, a given relationship can be increasing or decreasing (i.e., A increases B or A decreases B). Arc weights are typically defined on [−1,1] to represent direction (positive weights are reinforcing; negative are decreasing) and magnitude of influence (a weight of one means complete influence of one concept over another, whereas a weight of zero indicates no connection between two concepts).

Concepts can represent variables in a system (see, e.g., Tsadiras, 2008, pp. 3881–3882), with concept values defined on [0,1] as their value relative to the defined range for the given variable (i.e., 50% for a valve status means 50% open). Concepts can also represent system performance measures (see, e.g., Tsadiras, 2008, pp. 3882–3883), where causal relationships show the effect of increasing or decreasing a given performance measure on others. Carvalho (2013) elaborates on the meaning of a concept:

> Each concept represents the actors, entities and social, political, economic or abstract concepts that compose the system. Examples of concepts might be Inflation, the actions of an influent Politic, a Revolution, the Wealth of an individual or a nation, the Welfare of population, Road conditions, etc. Each concept is characterized by a value usually ranging from [0…1] or [−1…1] representing a normalized transformation from its real world value. (p. 8)

Mathematically, the concepts within a FCM can be represented by a matrix ***C***, the relative activation level of each concept can be represented by the matrix ***A***, and

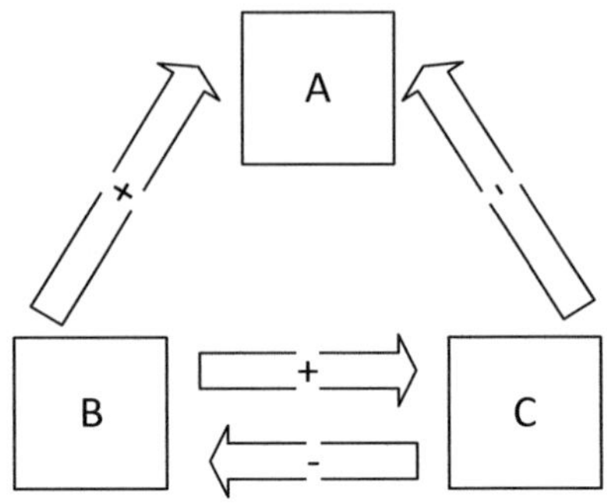

Fig. 5.3 Example FCM

Table 5.3 Example FCM matrix (W)

	A	B	C
A	0	0	0
B	+1	0	+1
C	−1	−1	0

W can represent the matrix of weights where each element w_{ij} represents the influence of concept i on j. An example of a three concept FCM is shown in Fig. 5.3.

In this example, there is positive causality from B to both A and C, and negative causality from C to both A and B. This same example is presented as an adjacency matrix in Table 5.3. Note that the row element imparts a causality on the column element (e.g., B imparts a positive causal relationship to A).

"An FCM can be considered as a type of recurrent artificial neural network" (Tsadiras, 2008, p. 3884). Thus, FCMs evolve over time (i.e., are dynamic) and can be analyzed relative to this evolution. We can use the matrix form of the example FCM (shown in Table 5.3) to study the evolution of this problem over time. Time step $t + 1$ can be evaluated using information from the previous time step, t, as shown in Eq. 5.1:

$$\boldsymbol{A}^{t+1} = f(\boldsymbol{A}^t \boldsymbol{W}) \tag{5.1}$$

where f is known as a transfer function used to evolve the FCM from one time stamp to the next. This transfer function typically takes on one of the following three forms (Tsadiras, 2008): (1) binary, (2) trivalent, or (3) sigmoid.

The binary function is shown in Eq. 5.2:

$$f_{\mathrm{bi}}(\boldsymbol{x}) = \begin{cases} 1, & x > 0, \\ 0, & x \leq 0. \end{cases} \tag{5.2}$$

This creates a two-state FCM. When the activation level of a concept, C_i, is 1, the concept is said to be activated (i.e., on), whereas a value of 0 indicates the concept is not activated (i.e., off).

The trivalent function is shown in Eq. 5.3:

$$f_{\text{tri}}(\boldsymbol{x}) = \begin{cases} 1, & x > 0, \\ 0, & x = 0, \\ -1, & x < 0. \end{cases} \tag{5.3}$$

This creates a three-state FCM. When the activation level of concept C_i equals 1, the concept is increasing, when the activation level equals −1, the concept is decreasing, and when the activation level equals 0, there is no change in the concept.

Finally, there is a sigmoid function in Eq. 5.4, with limits of [−1,1]:

$$f_{\text{sig}}(\boldsymbol{x}) = \tanh(\lambda x) = \frac{e^{\lambda x} - e^{-\lambda x}}{e^{\lambda x} + e^{-\lambda x}} \tag{5.4}$$

The activation level of a given concept can take any value over [−1,1]. Thus, a continuous FCM is created. In this formulation, λ represents a tuning parameter, "a constant value that indicates the function slope (degree of normalization) of the sigmoid functions. Each FCM designer specifies the value of λ" (Bueno & Salmeron, 2009, p. 5223). As λ increases to 10 or more, it approximates a trivalent function. For a small value of λ (e.g., 1 or 2), the function is near linear. Bueno and Salmeron (2009) advocate the use of a λ value of 5.

Tsadiras (2008) provides additional details on each of the three transfer functions as well as the following guidance on the appropriateness and implementation of each:

(1) Binary FCMs are suitable for highly qualitative problems where only representation of increase or stability of a concept is required.
(2) Trivalent FCMs are suitable for qualitative problems where representation of increase, decrease, or stability of a concept is required.
(3) Sigmoid FCMs are suitable for qualitative and quantitative problems where representation of a degree of increase, a degree of decrease, or stability of a concept is required and strategic planning scenarios are going to be introduced (p. 3894).

Additional guidance on transfer function choice is found in Bueno and Salmeron (2009). As a general rule, for all but the simplest of problems, binary FCMs do not offer a great deal of insight (although, as it will be shown later in this chapter, they are useful for FCM validation purposes). Trivalent and sigmoid FCMs are more useful. Sigmoid functions yield more detailed results but require more detailed input (i.e., nonbinary weights). Sigmoid functions were observed empirically by Bueno and Salmeron (2009) to be:

> A useful tool for decisional processes…the sigmoid function can be considered an excellent decision support tool within any scope. In a complex decisional environment the sigmoid function offers the possibility of attaining easily comparative analyses between scenarios that define decisional situations. These scenarios allow the obtaining of a future vision of

> the more suitable alternatives, and, therefore, the reaching of successful decisions. Therefore, the sigmoid function can define decisional scenarios for decision-makers of any decisional environment.
> Nevertheless, it has some disadvantages if it is compared with the other functions analyzed. First, decision-makers need an extensive number of interactions with the sigmoid function to reach the stable scenario, whereas with the other functions few interactions are needed. (p. 5228)

Thus, the utility of each function is dependent on the scenario and the desired output to be gained from the use of an FCM. It is worth making a few comments regarding the potential end states of each method. A binary FCM composed of n concepts has 2^n potential end states, representing each of the possible final configurations each FCM may exhibit. A trivalent FCM has 3^n potential end states, and a sigmoid function, with continuous variables, has an infinite number of final states.

Behavior of a FCM is dynamic, but deterministic. Dynamic behavior means that we must study the evolution of an FCM over time to gain the most meaningful insights. "Such systems are composed of a number of dynamic qualitative concepts interrelated in complex ways, usually including feedback links that propagate influences in complicated chains, that make reaching conclusions by simple structural analysis an utterly impossible task" (Carvalho, 2013, p. 6). Indeed, it is in this dynamic behavior where the true insight from a FCM lies:

> …one might question what is the added value of FCM in what concerns causality. The answer is quite simple: it resides in the fact that even if the direct relations between concepts is certain, the propagation of the effects through time is hard to understand or predict due to the feedback loops. Hence, in FCM one jumps the focus from the problem of finding and representing causality, to the problem of system dynamics and scenario simulation. (Carvalho, 2013, p. 11)

"System behavior is a function of both the system itself (which provides the internal structure that can allow for complex behaviors) and the environment (i.e., initial conditions) that the system is placed in" (Hester, 2016). A given system can exhibit simple, complicated, complex, and chaotic behavior when initialized with a different initial state vector, $\boldsymbol{A}^0$. This initial state vector can either represent the stakeholders' estimates of the values of the current system concepts or be based on a specific *what-if* scenario that we wish to explore. "After that, the concepts are free to interact. The activation level of each concept influences the other concepts according to the weight connections that exist between them" (Tsadiras, 2008, p. 3885).

Then, one of the following end states is reached as discussed by Hester (2016):

(1) Equilibrium is reached. Equilibrium is defined as where $\boldsymbol{A}^{t+1} = \boldsymbol{A}^t$ for all concepts. This behavior can be described as *complicated*. It is dynamic and more difficult to predict than *simple* behavior, which is a special case of equilibrium where $\boldsymbol{A}^t = \boldsymbol{A}^1 = \boldsymbol{A}^0$. In other words, simple behavior is static and

does not change after an initial disturbance. It is obvious and easily predicted by an analyst.

(2) Cyclic behavior is reached. This is described as *complex* behavior. This is defined as where $A^{t+\Delta t} = A^t$ for all concepts. "The system exhibits a periodic behavior where after a certain number of time steps, that is equal to the period [Δt] of the system, the system reaches the same state" (Tsadiras, 2008, p. 3885). It is dynamic and more difficult to predict than complicated behavior.

(3) The system exhibits *chaotic* behavior. In this case, there is no equilibrium reached and the scenario never demonstrates periodic behavior. By its very definition, truly chaotic behavior is not predictable.

The deterministic nature of an FCM is also important to note. Determinism means that for a given initial state vector, A^0, the FCM will always produce the same end state. If a FCM enters a state it has encountered previously, "the system will enter a closed orbit which will always repeat itself" (Tsadiras, 2008, p. 3885). Given the finite number of end states in both binary and trivalent cases, they cannot exhibit chaotic behavior, but only either reach equilibrium or exhibit cyclic behavior with a periodicity of at most 2^n states (in the case of binary functions) or 3^n states (in the case of trivalent functions). Continuous FCMs, however, due to their infinite number of potential end states, can exhibit chaotic behavior. Both the equilibrium point and the limit cycle behavior reveal hidden patterns encoded in the FCM (Kosko, 1988). Tsadiras (2008) suggests that encoding these patterns in the underlying structure of an FCM remains an open research question.

It is worth noting, finally, that given their deterministic and finite nature, the end states of binary and trivalent functions could be exhaustively enumerated, certainly with the assistance of a computer algorithm. This enumeration may lead to additional insight regarding the behaviors of the underlying system. Sigmoid functions are unable to enumerate completely, due to their continuous nature, but they can be probabilistically approximated.

5.5 A Framework for FCM Development

While there may be multiple manners in which a FCM can be constructed, tested, and utilized, the authors advocate the use the following six-step framework developed by Jetter and Kok (2014) for its simplicity and ease of deployment:

(1) Clarification of project objectives and information needs (Step 1),
(2) Plans for knowledge elicitation (Step 2),
(3) Knowledge capture (Step 3),
(4) FCM calibration (Step 4) and testing (Step 5), and
(5) Model use and interpretation (Step 6).

This general framework provides a straightforward and robust approach to utilizing FCM to increase the understanding of a mess. The following subsections detail the mechanics regarding each of these steps.

5.5.1 Step 1: Clarification of Project Objectives and Information Needs

Step 1 is, fundamentally, a problem structuring effort. What is the purpose of the model you are constructing? A model is built with a purpose, and its construction and use should reflect this purpose. Sterman (2000) offers the following insights regarding problem articulation as it relates to model formulation:

> The most important step in modeling is problem articulation. What is the issue the clients are most concerned with? What problem are they trying to address? What is the real problem, not just the symptom of difficulty? What is the purpose of the model?...Beware the analyst who proposes to model an entire business or social system rather than a problem. Every model is a representation of a system-a group of functionally interrelated elements forming a complex whole. But for a model to be useful, it must address a specific problem and must simplify rather than attempt to mirror an entire system in detail. (p. 89)

More succinctly, he offers, "Always model a problem. Never model a system" (Sterman, 2000, p. 90). The intent of his remarks are to clarify what was said at the outset of the chapter regarding models as a "purposeful abstraction of reality." Models exist for a specific reason and that reason is to explore a particular problem in a manner that is a necessary simplification of the real-world system it is representing.

Timing is an important element to consider when examining the purpose of a FCM. What is the intended model time frame (i.e., 6 months, 10 years)? Carvalho (2013) discusses the importance of time:

> It is important to notice that "time" should be considered essential when modeling a FCM, since the rate of change on a social system (or in fact in most real world systems) cannot be infinite; i.e., when simulating a FCM, one cannot assume that the value of a concept can change from its minimum to its maximum value on a single iteration unless this iteration represents a large enough amount of time. (pp. 8–9)

Carvalho and Tome (2000) provide further guidance on the selection of a time interval for a FCM:

> It is important to choose a base time interval (btime) to represent each iteration (1 day, 2 days, 1 week, 1 month, etc.). When defining the relations, btime must always be implicitly present. The rules that represent causal effects are tightly dependent on btime: If btime is 1 day, then rules expressing the effect of a Level in Inflation would most certainly indicate a very small change. If however btime is 1 year then the rules would have to indicate a larger variation. Btime is obviously dependent on the system we are representing and on the time gap we are hoping to analyze. However smaller btimes usually need more detailed and complex rule bases. (p. 411)

Proper problem articulation provides a purpose for each FCM. The intent of a problem-level FCM is to address the stated problem, whereas the intent of a mess-level FCM is to model the interaction effects among the problems in a holistic manner.

5.5.2 Step 2: Plans for Knowledge Elicitation

Since FCMs are primarily qualitative in nature, the knowledge used to construct them comes from expert elicitation. There are three options for this step (Jetter & Kok, 2014):

1. *The modeler is the expert*. This is the case with most academic literature but may be unrealistic for a real-world scenario, where multiple stakeholders are necessary. While this may be a necessary approach due to a lack of expert availability, it is advised against due to its reliance on a unitary perspective.
2. *The modeler surveys experts*. This can be done either on an individual basis or in a group setting. This is the preferred method of knowledge elicitation when available. Although it requires stakeholder commitment to complete, it will result in a representation of the problem that has the greatest degree of buy-in as the stakeholders were an integral part of its development. Time or resource constraints may prevent this method from being deployed.
3. *The modeler analyzes documents*. This involves the use of content analysis to infer cognitive maps from relevant documentation, i.e., scientific publications, technical reports, newspaper articles, and textbooks. This is the second-best option in the absence of the expert availability.

Additionally, these methods can be combined. For example, the modeler can develop an initial articulation of the problem and then present it to interested stakeholders for refinement. This can, if done correctly, reduce the required stakeholder involvement as it eliminates the need for all stakeholders to begin from a blank slate. It can also, however, bias them toward a singular representation of the problem, which may be problematic.

It is worth noting that the plan for knowledge elicitation may influence problem articulation. For example, a group of experts should be surveyed regarding their agreement with project objectives from Step 1. Failure to do so puts individuals at risk of committing the Type III error that was described in Chap. 1.

5.5.3 Step 3: Knowledge Capture

Once the problem has been agreed upon and the participants selected, the process of knowledge capture can begin. This can be either face to face or via written instruction. Coached map development may lead to imposition of facilitator worldviews, but may also preclude questions from experts. At a minimum, experts should be reminded that the maps they are creating are *causal* maps and not *correlation* maps. Thus, in an effort to avoid the Type VI error of an unsubstantiated inference (described in Chap. 1), experts should not overly prescribe causality among concepts. Asking the simple question, *does a change in A cause a change in B to occur*?, will assist in the development of appropriate causal linkages.

Information can be collected individually or in a group setting. Group elicitation may lead to groupthink, but individuals may also benefit from the presence of one another. "Group cognitive mapping furthermore has practical advantages: it requires fewer contact hours between interviewers and respondents and directly results in the integrated map needed for most FCMs…To balance the advantages and drawbacks of individual cognitive mapping and group meetings approaches can be mixed" (Jetter & Kok, 2014, pp. 50–51).

Combination of individual maps can be done in a straightforward manner by computing the average map (adding all stakeholder maps together and dividing by the number of stakeholders) (Kosko, 1988). Additionally, credibility weights can be assigned to experts (Taber, 1991; Taber & Siegel, 1987), or more advanced combination methods can be invoked (Kosko, 1988; Özesmi & Özesmi, 2004). "In participatory studies, however, that equally value the input of all respondents, it is not applied" (Jetter & Kok, 2014, p. 51). Given the focus in systemic decision making on the value of complementarity, it is advantageous that all opinions are equally considered. It is worth noting, however, that we do not advocate consensus as a necessary requirement for FCM development. Consensus is defined by Susskind (1999, p. 6) as follows:

> … reached when everyone agrees they can live with whatever is proposed after effort has been made to meet the interests of all stakeholding parties.…Participants in a consensus building process have both the right to expect that no one will ask them to undermine their interests and the responsibility to propose solutions that will meet everyone else's interests as well as their own.

Consensus is an unnecessarily restrictive requirement and problematic to achieve. Bueno and Salmeron (2009) elaborate on the topic within the framework of FCM development:

> The difficulty is in reaching a consented value not only for the causal weight but for the sign between the cognitive map relationships as well. This measure will be distinct with respect to the experts who assign the fuzzy numbers to each of the relationships. (p. 5222)

Integration of maps requires that concept names and definitions are standardized. "Most workshops devote about half of the total time on finalizing a commonly agreed upon list of concepts" (Jetter & Kok, 2014, p. 51). It should be recognized that some details may be lost to aggregation. For example, if two respondents assign equal weight to the same connection, one with a + sign and one with a – sign, the two will cancel one another out, thereby losing the important detail that both

Table 5.4 Facilitator FCM aggregation versus group FCM construction

Approach	Facilitator aggregation of individual models	Group FCM construction
FCM appropriation	Capturing knowledge	Stakeholder learning
Emphasis	Models as "artifacts for decision making or enhancing systemic understanding"	"Model and modeling process as a tool for social learning"
Research purpose	"Combining representations of stakeholder/expert mental models to (1) reduce uncertainty about a system and (2) compare mental models across groups"	"Community generated representation of knowledge used for planning and learning, often through participatory scenario development"

participants felt a connection was present. "Given the limited state-of-the-art, mathematical aggregation of individual maps, modeler-generated integration of all maps, and group generated cognitive maps all seem viable approaches to pooling the knowledge of individual respondents" (Jetter & Kok, 2014, p. 51). It may not always be advantageous to combine the perspectives, i.e., if the aim is to inform discussion. Gray, Zanre, and Gray (2014, p. 42) elaborate on the differences between facilitator aggregation and group FCM construction in Table 5.4.

Once concepts are agreed upon (if desired), then linkages between concepts are identified. Respondents then add signs (+ or −) to indicate the direction of causality. Once direction is specified, magnitude of weight can also be assigned. This can use a qualitative scale such as the Likert (1932)-type scale as shown in Table 5.5 to indicate the strength of connection, a simple positive or negative weight, or respondents can assign numerical weights in the range of [−1,1]. It is worth noting that FCM values are taken to be relative to one another, rather than absolute, so it is advisable that respondents use linguistic scales to avoid subconsciously quantifying relationships.

These maps are then translated into adjacency matrices for computation as FCMs. Several adjustments may need to be made (Jetter & Kok, 2014):

Table 5.5 Sample weight scale

Qualitative rating	Associated weight
High negative	−1.0
Medium negative	−0.5
Low negative	−0.25
No effect	0
Low positive	+0.25
Medium positive	+0.5
High positive	+1.0

- Elimination of causal links that exist only for definitional and not causal purposes.
- Removal of concepts with no "Out"-arrows, unless they are a target concept, that is, one that is a focus of the analysis. These links typically indicate poor map construction.
- Indication of conditional causality using activation function thresholds or using a nested FCM.
- Synchronization of time steps using intermediate concepts to break up long-term concepts. This may lead to redefinition of concepts.

This stage, as well as subsequent ones, may be aided by the use of software packages. Mental Modeler (Gray, Gray, Cox, & Henly-Shepard, 2013) and FCMapper (www.fcmappers.net) are two packages designed specifically for FCM modeling and assessment, although the authors advocate the use of Mental Modeler due to its graphical capabilities and ease of use.

5.5.4 Step 4: FCM Calibration and Step 5: Testing (Step 5)

Once a FCM has been constructed, it is necessary to calibrate it. It is extremely important to note that the goal of FCM development is not to create an objectively "correct" model, "but a useful and formal description of the perception of a group of people, such as subject matter experts or stakeholders, of the problem at hand" (Jetter & Kok, 2014, p. 54). In the absence of historical data against which to benchmark a model, three steps should be taken to calibrate and test it (Jetter & Kok, 2014):

(1) Calibrate the FCM with binary nodes and trivalent edges (Kosko, 1988). Test it against known or expected cases or generally understood phenomena. If the FCM does not perform as expected, it should be investigated for mistakes.
(2) Once it is calibrated, more sophisticated causal weights and activation functions can be used to gain more insight. Remember that FCMs are constructed to represent scenarios where there is a lack of concrete information, so trying to overspecify them may defeat their purpose.
(3) The model can then be tested against more complex scenarios. Disagreements may arise due to a dichotomy between expected and observed behaviors. This is an opportunity for insight into the expert's mental models and assumptions, or an opportunity to revise the model if necessary. The key is to have a discussion with experts as to the meaning of the results. Additional model tests can be undertaken to assess their validity.

It is noted that "FCM researchers have few approaches, other than trial-and-error, to know if their model will reach a stable state, how many stable states it has, how to select [transfer] functions, and how to deal with temporal aspects" (Jetter & Kok, 2014, p. 56). Sterman (2000) elaborates on this idea, specifically as it pertains to model validation:

> The word validation should be struck from the vocabulary of modelers. All models are wrong, so no models are valid or verifiable in the sense of establishing their truth. The question facing clients and modelers is never whether a model is true but whether it is useful. The choice is never whether to use a model. The only choice is which model to use. Selecting the most appropriate model is always a value judgment to be made by reference to the purpose. Without a clear understanding of the purpose for which the model is to be used, it is impossible to determine whether you should use it as a basis for action. (p. 890)

The goal, then, is to determine whether or not the perception is that the model is acting as expected. This, ultimately, is a subjective evaluation, as the model inputs are subjective and it is very qualitative in nature.

5.5.5 Step 6: Model Use and Interpretation

The model should spark debate and discussion and lead to further narratives being developed regarding expected system behaviors. Given the explored scenarios, they could also feed into decisions about future actions. Given the results of actions, models may be revised as appropriate. The notion of using FCMs as a decision support mechanism will be explored later in the book in Part III.

At a minimum, the model can be used to explore speculative, *what-if* scenarios. An initial scenario can be suggested, and the equilibrium resulting from this initial scenario can be explored. This may take the form, for example, of a temporary rise of a concept such as *S&P 500 Value* in a macroeconomic model of the economy. The interest of exploring such an initial perturbation may be to investigate what effects this change would have on the overall economy. This may lead to insight regarding emergent behavior and unexpected model outputs. Alternatively, a concept may be *clamped* or permanently set to a particular value. This may represent a scenario such as a permanent policy change. "The calculation is slightly different, if activation of concept C_1 is not a one-time impulse (e.g. an election, a natural disaster), but a change that lasts over extended periods of time (e.g. new tax laws). In this case, the concept is 'clamped' and always set back to its initial activation level…" (Jetter & Schweinfort, 2011, p. 55). Either scenario (an initial perturbation or a clamped variable), or a combination of them, will yield insight into the model's behaviors.

5.6 Example FCM Application

This example expands on the one developed in Hester, Akpinar-Elci, Shaeffer, and Shaeffer (2016). It centers on examination of the Ebola virus disease. The identified problem is how to reduce the incidence of the Ebola virus disease, viewed through a global health perspective. The developed FCM included economic, health care, and political ramifications, among others. Although this example focuses on the Ebola virus disease, the specific disease is not important; rather, the general case of communicable diseases is the intent of analysis. For contextual understanding, we may define communicable diseases and those that can be transmitted "from person to person by either direct or indirect methods. Direct transmission is either by direct physical contact or by means of droplet spread, such as by coughing or sneezing. Indirect transmission of an infectious agent is accomplished by some intermediary mechanism, such as transmission in contaminated water or by means of insects. A few communicable diseases are primarily diseases of animals and are transmitted to humans only incidentally" (Crowley, 2007, p. 149). Figure 5.4 shows a depiction of the Ebola virus disease mess.

Given that the focus of analysis is to determine ways in which we may reduce the incidence rate of Ebola virus disease, this becomes the central focus of the scenarios that we may choose to explore. For an initial inquiry, it may be worthwhile to simply explore the question, *What if we could reduce the Incidence rate of Ebola virus disease*? Thus, absent a mechanism for doing so, we wish to explore what the effects would be on the overall FCM if we initially set the concept

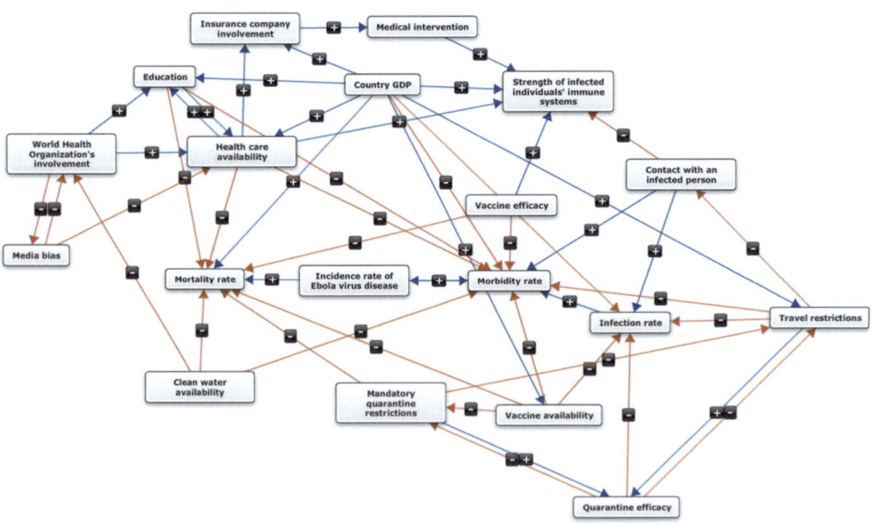

Fig. 5.4 Ebola virus disease depiction (adapted from Hester et al., 2016)

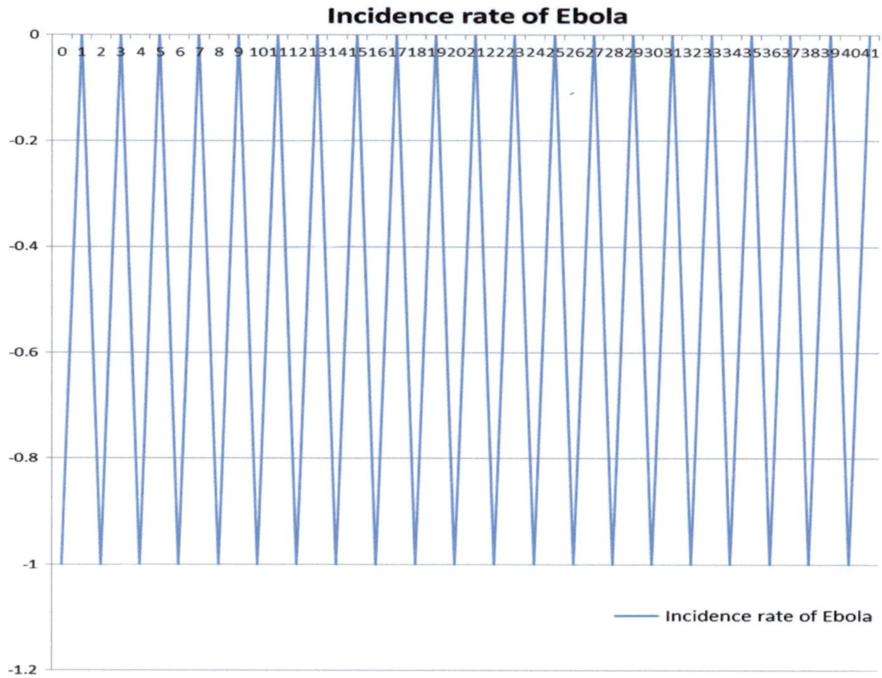

Fig. 5.5 Effect of initial reduction in incidence rate of Ebola

Incidence rate of Ebola virus disease to −1 (understanding that if we were able to take such an action, we would not have a mess to begin with). This result is shown in Fig. 5.5. This scenario exhibits complex behavior. Despite initially reducing the concept, *Incidence rate of Ebola* exhibits periodic behavior, oscillating between concept reduction and no change. This tells us that an initial reduction of the incidence rate of Ebola would not be a sustainable solution and would require additional intervention for longevity.

The next experiment results from a qualitative investigation of Fig. 5.4. In it, the *World Health Organization's involvement* appears to be a driving concept, with many other concepts linked to it. Practically, influence of this concept also seems achievable as getting the World Health Organization (WHO) more involved simply requires their buy-in and minimal resources to be committed, which is much simpler than, for example, improving *Vaccine efficacy* or *Clean water availability*. Figure 5.6 shows the effects on *Incidence rate of Ebola virus disease* from initially setting *World Health Organization's involvement* to +1.

This scenario appears to exhibit complicated behavior. While the results of Fig. 5.6 appear promising, an examination of the effects of this scenario on all of

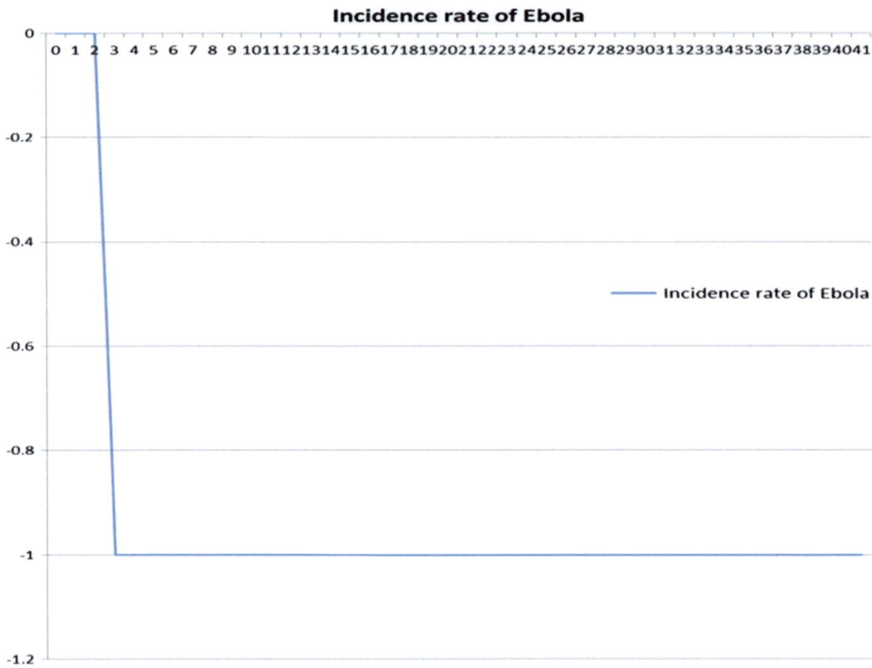

Fig. 5.6 Effect of initial reduction in World Health Organization's involvement on incidence rate of Ebola

the concepts reveals additional insights, as shown in Fig. 5.7, which reveals complex behavior. This reveals that both *media bias* and *WHO involvement* are cyclical. *WHO involvement* increases (to +1), followed by a drop in *media bias* (to −1) and *WHO involvement* (to 0) the following period, followed by an increase in *WHO involvement* increases (to +1) and *media bias* (to 0). The pattern repeats with a periodicity of 2.

If the scenario depicted in Fig. 5.7 is acceptable, then we have found an acceptable mechanism for reducing the *Incidence rate of Ebola*. If, however, a more stable scenario is desired, then we can explore additional options. Given that *media bias* appears to be a trigger for the oscillating behavior, we should explore an initial increase of *WHO involvement* and an initial decrease of *media bias*. This scenario is shown in Fig. 5.8.

The results of this scenario are complicated; the scenario shown in Fig. 5.8 is stable for all concepts after period 2. It results in an increase in the concepts of *WHO involvement*, *Education*, *Insurance company involvement*, *Medical intervention*, *Strength of infected individuals' immune systems*, and *Healthcare availability*. The scenario also results in the reduction of the concepts of *Media bias*,

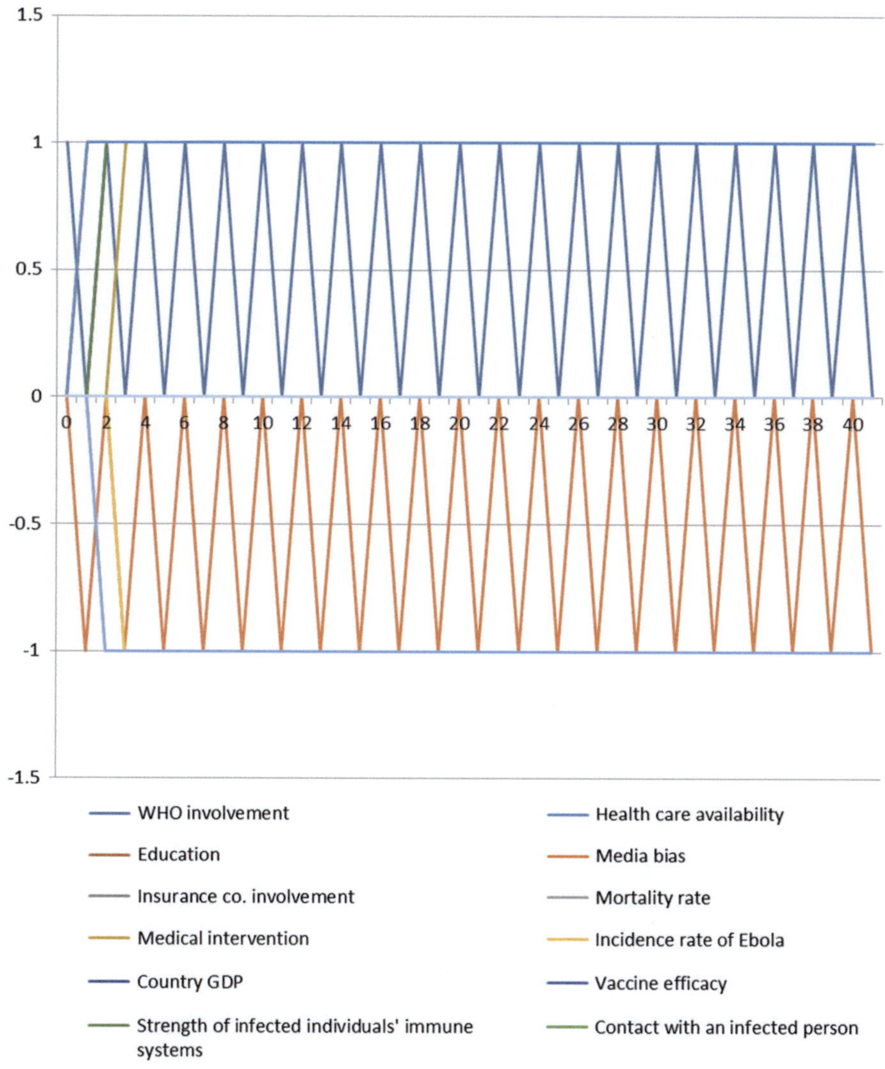

Fig. 5.7 Effect of initial increase in World Health Organization's involvement on all concepts

Mortality rate, *Morbidity rate*, and, most importantly, *Incidence rate of Ebola*. The remaining concepts had no change. If we believe that we have no control over the media and thus cannot influence *Media bias*, then we can explore an additional scenario, that of clamping *WHO involvement* at +1. That is, we can secure a long-term commitment from the WHO for involvement in the fight against the Ebola virus disease. This result is shown in Fig. 5.9.

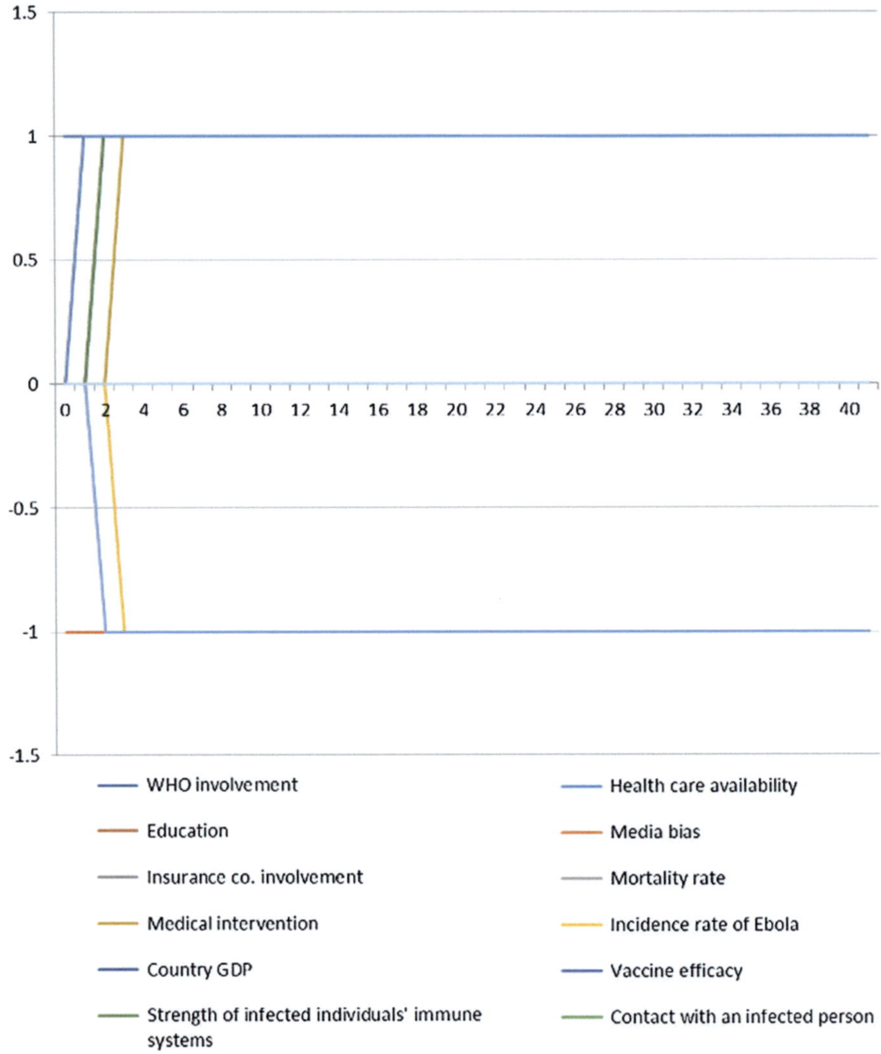

Fig. 5.8 Effect of initial reduction in World Health Organization's involvement and initial decrease in media bias on all concepts

The scenario shown in Fig. 5.9 is complicated as well, but it may be characterized as the most desirable. It results in the same equilibrium as the one depicted in Fig. 5.8, yet it requires no initial change in *Media bias*, which may be problematic to achieve. Thus, with this simple set of experiments, this section has demonstrated the utility of fuzzy cognitive maps in assessing a complex system.

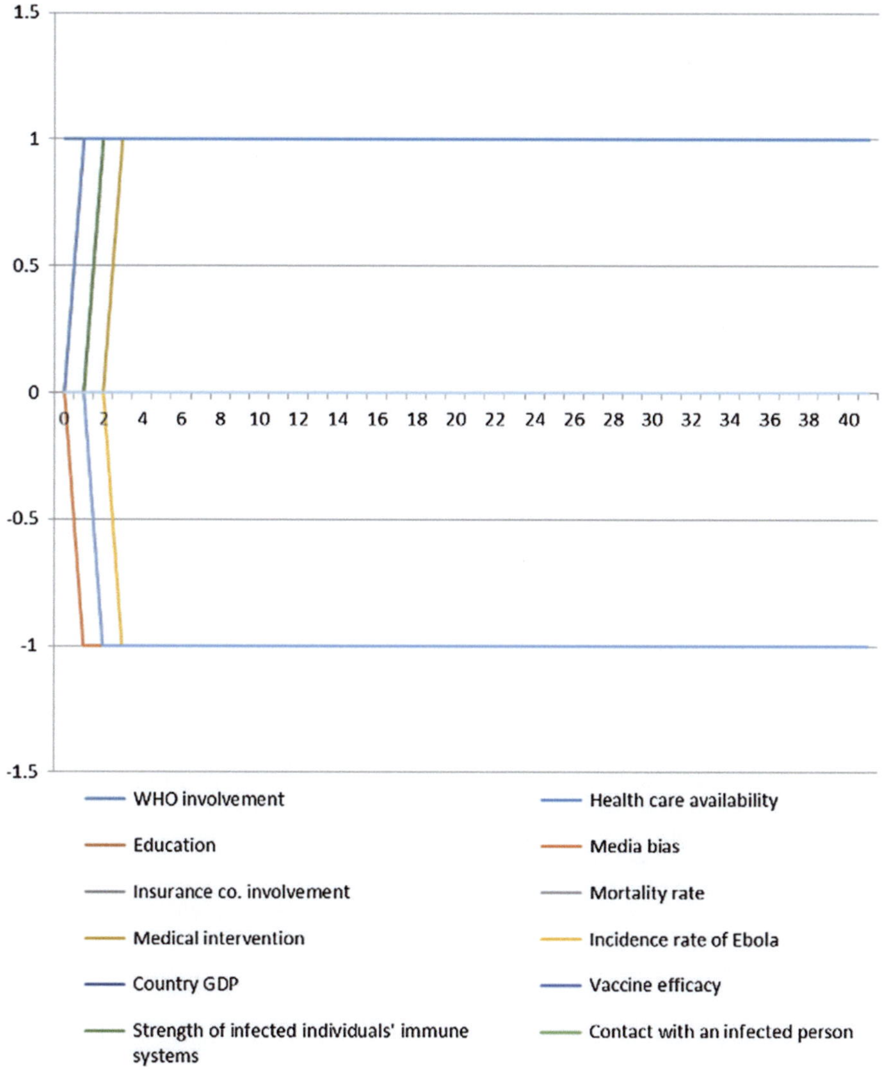

Fig. 5.9 Effect of clamped increase in World Health Organization's involvement on all concepts

5.7 Summary

This chapter introduced the reader to complex systems modeling methods. Specifically, a comparison of available methods was presented which outlined the appropriateness of fuzzy cognitive mapping for understanding and investigating complex systems. Finally, a framework was presented for the development and use of a fuzzy cognitive map to explore these problems. Efficacy of FCM modeling was demonstrated on an example problem. While it may be unclear at this point to the reader, ensuing chapters will provide additional details regarding the development and use of FCMs, as a mechanism for thinking about, acting on, and observing messes.

After reading this chapter, the reader should

1. Be able to identify methods available for complex systems modeling;
2. Understand the appropriateness of fuzzy cognitive mapping for representing complex systems; and
3. Understand how to construct and use a fuzzy cognitive map.

References

Ackermann, F. (2012). Problem structuring methods 'in the Dock': Arguing the case for Soft OR. *European Journal of Operational Research, 219,* 652–658.

Amer, M., Jetter, A. J., & Daim, T. (2011). Development of fuzzy cognitive map (FCM)-based scenarios for wind energy. *International Journal of Energy Sector Management, 5*(4), 564–584.

Axelrod, R. (1976). *Structure of decision: The cognitive maps of political elites*. Princeton, NJ: Princeton University Press.

Banerjee, A., Chitnis, U. B., Jadhav, S. L., Bhawalkar, J. S., & Chaudhury, S. (2009). Hypothesis testing, type I and type II errors. *Industrial Psychiatry Journal, 182*(2), 127–131.

Bueno, S., & Salmeron, J. L. (2009). Benchmarking main activation functions in fuzzy cognitive maps. *Expert Systems with Applications, 36,* 5221–5229.

Carvalho, J. P. (2013). On the semantics and the use of fuzzy cognitive maps and dynamic cognitive maps in social sciences. *Fuzzy Sets and Systems, 214,* 6–19.

Carvalho, J. P., & Tome, J. A. B. (2000, 2000). *Rule based fuzzy cognitive maps-qualitative systems dynamics*. Paper presented at the Fuzzy Information Processing Society, 2000. NAFIPS. 19th International Conference of the North American.

Checkland, P. B. (1999). *Systems thinking, systems practice*. New York: Wiley.

Checkland, P. B., & Scholes, J. (1999). *Soft Systems Methodology in action*. Chichester, UK: John Wiley & Sons.

Crowley, L. V. (2007). *An introduction to human disease: Pathology and pathophysiology correlations* (7th ed.). Sudbury, MA: Jones and Bartlett Publishers.

Eden, C. (1992). On the nature of cognitive maps. *Journal of Management Studies, 29*(3), 261–265.

Eden, C., & Ackermann, F. (1998). *Making strategy: The journey of strategic management*. London: Sage.

Forrester, J. W. (1961). *Industrial dynamics*. Cambridge, MA: MIT Press.

Friend, J. K., & Hickling, A. (1987). *Planning under pressure: The Strategic Choice Approach*. Oxford, UK: Pergamon.

Friend, J. K., & Jessop, W. N. (1977). *Local government and Strategic Choice* (2nd ed.). Oxford, UK: Pergamon.

Gray, S. A., Gray, S., Cox, L. J., & Henly-Shepard, S. (2013). *Mental modeler: A fuzzy-logic cognitive mapping modeling tool for adaptive environmental management.* Paper presented at the 2013 46th Hawaii International Conference on System Sciences (HICSS), Wailea, Maui, HI.

Gray, S. A., Zanre, E., & Gray, S. R. J. (2014). Fuzzy cognitive maps as representations of mental models and group beliefs. In E. I. Papageorgiou (Ed.), *Fuzzy cognitive maps for applied sciences and engineering*. Berlin: Springer.

Hester, P. T. (2016). *Unraveling complexity*. Paper presented at the 2016 Industrial and Systems Engineering Research Conference, Norcross, GA.

Hester, P. T., Akpinar-Elci, M., Shaeffer, J. M., & Shaeffer, M. B. (2016). On complementarity and the need for a transdisciplinary perspective in addressing emerging global health issues. In F. J. Kahlen, S. Flumerfelt, & A. Alves (Eds.), *Transdisciplinary perspectives on complex systems: New findings and approaches* (pp. 193–212). Cham, Switzerland: Springer International.

Hester, P. T., & Tolk, A. (2010). *Applying methods of the M&S spectrum for complex systems engineering*. Paper presented at the Proceedings of the 2010 Spring Simulation Multiconference, Book 5—EAIA, Vista, CA.

Heyer, R. (2004). *Understanding soft operations research: The methods, their application and its future in the defence setting (DSTO-GD-0411)*. Edinburgh, Australia: DSTO Information Sciences Laboratory.

Jetter, A. J. (2006). *Fuzzy cognitive maps for engineering and technology management: What works in practice*. Paper presented at the PICMET 2006, Istanbul, Turkey.

Jetter, A. J., & Kok, K. (2014). Fuzzy Cognitive Maps for futures studies—A methodological assessment of concepts and methods. *Futures, 61,* 45–57.

Jetter, A. J., & Schweinfort, W. (2011). Building scenarios with Fuzzy Cognitive Maps: An exploratory study of solar energy. *Futures, 43,* 52–66.

Kosko, B. (1986). Fuzzy cognitive maps. *International Journal of Man-Machine Studies, 24,* 65–75.

Kosko, B. (1988). Hidden patterns in combined and adaptive knowledge networks. *International Journal of Approximate Reasoning, 2,* 377–393.

Law, A. M., & Kelton, W. D. (2000). *Simulation modeling and analysis* (3rd ed.). New York: McGraw Hill.

Likert, R. A. (1932). A technique for the measurement of attitudes. *Archives of Psychology, 23* (140), 1–55.

Maria, A. (1997). *Introduction to modeling and simulation*. Paper presented at the Proceedings of the 29th Conference on Winter simulation.

Mingers, J. (2011). Soft OR comes of age-But not everywhere! *Omega, 39,* 729–741.

Özesmi, U., & Özesmi, S. L. (2004). Ecological models based on people's knowledge: A multi-step fuzzy cognitive mapping approach. *Ecological Modeling, 176,* 43–64.

Papageorgiou, E. I., & Salmeron, J. L. (2013). A review of fuzzy cognitive map research at the last decade. *IEEE Transactions on Fuzzy Systems, 21*(1), 66–79.

Sterman, J. D. (2000). *Business dynamics: Systems thinking and modeling for a complex world*. Boston: McGraw-Hill Higher Education.

Susskind, L. (1999). An alternative to Robert's Rules of Order for groups, organizations, and ad hoc assemblies that want to operate by consensus. In L. Susskind, S. McKearnan, & J. Thomas-Larmer (Eds.), *The consensus building handbook: A comprehensive guide to reaching agreement*. Thousand Oaks, CA: Sage Publications.

Taber, R. (1991). Knowledge processing with fuzzy cognitive maps. *Expert Systems with Applications, 1,* 83–87.

Taber, R., & Siegel, M. (1987). *Estimation of expert credibility weights using fuzzy cognitive maps*. Paper presented at the Proceedings of the IEEE First International Conference on Neural Nets, San Diego, CA.

Tsadiras, A. K. (2008). Comparing the inference capabilities of binary, trivalent and sigmoid fuzzy cognitive maps. *Information Sciences, 178,* 3880–3894.

Yilmaz, L., & Ören, T. (Eds.). (2009). *Agent-directed simulation and systems engineering*. Weinheim, Germany: Wiley-VCH.

Zeigler, B., Praehofer, H., & Kim, T. G. (2000). *Theory of modeling and simulation: Integrating discrete event and continuous complex dynamic systems* (2nd ed.). New York: Academic Press.

Part II
Thinking Systemically

A Multi-Methodology for Systemic Thinking

With Part II of the text, we have now transitioned from the structure to think phase in the systemic decision making process (as shown below (Fig. 1), repeated from Chap. 3).

The key to *thinking systemically* is consideration of the "5 W's and How?" That is, who, what, why, where, when, and how? The relevance of each is explained below.

- *Who* is relevant to understanding our mess? *Who* concerns holistic consideration for the stakeholders involved in a situation. Stakeholder analysis and management is discussed in detail in Chap. 6.
- *What* are we trying to achieve in addressing our mess? What are the objectives we wish to achieve? These and other questions are discussed in Chap. 7.
- *Why* are we interested in this mess? We all only have 24 hours in a day with which to expend our resources. Why does this mess demand our resources and efforts? What motivations exist for our involvement in this mess? These questions are discussed in Chap. 8.

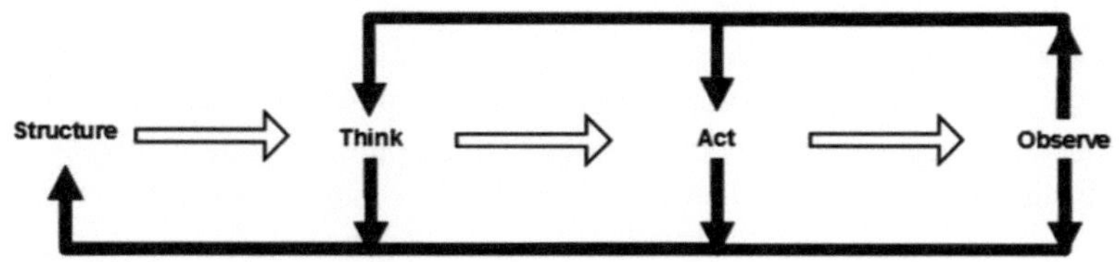

Fig. 1 Systemic decision making process with feedback

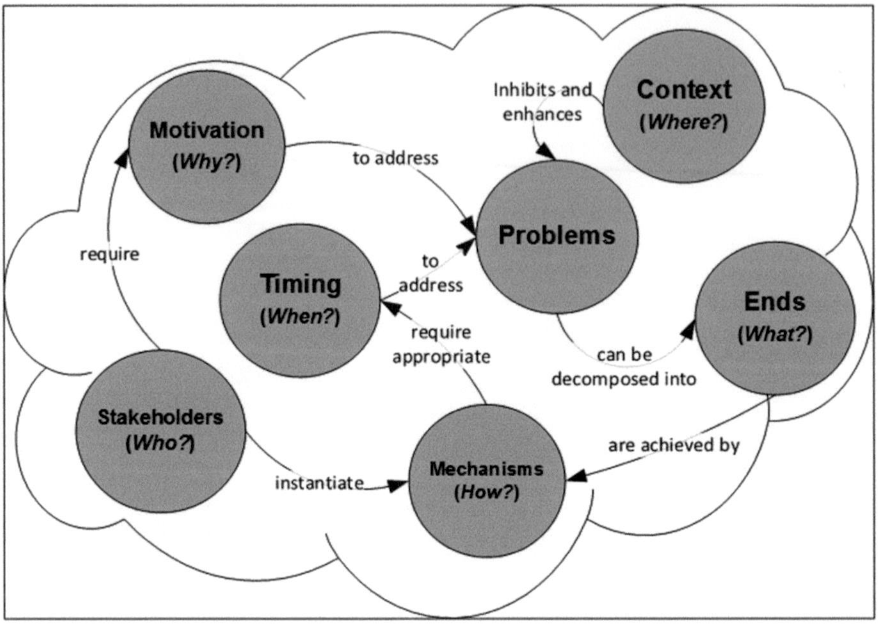

Fig. 2 Anatomy of a mess

- *Where* does our situation reside? What are the characteristics of the context of our mess? Where are the boundaries on our mess? Attention is given to these elements in Chap. 9.
- *How* do we achieve improved understanding and achieve the objectives of our mess? This question discusses mechanisms for improving our mess. How do we deploy mechanisms in order to achieve our aims? This is the focus of Chap. 10.
- *When* do we want to have increased mess understanding by? This question explores concerns related to timing. When should we intervene in a system to create the largest impact? These questions are addressed in Chap. 11.

Attempting to answer these questions forms the methodology for systemic thinking developed through Part II of this text. Figure 2 illustrates the interaction of these questions with one another within a mess, depicted as a cloud. As the figure demonstrates, each of these perspectives is inextricably linked to each other.

While this figure seems innocent enough, one could imagine it increasing substantially in complexity if we were to decompose a mess as shown in Fig. 3. We

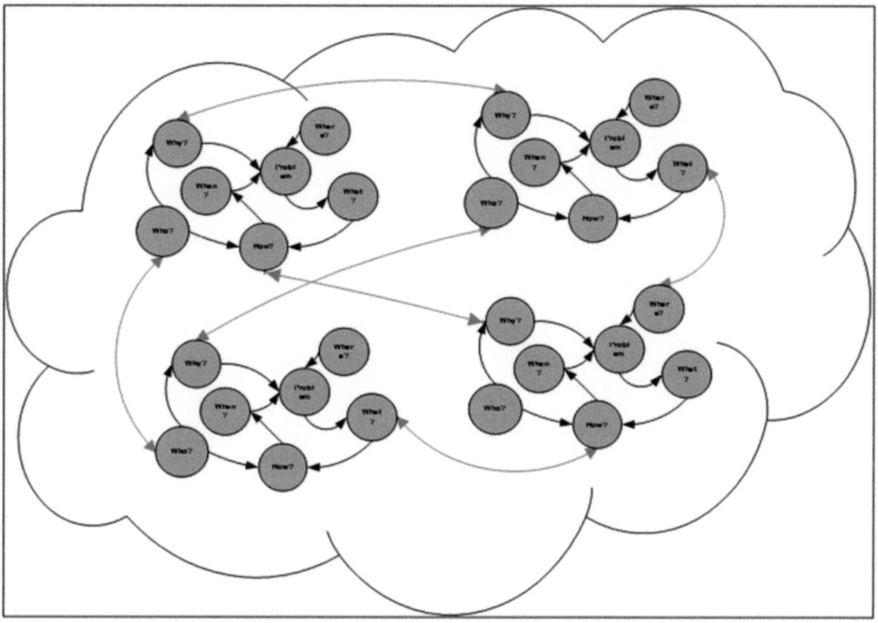

Fig. 3 Relationship among several problems and systemic thinking elements

have to account for the relationships between elements, e.g., the resources of one problem being tied to those of another. In these interactions and conflicts, our mess truly arises.

Chapter 6
The *Who* of Systemic Thinking

Abstract The main focus of the *who* question of systemic thinking is on the stakeholders associated with our mess. This chapter discusses our approach for the analysis and management of stakeholders. This introduction provides a brief background of stakeholder analysis and an introduction to our approach to stakeholder analysis and management, which is then followed by a detailed discussion of each of these steps. Finally, a framework for stakeholder analysis and management is presented and demonstrated.

6.1 Stakeholder Analysis

Study of the individuals and organizations involved in our mess is critical to understanding (and influencing it). There are two competing theories as to how to undertake this analysis, *shareholder theory* and *stakeholder theory*. Shareholder theory, or the theory that corporations are strictly beholden to their shareholders and thus, driven entirely by financial objectives, was championed by Friedman (1962). Seen by many as too myopic a viewpoint, this perspective was later broadened to include all stakeholders with the development of R. Edward's stakeholder theory (Freeman, 1984). Another way to view this expansion is to understand that value, broadly defined, had expanded in scope from a purely financial perspective to one that is more inclusive. In fact, Freeman's view was founded in corporate social responsibility, coupled with financial responsibility, as complementary perspectives to consider in running a business. "Stakeholder analysis was first explored by Freeman (1984) as a methodology to assist business organization leadership with their strategic management functions. Stakeholder analysis has since expanded beyond the corporate arena" (Hester & Adams, 2013, p. 337). Stakeholder analysis is now considered an essential part of many complex problem solving endeavors (Hester & Adams, 2013).

Shareholder theory is singularly focused on maximization of return on investment or ROI. Stakeholder theory, on the other hand, is focused on maximizing *value* to stakeholders. As we have shifted from a shareholder-driven perspective in

P.T. Hester and K.M. Adams, *Systemic Decision Making*, Topics in Safety, Risk, Reliability and Quality 33, DOI 10.1007/978-3-319-54672-8_6

which maximizing value = maximizing ROI, the goal of maximizing value for stakeholders has grown more complicated. We must now widen our aperture and appreciate that many different, and possibly competing, stakeholders can derail or enhance our system's goals. Thus, we must appreciate the richness of value representation to a diverse stakeholder body. While maximizing ROI may be synonymous with maximizing value to some stakeholders, it may be drastically different for others. The notion of value and its ties to personal objectives is explored more in depth in Chap. 7.

So, what exactly is a stakeholder? There are many perspectives on this question. Friedman and Miles (2002) cite 75 different sources offering individual views or adoptions on what a stakeholder is. They also cite the statistic of 100,000 references to be found in Google Scholar for a simple search of the term *stakeholder*. One of the earliest and broadest definitions of a stakeholder comes from Freeman (1984), who defined a stakeholder as someone who "can affect or is affected by the achievement of the organization's objectives" (p. 46). Mitchell, Agle, and Wood (1997) expand on these notions, questioning, … "who (or what) are the stakeholders of the firm? And to whom (or what) do managers pay attention?" (p. 853).

What about the perspective of value? Adopting Freeman's (1984) definition, we can say that stakeholders are those individuals or organizations whose value is affected by the achievement of the organization's objectives. Hester and Adams (2013) offer a big picture view of stakeholders:

> Stakeholders exist at the center of any complex problem solving effort and holistic consideration of them is a key element of analyzing a problem systemically. Stakeholders are the customers, users, clients, suppliers, employees, regulators, and team members of a system. They fund a system, design it, build it, operate it, maintain it, and dispose of it. Each stakeholder contributes their own value-added perspective, as described by the systems principle known as complementarity. (p. 337)

Thus, stakeholders are far reaching and affect every element of our organization's goals. To that end, we must analyze and manage them holistically in order to improve our mess understanding and doing so can invoke a number of different approaches based on the underlying theory being utilized. Friedman and Miles (2002) discuss the differing stakeholder theory classes as follows:

- *Normative* stakeholder theory which describes how managers and stakeholders should act based on ethical principles.
- *Descriptive* stakeholder theory describes how managers and stakeholders actually behave.
- *Instrumental* stakeholder theory describes how managers should act if they wish to further their own interests and the interests of the organization, typically viewed as profit maximization.

Normative stakeholder theory is interesting but not the focus of the remainder of this chapter. Descriptive stakeholder theory invokes elements such as human psychology and organizational behavior, which, while also interesting, are not particularly relevant to the emphasis of this chapter. Instead, the proposed approach will focus on instrumental stakeholder theory in an effort to provide readers a methodology by which to advance their own interests and make better decisions in the context of a stakeholder-rich environment. To support these objectives, we propose the following six-step process for stakeholder analysis and management:

1. Brainstorm stakeholders
2. Classify stakeholders
3. Evaluate stakeholder attitudes
4. Map stakeholder objectives
5. Determine stakeholder engagement priority
6. Develop a stakeholder management plan.

This is followed by an implicit 7th step, manage stakeholders. Much like other elements of systemic decision making, stakeholder analysis and management is an iterative process as depicted in Fig. 6.1.

Thus, while we begin with brainstorming, as systemic decision makers, we recognize that we will likely have to revisit our steps as our understanding of our problem evolves. The following sections provide details regarding each of the six steps, and a framework for undertaking stakeholder analysis and management, which is demonstrated on a simple example concerning real estate rezoning, which will be carried throughout the remainder of the text.

Fig. 6.1 Stakeholder analysis and management process

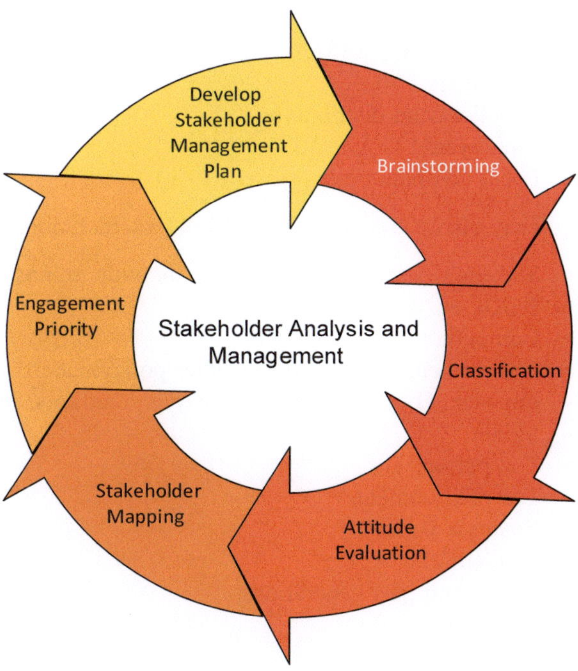

6.2 Brainstorm Stakeholders

The first step necessary for stakeholder analysis is arguably the most straightforward, that is, identifying the stakeholders relevant to the problem being analyzed and speculating as to their desires. It should be noted that the issue of which was came first, the stakeholder or the problem, is a classic chicken-or-egg issue. We must have some notion of our problem before we can brainstorm who might be relevant to our systemic decision making effort; however, we need those very stakeholders to help us clearly structure (and potentially later restructure) our problem. Thus, we must, in all but the simplest of cases, start with an initial problem formulation, perhaps with a subset of stakeholders, and iterate on both stakeholders and problem definition (as well as our context). This naturally leads to the question of who should be considered as a stakeholder for our problem. While the notion of a stakeholder is fairly ubiquitous, we will show throughout the course of this chapter that analysis of them is anything but trivial.

Given Freeman's (1984) seminal stakeholder definition and Mitchell et al. (1997) emphasis on managerial attention, we must consider (1) how to identify stakeholders and (2) how to engage these stakeholders in support of our organizational objectives. These two elements are crucial to effective stakeholder analysis and management.

Maintaining a problem-centric posture on our effort, we focus on the question of *who can affect or is affected by the problem solution*. But where do we start in generating a comprehensive list of possible stakeholders to answer such a broad question? Friedman and Miles (2002) provide the following common list of stakeholders to serve as a sufficient starting point:

- Shareholders
- Customers
- Suppliers and distributors
- Employees
- Local communities.

They also add additional stakeholders, including the following:

- Stakeholder representatives such as trade unions or trade associations
- NGOs or "activists"
- Competitors
- Governments, regulators, or other policymakers
- Financiers beyond stockholders (e.g., creditors, bondholders, debt providers)
- Media
- The public
- The environment
- Business partners
- Academics
- Future and past generations
- Archetypes.

With this frame of reference in mind, we can see why stakeholder analysis is a crucial element in systemic decision making. Stakeholders influence every aspect of our problem. The choice of Freeman's definition, admittedly an intentionally broad definition, is purposeful. Systemic decision making involves taking a broad perspective on a problem and, in the case of stakeholders, we ought to err on the side of inclusion rather than exclusion. Step 1 of the stakeholder analysis process truly is a brainstorming exercise. At this point, it is up to the systems practitioner and other identified participants to brainstorm answers to a question form of Freeman's notion of stakeholders, that is, *who can affect or is affected by the problem solution*? This list may include any or all of the list suggested by Friedman and Miles (2002). The next question we must ask ourselves is *what does the stakeholder want as a result of problem resolution?* Articulation of a stakeholder desire is a simple narrative summarizing what a stakeholder may wish to achieve as the result of a successful problem resolution. This allows us to brainstorm what the stakeholder wants from the intervention or, if possible, simply ask the stakeholder about their desires with respect to the problem (this of course is the most straightforward manner to obtain this information but it may not be feasible or desirable). This should be written as a simple statement of stakeholder desire, including a verb and object. For example, we may wish to *maximize safety*, *mitigate environmental impact*, or *maximize ROI*. It may be necessary to ask *why* to understand the fundamental desires of our stakeholders. A stakeholder expressing a desire to see a competitor fail may really be seeking to advance his or her own interests (e.g., financial return), which do not necessarily come at the expense at a competitor (e.g., by growing the market, each company may flourish). It is worth noting that the focus is on what a stakeholder *wants* and not what they *need* due to the *principle of suboptimization* (Hitch, 1953); that is, everyone will not get what they want in order for the problem to be resolved in the most effective manner.

The output of the brainstorming step is simply a list of individuals and groups that *may* be considered as stakeholders and their desires. The following is an example list of stakeholders and their associated expectations that might be generated by a real estate development company after they have been awarded a contract for a new commercial real estate development:

1. The real estate developer wants financial gain.
2. City council wants to be reelected.
3. State government wants tax revenue.
4. Zoning commission wants compliance from any new development.
5. Tenants of the proposed development want a nice place to live at an affordable price.
6. Customers of proposed commercial entities want attractive shopping.
7. Environmentalists want a development with minimal environmental impact.
8. Rival real estate developers want the development to fail.
9. Safety personnel want compliance of the design with ADA standards.
10. Tourists want additional attractions to consider during their visit.

11. The Chamber of Commerce wants additional members.
12. and so on…

It is clear that this list can grow quite large rather rapidly. The key to this step is to capture all of these entities in Step 1, without regarding for classification, attitude, or relationship of these stakeholders in any manner. Consideration for these elements will be accounted for in subsequent steps of the stakeholder analysis process. If we think that they may affect or be affected by the problem, then they should be included as potential stakeholders.

6.3 Classify Stakeholders

As we complete Step 1, we have a potentially overwhelming list of stakeholders to consider during our stakeholder analysis and management effort. In order to begin to make sense of this list, we must classify these stakeholders. To do so, we draw from Mitchell et al. (1997), who developed a typology in order to enable organizations to analyze and decide which stakeholders demanded the greatest organizational attention. Their typology specifies three key stakeholder attributes: (1) power; (2) legitimacy; and (3) urgency. These terms are defined in Table 6.1 in terms of their sources and the definitions provided for them by Mitchell et al. (1997).

For each stakeholder, one should answer the question of whether or not each attribute is exhibited by the stakeholder on the range [0,1], with 0 being a complete lack of attribute in question, and 1 being the highest possible value. We can then go on to define a combined measure, $Prominence_i$ of the ith stakeholder as follows:

$$Prominence_i = [P_i + L_i + U_i]/3 \tag{6.1}$$

where P is Power, defined on [0,1]; L is Legitimacy, defined on [0,1]; and U is Urgency, defined on [0,1].

Table 6.1 Stakeholder attribute definitions

Attribute	Definition	Sources
Power	"A relationship among social actors in which one social actor, A, can get another social actor, B, to do something that B would not" (Mitchell et al., 1997, p. 869)	Dahl (1957), Pfeffer (1981), Weber (1947)
Legitimacy	"A generalized perception or assumption that the actions of an entity are desirable, proper, or appropriate within some socially constructed system of norms, values, beliefs, definitions" (Mitchell et al., 1997, p. 869)	Suchman (1995), Weber (1947)
Urgency	"The degree to which stakeholder claims call for immediate attention" (Mitchell et al., 1997, p. 869)	Mitchell et al. (1997)

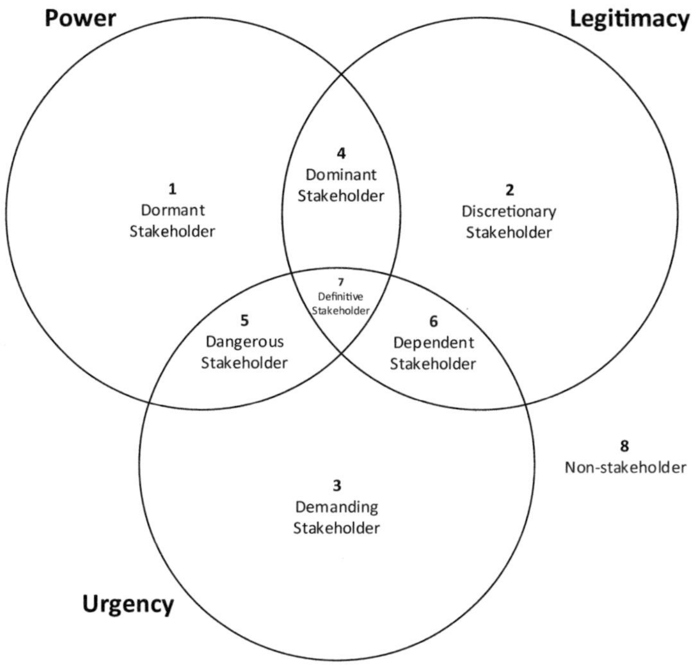

Fig. 6.2 Stakeholder typology, adapted from Mitchell et al. (1997)

Prominence represents a relative level of importance of each stakeholder to a given problem. The number and type of attributes possessed help to define the class for each stakeholder. Mitchell et al. (1997) go on to classify each of the eight possible combinations of these attributes as shown in Fig. 6.2. This graphic should be interpreted as intersecting regions indicate any presence of the contributing attributes; however, the stronger an attribute's presence, the stronger the shared category. For example, a stakeholder who has a P of 0.3, L of 0.2, and U of 0 can be categorized as Dominant; however, a different stakeholder with a P of 0.5, L of 0.7, and U of 0 would also be dominant, although in this case, the attributes are more strongly possessed, so we may say that this stakeholder is *more dominant*.

Further, these stakeholders can be classified in terms of the number of attributes they exhibit; thus, any given stakeholder classification contains one or more class of stakeholders. Individuals who exhibit none of the attributes are considered to be *Nonstakeholders*. Stakeholders exhibiting any one of power, legitimacy, or urgency are classified as *Latent* (either dormant, discretionary, or demanding). Latent stakeholders have little expectation for influence on an associated system, and "managers may not even go so far as to recognize those stakeholders' existence" (Mitchell et al., 1997, p. 874). Stakeholders exhibiting any two attributes can be classified as *Expectant* (dominant, dangerous, or dependent), individuals who "are seen as 'expecting something,' because the combination of two attributes leads the

Table 6.2 Stakeholder class, attributes, and classifications

Stakeholder class	Stakeholder attribute			Stakeholder classification
	Power	Legitimacy	Urgency	
Dormant	Yes	No	No	Latent
Discretionary	No	Yes	No	
Demanding	No	No	Yes	
Dominant	Yes	Yes	No	Expectant
Dangerous	Yes	No	Yes	
Dependent	No	Yes	Yes	
Definitive	Yes	Yes	Yes	Definitive
Nonstakeholder	No	No	No	Undefined

stakeholder to an active versus a passive stance, with a corresponding increase in firm responsiveness to the stakeholder's interests" (Mitchell et al., 1997, p. 876). Those stakeholders classified as latent or expectant may be thought of as so-called secondary stakeholders in Clarkson's (1995) typology, stakeholders on whom the "corporation is not dependent for its survival…Such groups, however, can cause significant damage to a corporation" (p. 107). Finally, *Definitive* stakeholders exhibit all three stakeholder attributes. With these individuals, "managers have a clear and immediate mandate to attend to and give priority to that stakeholder's claim" (Mitchell et al., 1997, p. 878). Definitive stakeholders are akin to what Clarkson (1995) calls *primary stakeholders*, describing them as "one without whose continuing participation the corporation cannot survive…" (p. 106). Table 6.2 illustrates stakeholder class, attributes, and classification as they relate to one another.

While this is a useful typology and Mitchell et al. (1997) make some initial recommendations regarding actions to deal with stakeholders based on their classification, we contend that it is insufficient. Their typology fails to account for the underlying attitude of the stakeholder, to which we now turn our attention.

6.4 Evaluate Stakeholder Attitudes

As we transition to Step 3 of the stakeholder analysis process, we have brainstormed our stakeholders and classified them according to their prominence within the context of the problem we are addressing. A strategy for engaging stakeholders based solely on their relative classification is insufficient as it does not account for stakeholder support or opposition to a particular endeavor. For example, if a stakeholder is supportive of a project, while they may not be classified as definitive, it still may be advantageous for us to engage them in developing strategies for dealing with a complex problem. Thus, it is imperative that we evaluate the attitude of our stakeholders with respect to our particular effort. For this classification, the authors draw on work by Savage, Nix, Whitehead, and Blair (1991), who categorize

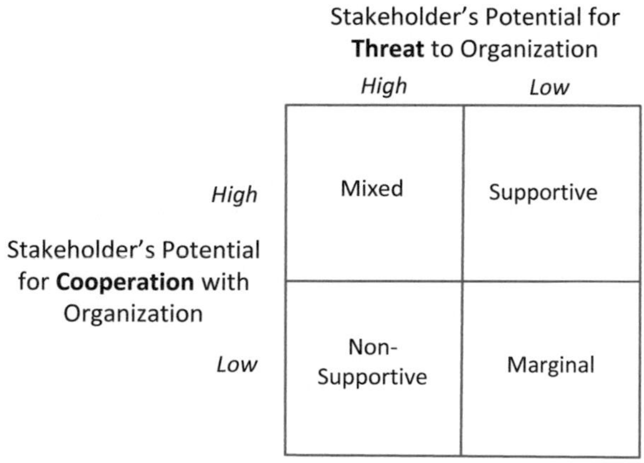

Fig. 6.3 Stakeholder attitude characterization, adapted from Savage et al. (1991)

stakeholder attitude according to two characteristics: (1) potential for threat and (2) potential for cooperation, as shown in Fig. 6.3.

Savage et al. (1991) propose four strategies for dealing with stakeholders of varying attitudes as follows:

1. *Involve*: Leverage key relationships and network, possibly engage in an active champion role.
2. *Collaborate*: Enter strategic alliances or partnerships, educate if necessary.
3. *Defend*: Move toward reducing dependency on stakeholder.
4. *Monitor*: Gather information and observe.

To this set of four strategies, we add the strategy of *no action*. As we will show in the ensuing discussion, this is a valid approach for particular stakeholder classification and attitudes. Figure 6.4 shows all of these strategies in what Hester, Bradley, and Adams (2012) term a *continuum of stakeholder involvement*.

The continuum of stakeholder involvement shows the strategies available for an organization to use when dealing with a stakeholder. As the strategies progress from left to right, stakeholders become more involved, thereby requiring substantially more resources at every step, thus, *monitor* is more resource intensive than *no action*, *defend* is more resource intensive than *monitor*, and so on. Savage et al. (1991) propose the following strategies for their four stakeholder types:

- *Involve* supportive stakeholders
- *Collaborate* with mixed stakeholders

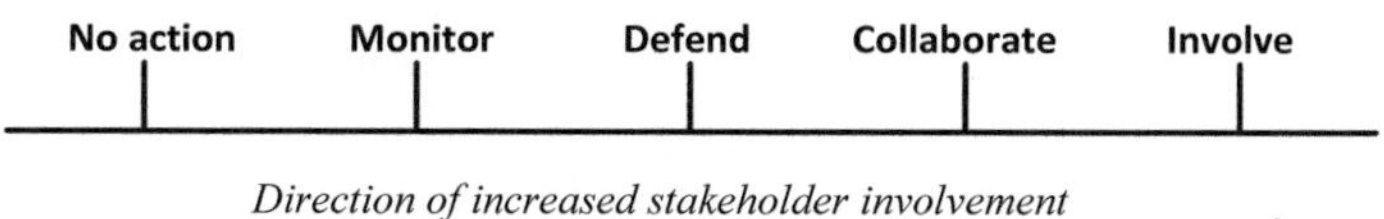

Fig. 6.4 Continuum of stakeholder involvement, adapted from Hester et al. (2012)

- *Defend* against nonsupportive stakeholders
- *Monitor* marginal stakeholders.

Aligning the appropriate strategy with a stakeholder's attitude toward a problem is critically important. Expending too many resources on a stakeholder is at best a resource waste and at worst a risk. We risk alienating that particular stakeholder and turning their attitude into one that is in opposition to our endeavor. Thus, if we involve a nonsupportive stakeholder, they will consume resources which are better spent on stakeholders who may otherwise have supported our effort. Conversely, spending insufficient resources on a stakeholder means that we have wasted an opportunity. Merely collaborating with a supportive stakeholder means that we have potentially missed out on an opportunity to involve them in the solution process.

Savage et al. (1991) devote specific attention to the dangers of the collaborate strategy. Collaborating with a mixed stakeholder can result in either a positive outcome (they become supportive) or a negative one (they become nonsupportive). Thus, once again with an eye toward resource conservation, we must be careful as to which stakeholders we choose to engage with and to what extent. While offering an additional stepping stone toward a complete set of stakeholder strategies, we must point out a deficiency of the approach developed by Savage et al. (1991), namely that it doesn't account for the relative importance of the stakeholder. Using the typology of Mitchell et al. (1997), we understand the importance of investing more heavily in ensuring that definitive stakeholders (e.g., those with power, legitimacy, and urgency) maintain a supportive attitude toward our endeavor. Thus, both approaches provide insights into the stakeholder problem, yet neither paints a complete picture. For a more comprehensive approach to dealing with stakeholders, we can utilize the concept of a Power-Interest grid, a common stakeholder analysis technique which plots stakeholder *Power* versus *Interest* in order to consider both elements as they relate to an engagement strategy. The Power-Interest grid approach, developed by Mendelow (1991), is shown in Fig. 6.5, complete with stakeholder categories from Eden and Ackermann (1998).

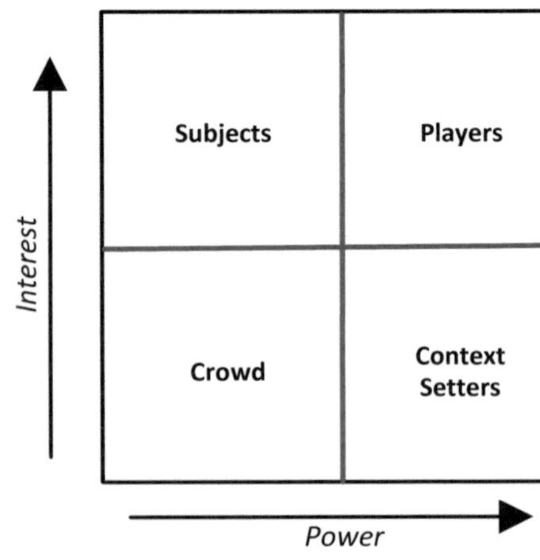

Fig. 6.5 Power-Interest grid, adapted from Eden and Ackermann (1998)

We can adapt the Power-Interest grid approach using Prominence, as defined in the previous section, as a proxy measurement for Power and Support, as defined below using terms from Savage et al. (1991), as a proxy measurement for Interest. We can calculate a stakeholder's support for a given problem as follows:

$$Support_i = C_i - T_i \text{ on } [-1,1] \tag{6.2}$$

where i represents the ith stakeholder, T_i is potential for threat, defined on [0,1], and C_i is potential for cooperation, defined on [0,1], with 0 being a complete lack of the particular attribute, and 1 being the highest possible value for both T_i and C_i.

Interest, as it is conceptualized by Mendelow (1991) and Eden and Ackermann (1998), is simply the magnitude of $Support_i$. The suggested use of *Support*, vice *Interest*, is purposeful. *Interest* is devoid of direction; thus, an individual can be interested in our project but only because they wish to see it fail. Conversely, they may be interested in our project as an active champion. Given the insights of Savage et al. (1991), it is clear that direction of support will have a bearing on the strategy we choose to engage a stakeholder. Power-Interest grids can be adapted to account for support and prominence, and to reflect appropriate stakeholder strategies, as shown in the adapted Power-Interest grid in Fig. 6.6.

While Fig. 6.6 shows crisp separation between categories, the reality is that category membership is fuzzy. Thus, this grid is intended merely as a guideline to

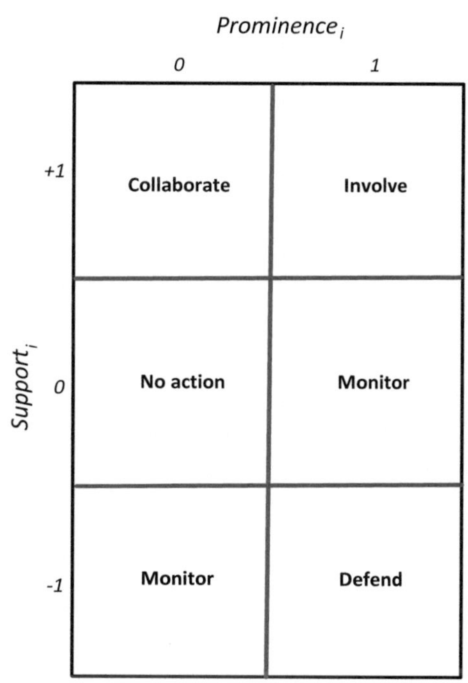

Fig. 6.6 Prominence-support grid, including stakeholder strategies

readers. In accordance with Savage et al. (1991) and Mitchell et al. (1997), we can identify five strategies corresponding to the cells shown in Fig. 6.6 as follows:

- *Involve* supportive, prominent stakeholders
- *Collaborate* with supportive, less prominent stakeholders
- *Defend* against nonsupportive, prominent stakeholders
- *Monitor* neutral, prominent, and nonsupportive, less prominent stakeholders
- *Take no action* pertaining to neutral, less prominent stakeholders.

The goal of each of these strategies is to ensure all active stakeholders (latent, expectant, and definitive) are supportive and to increase the prominence of supportive stakeholders. Figure 6.7 illustrates the outcome when implementing the strategies based on Fig. 6.6.

Examination of Fig. 6.7 provides some insight regarding stakeholder treatment. We would like to secure all stakeholders as supportive. Of course, this becomes a resource constraint issue as engagement of stakeholders is a resource-intensive process that is not without risk. To this end, we must engage stakeholders in an effort to maximize our resources. However, this entire analysis supposes that stakeholders exist in isolation, which we know not to be the case. In an effort to understand stakeholder interactions (and their effect on the prioritization of our actions), we now turn to the idea of mapping stakeholder objectives.

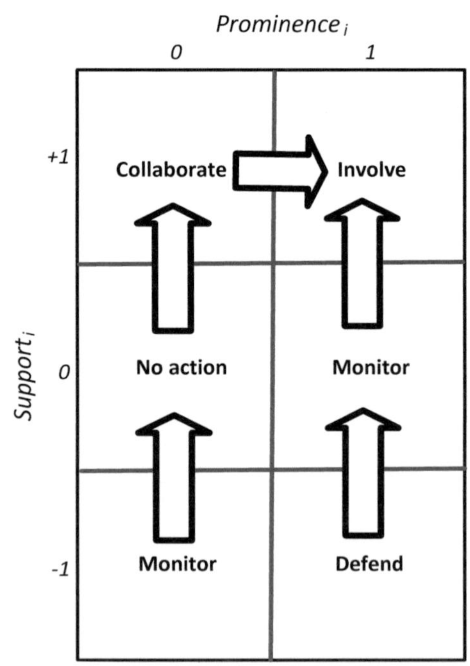

Fig. 6.7 Transformation of stakeholders

6.5 Map Stakeholder Objectives

At this point, we have brainstormed appropriate stakeholders, and determined their prominence and support. However, we lack the ability to prioritize our efforts regarding stakeholder engagement. This is crucial to our endeavor as we must focus our stakeholder management efforts on the stakeholders who can affect the largest amount of change. In order to determine engagement priority, we must first think about our stakeholders in relation to one another. We must complete the fourth step in the stakeholder analysis process, mapping our stakeholder's objectives.

Since the *darkness principle* (Cilliers, 1998) informs us we are not capable of complete knowledge of a mess, we must consider multiple perspectives (i.e., stakeholders) and their relation to one another. Our suggested mechanism for capturing these relationships is with a network-based representation of stakeholders and their relationships. Nodes within a network may be thought to represent stakeholders (and their objectives), while a connection between two nodes indicates a causal influence between the two nodes. More specifically, a directed graph can be constructed, where the directionality of arrows between nodes may represent the direction of influence exerted by one stakeholder on another (e.g., the CEO of a company, whose goal is to maximize company profits, may exert influence over the company janitor, whose goal is to keep his job, and this influence is likely not symmetric, thus in this case their relationship is unidirectional), as well as the magnitude and direction of this influence on [−1,+1], in keeping with FCM guidelines discussed in Chap. 5. A depiction of this relationship is shown in Fig. 6.8.

Thus, we should create a concept for each stakeholder and their associated objective (from Step 1) and identify any causal linkages between these objectives. This allows for a more holistic perspective of our stakeholders and their relationships between one another. When we proceed to the next stage of stakeholder analysis, this will help us prioritize our efforts in seeking resolution to our mess.

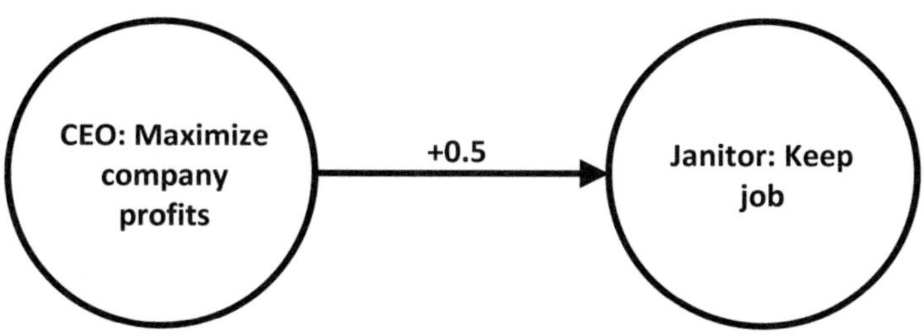

Fig. 6.8 Illustration of causal influence

6.6 Determine Stakeholder Engagement Priority

At this point, we have brainstormed appropriate stakeholders, determined their attitude and classification, and mapped them. The fifth step in the stakeholder analysis process is to determine the priority with which we should engage stakeholders to gain increased understanding about our problem. In order to fully capture the relationship between stakeholders, we can explore various notions of what is termed node centrality (Bavelas, 1948; Bonacich, 1987; Borgatti, 2005; Borgatti, Carley, & Krackhardt, 2006; Freeman, 1979). Centrality is a measure of determining the importance of a node within a network. Table 6.3 is a list of three formalized measures of centrality as formalized by Freeman (1979).

There are several issues with the measures present in Table 6.3. Directed graphs are problematic to assess using the *closeness* measure as many nodes in a directed graph may be unconnected with one another (i.e., we cannot travel from node A to node B). Further, most networks have a large proportion of nonshortest-path nodes that therefore are each equally determined to have zero *betweenness*, and thus, no influence on the network. Finally, the measures in Table 6.3 were intended only for binary networks, i.e., those with arcs whose values are either one of zero. This is problematic as stakeholders are likely to have varying degrees of influence on one another and thus, a more sophisticated measure is necessary. Barrat, Barthélemy, Pastor-Satorras, and Vespignani (2004), Brandes (2001), and Newman (2001) attempted to generalize the work of Freeman (1979) to weighted networks, but their work focused on weighted arcs and not on the number of connections of a particular node.

If we explore *degree*, recent research has provided adequate evolution to consider its use in a directed graph. Freeman's original notion of degree can be defined using nomenclature from Opsahl, Agneessens, and Skvoretz (2010) as follows:

Table 6.3 Freeman's measures of centrality (Freeman, 1979)

Measure of centrality	Description	Comments
Degree	The number of nodes that a given node is adjacent to	While this is a simple, and therefore appealing, measure, it lacks the ability to account for the relative importance of the nodes to which a given node is connected to
Closeness	The inverse sum of shortest distances to all nodes from a given node	This has problems when networks have unconnected nodes, a problem that is of particular concern in a directed graph, where connections may not be symmetric
Betweenness	The degree to which a node lies on a shortest path between any other two nodes	Its appearance along a shortest path indicates that the node acts as a conduit for information flow, and thus, is an important contributor to network information transfer

$$k_i = C_D(i) = \sum_{j}^{N} x_{ij} \tag{6.3}$$

where C_D is the degree centrality, i is the node of interest, j represents all other nodes, N is the total number of nodes, and x_{ij} is the adjacency matrix, defined as 1 if an arc exists between i and j, and 0 otherwise.

Degree has generally been revised (Barrat et al., 2004; Opsahl et al., 2010; Opsahl, Colizza, Panzarasa, & Ramasco, 2008) for weighted networks as the sum of arc weights and redefined as *strength* as follows:

$$s_i = C_D^W(i) = \sum_{j}^{N} w_{ij} \tag{6.4}$$

where C_D^W is the weighted degree centrality and w_{ij} is the weighted adjacency matrix, defined as the weight of the connection between i and j (>0) if i is connected to j, and 0 otherwise. This weight is an assessment of the strength of causal influence between concepts, defined on [−1,1].

A further complication is the presence of both positive and negative weights. Thus, in order to calculate strength properly, we define a new term, $s*$, which calculates strength based only on the magnitude of influences as follows:

$$s_i^* = \sum_{j}^{N} |w_{ij}| \tag{6.5}$$

This measure of influence can be conceptualized as a proxy for the *communication principle* (Shannon 1948a, b); are cited in the text but not provided in the reference list. Please provide the respective references in the list or delete these citations." –>nnon 1948a, b); i.e., if a strong influence exists between two stakeholders, then a strong communication channel can be thought to exist between the two, whereas the absence of influence is an indicator of poor communication. Two additional elements are worth noting for this assessment. The first element is that the relationships are likely not to demonstrate symmetric behavior. That is, the CEO discussed in Fig. 6.8 likely has a high influence on the Janitor, yet the feeling is likely not to be mutual. Further, we can think of entities that exhibit no influence on one another as not having a linkage between them. Thus, in the network depiction of the problem, no arc exists between any stakeholders who have no influence between them (i.e., $w_{ij} = 0$).

Simply evaluating their strength, however, is insufficient. "Since degree and strength can be both indicators of the level of involvement of a node in the surrounding network, it is important to incorporate both these measures when studying the centrality of a node" (Opsahl et al., 2010, p. 246). Based on this assertion, Opsahl et al. (2010) developed a measure which combines degree and strength as follows (note, this measure has been modified to use $s*$):

$$C_D^{W\alpha}(i) = k_i\left(\frac{s_i^*}{k_i}\right)^{\alpha} = k_i^{(1-\alpha)}(s_i^*)^{\alpha} \quad (6.6)$$

where α is a positive tuning parameter used to adjust the relative importance of degree and strength. If $\alpha = 0$, the measure reduces to degree, as shown in Eq. 6.3. If $\alpha = 1$, the measure reduces to strength, as shown in Eq. 6.4. We suggest adopting an α of 0.5 for the purposes of this analysis, thereby ensuring that the effect of both strength and degree are accounted for.

Use of this measure is complicated somewhat by the fact that our stakeholder network is directed. Opsahl et al. (2010) elaborate on this issue as follows:

> Directed networks add complexity to degree as two additional aspects of a node's involvement are possible to identify. The activity of a node, or its gregariousness, can be quantified by the number of ties that originate from a node, k^{out}. While the number of ties that are directed toward a node, k^{in}, is a proxy of its popularity. Moreover, since not all ties are not necessarily reciprocated, k^{out} is not always equal to k^{in}. For a weighted network, s^{out} and s^{in} can be defined as the total weight attached to the outgoing and incoming ties, respectively. However, these two measures have the same limitation as s in that they do not take into account the number of ties. (p. 247)

Opshal et al. (2010) go on to define *activity* and *popularity*, respectively, as follows (note again, these measures are modified to use s^*):

$$Activity(i) = C_{D-\text{out}}^{W\alpha}(i) = k_i^{\text{out}}\left(\frac{s_i^{*-\text{out}}}{k_i^{\text{out}}}\right)^{\alpha} \quad (6.7)$$

$$Popularity(i) = C_D^{W\alpha}(i) = k_i^{\text{in}}\left(\frac{s_i^{*-\text{in}}}{k_i^{\text{in}}}\right)^{\alpha} \quad (6.8)$$

Activity is a measure of the amount of reach that a stakeholder has in a network. It is a function of both the number of outgoing connections and the strength of these connections. Individuals with high activity are seen as highly connected and therefore important because their perspective carries a great deal of weight within the network. Recall that the *principle of redundancy of potential command* (McCulloch, 1959) informs us that "power resides where information resides" (Adams, 2011, p. 151). Those individuals with high activity are perceived to have power in our stakeholder network. They can disseminate information rapidly to many individuals. Thus, even though they may not be the CEO of an organization, their connectedness affords them power.

Popularity can be conceptualized of as the inverse of the ease with which someone is able to be influenced. That is to say, those with high popularity have a high number of incoming perspectives and are difficult to influence as a result. Those with low popularity have a small number of incoming perspectives and should be easier to influence with less dissenting opinions to deal with. Popularity considers both the number of incoming connections and the strength of those connections.

Table 6.4 Intersection of popularity and activity

		Popularity	
		Low	High
Activity	High	Important and easy to influence	Important but hard to influence
	Low	Not important but easy to influence	Not important and hard to influence

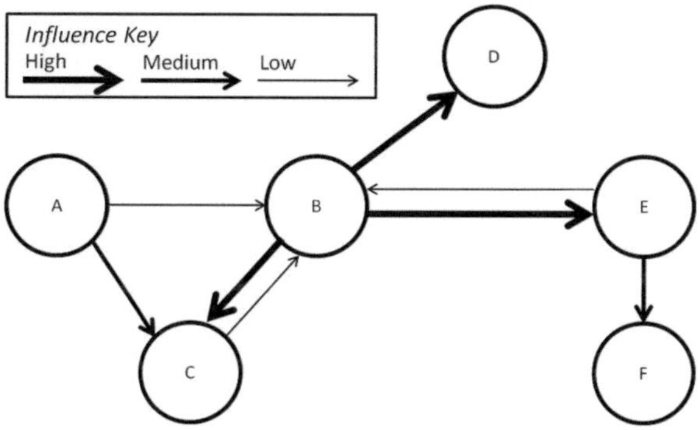

Fig. 6.9 Illustrative influence network

In terms of engaging our stakeholders, we must consider both their popularity and their activity. We want to influence individuals that are easy to influence, but that are important. The relationship of these two elements is important and is shown in Table 6.4.

A simple illustrative example demonstrates the calculation of activity and popularity and how we would use these characteristics to prioritize our stakeholders. We adopt an illustrative example provided by Opsahl et al. (2010) and shown in Fig. 6.9 with directionality added and high influence defined as a weight of 1, medium influence a weight of 0.5, and low influence a weight of 0.25, all positive for simplicity's sake. Note all causal influences in this network are positive.

Table 6.5 illustrates the Popularity and Activity results for this network, including the supporting calculations necessary for strength and degree.

Examination of Table 6.5 shows that the most active node is B. This makes sense as B has more outgoing influences than any other node and these are all rated as high. Further examination shows that the least popular (i.e., easiest to influence)

Table 6.5 Illustrative network characteristics

Node	k_i^{in}	k_i^{out}	s_i^{*-in}	s_i^{*-out}	Popularity	Activity
A	0	2	0	0.75	0.0	1.2
B	3	3	0.75	3	1.5	3.0
C	2	1	1.5	0.25	1.7	0.5
D	1	0	1	0	1.0	0.0
E	1	2	1	0.75	1.0	1.2
F	1	0	0.5	0	0.7	0.0

Table 6.6 Illustrative prioritization of stakeholders

Node	Popularity	Activity	Engagement priority
B	1.5	3	1
A	0	1.2	2
E	1	1.2	3
C	1.7	0.5	4
F	0.7	0	5
D	1	0	6

node is node A. This also makes sense as it has no incoming influences and therefore, no outside detracting opinions to contend with. Accounting for popularity and activity to determine stakeholder engagement priority should be done with an eye for accomplishing the movement of all stakeholders toward a supportive role (as shown in Fig. 6.7). It is our belief that, in order to do this, all stakeholders should be sorted by activity first (in descending order), and then, if multiple individuals share the same activity level, by popularity (in ascending order). This order reflects the order in which stakeholders should be engaged in support of an effort. Table 6.6 illustrates the prioritization values for the illustrative example.

One final element should be considered in engaging stakeholders. Each of the stakeholders A-F has a unique strategy associated with it, defined by the taxonomy shown in Fig. 6.6. Stakeholders with a more involved strategy (i.e., involve or collaborate) will require more resources to engage than a stakeholder demanding a more passive strategy (i.e., defend, monitor, or no action). This is a problem for us as we struggle with how to dispatch our scarce resources as we likely will have less resources than we have stakeholders. Resources must be utilized in a manner which gives us the most *bang for the buck*, a measure consistent with the approach presented here.

Before moving on the next step of the stakeholder analysis process, we would be remiss in not pointing out that, while we believe our first order approach to engagement priority is sufficient, we have also developed a higher order approach involving Leontief (1951) input-output modeling; the reader is referred to Hester and Adams (2013) for details of this approach. The approach presented in this book is intended to provide the reader with an approachable method for determining stakeholder priority without sacrificing resultant method insight. We believe the presented approach does just that.

6.7 Develop a Stakeholder Management Plan

At this point in the stakeholder analysis process, we have brainstormed stakeholders, classified them, determined their level of support, and mapped their objectives. The sixth step is the development of a Stakeholder Management Plan (SMP). The SMP allows us to track stakeholders and maintain a plan for dispatching resources to secure and maintain a stakeholder's support for our effort. At a minimum, a SMP should include the following:

Table 6.7 Construct for a stakeholder management plan (SMP)

Stakeholder name	Wants	Prominence	Support	Priority of engagement	Strategy

- Stakeholder name/identifier (from Step 1)
- Stakeholder wants (from Step 1)
- Stakeholder prominence (from Step 2)
- Stakeholder support (from Step 3)
- Stakeholder engagement priority (from Step 5)
- Strategy (defend, collaborate, etc.) for dealing with stakeholder, based on their prominence and interest (from Step 3)
- Method for engagement (e-mails, in-person, etc.)
- Frequency of engagement (e.g., monthly, weekly)
- Responsible party who pursues the identified strategy
- Notes that are necessary for housekeeping purposes (call before showing up to office, prefers early morning, etc.).

Table 6.7 is a generic construct for a SMP. Several columns have been eliminated for ease of reading, namely the method for engagement, frequency of engagement, responsible party, and notes.

Once a stakeholder management plan is generated, stakeholders should be sorted by their priority of engagement. This presents a ranking of the order in which stakeholders should be engaged. Recalling that the strategy for engagement is determined as a function of both classification and attitude, this provides a first pass at what level of involvement we should wish to afford a particular stakeholder. We wish to heavily involve those stakeholders that are both prominent and supportive. However, in most complex problems the myriad number of stakeholders involved will likely result in redundant engagement strategies across stakeholders. For example, multiple individuals will be assigned the strategy of *Involve*. Thus, stakeholder activity and popularity are used to determine engagement priority.

6.8 Manage Stakeholders

Once a stakeholder management plan has been generated, the organization is charged with executing it. That is to say, we must *follow through* on the strategies outlined by the SMP. The stakeholder analysis process does not end here, however. Thus, after establishing a SMP, we may wish to revisit our brainstorming exercise to identify stakeholders, perhaps streamlining our list as our knowledge gained from the process informs us that many of our previously identified stakeholders are no longer relevant to the problem at hand. Given its recursive and iterative nature, the process will necessarily continue throughout the resolution of our problem.

In each of the chapters discussing the six systemic thinking perspectives, a framework is provided to assist the reader in understanding which steps must be followed to sufficiently address the perspective as it pertains to a mess and its constituent problems. The first of these frameworks is provided in the following section.

6.9 Framework for Addressing *Who* in Messes and Problems

Undertaking a stakeholder analysis requires an individual to complete the six-step process outlined in this chapter as it pertains to an identified problem, namely

1. Brainstorm stakeholders
2. Classify stakeholders
3. Evaluate stakeholder attitudes
4. Map stakeholder objectives in a FCM
5. Determine stakeholder engagement priority
6. Develop a stakeholder management plan.

Each of these six steps is required to completely account for stakeholders in our messes and constituent problems. The following section demonstrates each step on an example problem.

6.10 Example Problem

The problem introduced in this section will be analyzed throughout the remainder of this text. It represents a more comprehensive examination of the problem discussed briefly in Hester et al. (2012). In this example, a local real estate developer sought to rezone portions of an upscale, single family home residential neighborhood. The impetus for this intended rezoning was the Great Recession during the late 2000s and early 2010s, which caused a decrease in the purchasing power of potential homebuyers. In order to recoup their investment in land which was suddenly no longer profitable, the developer aimed to build condominiums, which required that they rezone the land, necessitating approval from the city council. Viewing the change as undesirable largely from a financial standpoint, a group of nine local communities opposed the rezoning process and fought adamantly to prevent it. The intended rezoning needed to take into account the values of important stakeholders (e.g., neighbors, local government) in order to ensure project success.

The example is being discussed from the perspective of the developer, who is seeking to determine which stakeholders they will need to garner support from. The developer has been included as a stakeholder in the analysis in order to understand their relationship to other relevant stakeholders.

6.10.1 Example Stakeholder Brainstorming

Brainstorming stakeholders for the rezoning problem yields the following stakeholders and their associated wants as follows:

1. The real estate developer *wants* financial gain from the project.
2. Nine local communities *want* to maintain their property values.
3. Local media *want* news stories that sell.
4. City Staff *wants* minimal disruption.
5. City Planning Commission *wants* compliance with regulations.
6. City Council *wants* to be reelected.

While many more individuals and groups could be added into the analysis, it is thought that the initial stakeholder analysis should include, at a minimum, these six entities and their associated desires.

6.10.2 Example Stakeholder Classification

Table 6.8 shows evaluations of the attributes and class for each of the stakeholders identified in the previous section. They have been sorted according to decreasing order of prominence.

Clearly, the two most prominent stakeholders are the real estate developer and the local community affected by the developers' efforts. This is fairly intuitive as both of these groups possess all three attributes of power, legitimacy, and urgency. Moving to the next tier, the City Planning Commission and the City Council, both have power and legitimacy, but they are unlikely to possess the urgency to place a

Table 6.8 Example stakeholder classification

Stakeholder	Stakeholder attribute			Prominence
	Power	Legitimacy	Urgency	
The real estate developer	1	1	1	1.0
Nine local communities	1	1	1	1.0
City Planning Commission	1	1	0	0.67
City Council	1	1	0	0.67
Local media	0	1	0	0.33
City Staff	0	1	0	0.33

priority on the execution of this particular project due to other commitments. Finally, the local media and assorted city staff have legitimacy in that they should be involved in the planning process, but they have neither power nor urgency; they cannot directly influence the other members of the problem and they don't appear on the surface to have the urgency to see the project's execution occur.

6.10.3 Example Stakeholder Attitude Evaluation

Table 6.9 shows evaluations of the potential for threat and potential for cooperation for each of the stakeholders identified in the previous section. These two parameters provide an identification of the attitude of each stakeholder. They have been sorted in decreasing order of support according to their assigned stakeholder attitude.

Both the real estate developer and city staff are seen as supportive of this effort. The developer's support is obvious, while perception of the city staff as supportive comes from their unwillingness to object to the project's development. The City Planning Commission, City Council, and local media all have a high potential for cooperation as they would like to see the project succeed, but their high potential for threat demonstrates their unwillingness to be a champion for project success at the cost of their more prominent desires. Thus, these three stakeholder groups possess a mixed attitude. Finally, the nine local communities pose a high potential for threat and a low potential for cooperation. They have a vested interest in seeing the project fail as they are opposed to it on fundamental grounds (i.e., they believe it is likely to reduce their property values). They are therefore nonsupportive of the effort.

6.10.4 Example Stakeholder Objective Mapping

With classification and attitude defined in the previous two sections, Fig. 6.10 shows a stakeholder objective map (an FCM), including the influence (direction and magnitude) for all identified stakeholders involved in the problem. The thicker the line, the stronger the causal influence.

Table 6.9 Example stakeholder attitude evaluation

Stakeholder	Potential for threat	Potential for cooperation	Support
The real estate developer	0	1	1
City Staff	0	1	1
City Planning Commission	1	1	0
City Council	1	1	0
Local media	1	1	0
Nine local communities	1	0	−1

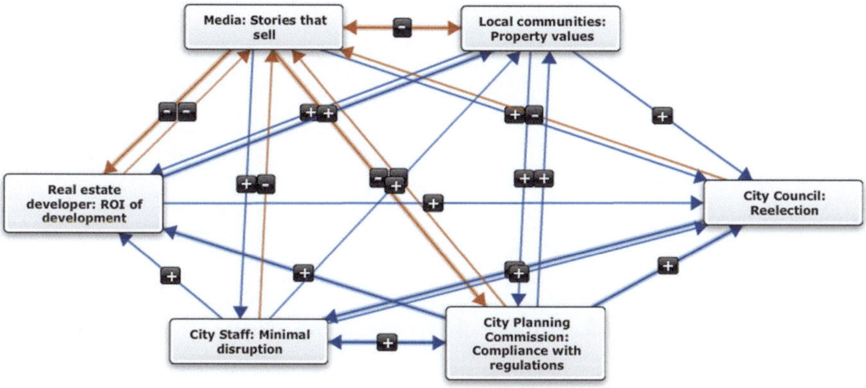

Fig. 6.10 Stakeholder relationship map

After examination of this relationship map, it is clear that there are a number of complicated connections at play in this problem.

6.10.5 *Example Stakeholder Engagement Priority*

In order to calculate the stakeholder engagement priority for all the stakeholders in the real estate development project, we need to calculate k_i^{in}, k_i^{out}, s_i^{*-in}, s_i^{*-out}, *Popularity*, and *Activity*, in accordance with earlier equations. These results are shown in Table 6.10.

We then sort the stakeholders by activity first (in descending order), and then, by popularity (in ascending order). Table 6.11 illustrates the order in which stakeholders should be engaged in support of this effort.

It is clear that the nine local communities should be prioritized in terms of their engagement in the development project. This makes intuitive sense given the stakeholder relationships shown in Fig. 6.10. On the other end of the spectrum, the city staff should be the final entity engaged. They have no influence on any other stakeholder and, thus, should be given a low priority in terms of their engagement.

Table 6.10 Real estate network characteristics

Stakeholder	k_i^{in}	k_i^{out}	s_i^{*-in}	s_i^{*-out}	Popularity	Activity
The real estate developer	3	3	2	0.75	2.45	1.50
City Staff	2	0	0.75	0	1.22	0.00
City Planning Commission	3	2	1	1.5	1.73	1.73
City Council	4	3	1.5	1.25	2.45	1.94
Local media	1	2	0.25	0.5	0.50	1.00
Nine local communities	1	4	0.25	2	0.50	2.83

Table 6.11 Real estate stakeholder prioritization

Stakeholder	Activity	Popularity	Engagement priority
Nine local communities	2.83	0.50	1
City Council	1.94	2.45	2
City Planning Commission	1.73	1.73	3
The real estate developer	1.50	2.45	4
Local media	1.00	0.50	5
City Staff	0.00	1.22	6

6.10.6 Example Stakeholder Management Plan

The final step in analyzing this example is to develop a stakeholder management plan. An example stakeholder management plan is shown below in Table 6.12. Two elements should be noted. Just like in Table 6.7, several columns have been eliminated for ease of reading, namely the method for engagement, frequency of engagement, responsible party, and notes. Second, as this stakeholder assessment is being performed by the real estate developer, their priority of engagement is a nonissue. They are inherently a part of the stakeholder management process. Thus, although they are both prominent and supportive, they are moved to the bottom of the list.

Using information gained by holistically considering our mess, we can identify priorities and manage our stakeholders. What is clear at this stage is that the strategy we employ varies greatly based on the stakeholder we are considering. It is very important, for example, for the real estate develop to defend against the nine local communities, rather than ignoring them. In order to do so, they should consider the wants of the communities (property values and quality of life). This is directly counter to their chosen strategy of simply ignoring the communities. Had they undertaken a thorough stakeholder analysis, they might have saved themselves from the eventual failure of their project. Unfortunately for them, they did not (Hester et al., 2012).

Table 6.12 Example stakeholder management plan

Stakeholder name	Wants	Prominence	Support	Priority of engagement	Strategy
Nine local communities	Property values and quality of life	1	−1	1	Defend
City Council	Re-election	0.67	0	2	Monitor
City Planning Commission	Regulation compliance	0.67	0	3	Monitor
Local media	Stories that sell	0.33	0	4	No action
City Staff	Minimal disruption	0.33	1	5	Collaborate
The real estate developer	Financial gain	1	1	n/a	Involve

6.11 Summary

Because stakeholders exist at the center of all systems problems and serve as the principal contributors to the solution of these problems, we must formally address them as part of the solution to any systems problem. In this chapter, we developed a six-step approach to stakeholder analysis and management. This approach includes identification of stakeholders, classification of these stakeholders, assessment of their attitude, calculation of their engagement priority, developing a plan for managing them, and carrying out the plan (i.e., managing them). This comprehensive technique is an important discriminator enabling systems practitioners with an effective method for dealing with stakeholders appropriately.

After reading this chapter, the reader should be able to:

1. Identify and classify stakeholders for a problem;
2. Evaluate stakeholder attitudes;
3. Map stakeholder objectives;
4. Calculate stakeholder engagement priority; and
5. Develop a stakeholder management plan.

References

Adams, K. M. (2011). Systems principles: Foundation for the SoSE methodology. *International Journal of System of Systems Engineering, 2*(2/3), 120–155. doi:10.1504/IJSSE.2011.040550.

Barrat, A., Barthélemy, M., Pastor-Satorras, R., & Vespignani, A. (2004). The architecture of complex weighted networks. *Proceedings of the National Academy of Sciences of the United States of America, 101*(11), 3747–3752. doi:10.1073/pnas.0400087101.

Bavelas, A. (1948). A mathematical model for group structures. *Applied Anthropology, 7*(3), 16–30.

Bonacich, P. (1987). Power and centrality: A family of measures. *American Journal of Sociology, 92*(5), 1170–1182.

Borgatti, S. P. (2005). Centrality and network flow. *Social Networks, 27,* 55–71.

Borgatti, S. P., Carley, K. M., & Krackhardt, D. (2006). On the robustness of centrality measures under conditions of imperfect data. *Social Networks, 28,* 124–136.

Brandes, U. (2001). A faster algorithm for betweenness centrality. *Journal of Mathematical Sociology, 25*(2), 163–177.

Cilliers, P. (1998). *Complexity and postmodernism: Understand complex systems*. New York: Routledge.

Clarkson, M. B. E. (1995). A stakeholder framework for analyzing and evaluating corporate social performance. *Academy of Management Review, 20*(1), 92–117.

Dahl, R. A. (1957). The concept of power. *Behavioral Science, 2*(3), 201–215.

Eden, C., & Ackermann, F. (1998). *Making strategy: The journey of strategic management*. London: Sage.

Freeman, L. C. (1979). Centrality in social networks: Conceptual clarification. *Social Networks, 1* (3), 215–239. doi:10.1016/0378-8733(78)90021-7.

Freeman, R. E. (1984). *Strategic management: A stakeholder approach*. Boston: Pitman.

Friedman, A. L., & Miles, S. (2002). Developing stakeholder theory. *Journal of Management Studies, 39*(1), 1–21.

Friedman, M. (1962). *Capitalism and freedom*. Chicago, IL: University of Chicago Press.

Hester, P. T., & Adams, K. M. (2013). Determining stakeholder influence using input-output modeling. *Procedia Computer Science, 20,* 337–341.

Hester, P. T., Bradley, J. M., & Adams, K. M. (2012). Stakeholders in systems problems. *International Journal of System of Systems Engineering, 3*(3/4), 225–232.

Hitch, C. J. (1953). Sub-optimization in operations problems. *Journal of the Operations Research Society of America, 1*(3), 87–99.

Leontief, W. W. (1951). Input-output economics. *Scientific American, 185*(4), 15–21.

McCulloch, W. S. (1959). *Embodiments of mind*. Cambridge, MA: MIT Press.

Mendelow, A. (1991). *Stakeholder mapping*. Paper presented at the Proceedings of the 2nd International Conference on Information Systems, Cambridge, MA.

Mitchell, R. K., Agle, B. R., & Wood, D. J. (1997). Toward a theory of stakeholder identification and salience: Defining the principle of who and what really counts. *Academy of Management Review, 22*(4), 853–886. doi:10.5465/amr.1997.9711022105.

Newman, M. E. J. (2001). Scientific collaboration networks. I. Network construction and fundamental results. *Physical Review E, 64*(1), 0161311–0161318.

Opsahl, T., Agneessens, F., & Skvoretz, J. (2010). Node centrality in weighted networks: Generalizing degree and shortest paths. *Social Networks, 32*(3), 245–251. doi:10.1016/j.socnet.2010.03.006.

Opsahl, T., Colizza, V., Panzarasa, P., & Ramasco, J. J. (2008). Prominence and control: The weighted rich-club effect. *Physical Review Letters, 101*(16), 1687021–1167024.

Pfeffer, J. (1981). *Power in organizations*. Marshfield, MA: Pitman.

Savage, G. T., Nix, T. W., Whitehead, C. J., & Blair, J. D. (1991). Strategies for assessing and managing organizational stakeholders. *The Executive, 5*(2), 61–75.

Shannon, C. E. (1948a). A Mathematical Theory of Communication, Part 1. *Bell System Technical Journal, 27*(3), 379–423.

Shannon, C. E. (1948b). A Mathematical Theory of Communication, Part 2. *Bell System Technical Journal, 27*(4), 623–656.

Suchman, M. C. (1995). Managing legitimacy: Strategic and institutional approaches. *The Academy of Management Review, 20*(3), 571–610. doi:10.2307/258788.

Weber, M. (1947). *The theory of social and economic organization*. New York: Free Press.

Chapter 7
The *What* of Systemic Thinking

Abstract The main focus of the *what* question of systemic thinking is on the articulation and organization of the objectives of the problem that we are trying to gain increased understanding of. Given that a mess is a system of problems as we described it in Chap. 2, we take the occasion in this chapter to dissect a given problem into its basic elements in order to gain further insight regarding its parent mess. This chapter builds on the stakeholder analysis undertaken in the previous chapter. The chapter begins by discussing the anatomy of a problem. Then, the importance of objectives is discussed. Next, we address objective articulation. We then distinguish between fundamental and means objectives and discuss how to organize them to increase our understanding. Finally, a framework for addressing the *what* question is presented and this framework is demonstrated on the real estate problem introduced in Chap. 6.

7.1 Anatomy of a Problem

There are many ways in which a problem may be decomposed into its constituent elements. Each is value-added and can be used adequately for this discussion. We choose to align our approach with decision analysis for its sustained and successful effort in this field. For the sake of simplicity, we begin by introducing the standard terminology of *problem*, *objective*, *attribute*, and *alternatives* (e.g., Hammond, Keeney, & Raiffa, 2002; Keeney, 1992; Keeney & Raiffa, 1976). These terms, and their definitions as they are used in this text, are found in Table 7.1.

Figure 7.1 shows a notional example of the relationship between the elements shown in Table 7.1. A few elements are worth noting regarding the depicted relationships. A given mess involves multiple interconnected problems, which may share one or more of the same objectives (e.g., an objective of profit maximization may be shared across multiple problems). This is complicated by the presence of conflict between problems. For example, two competing businesses that both wish to maximize profits may wish to do so at the expense of one another. Also, more than one objective may be evaluated using the same attribute. For example,

P.T. Hester and K.M. Adams, *Systemic Decision Making*, Topics in Safety, Risk, Reliability and Quality 33, DOI 10.1007/978-3-319-54672-8_7

Table 7.1 Problem terminology

Term	Definition
Problem	An undesirable situation without a clear resolution that an individual or group wishes to see resolved (see Chap. 2 for further discussion)
Objective	"A statement of something that one desires to achieve" (Keeney, 1992, p. 34)
Attribute	A scale used to measure achievement of a fundamental objective
Alternative	A potential course of action (alternatives are revisited in Chap. 13)

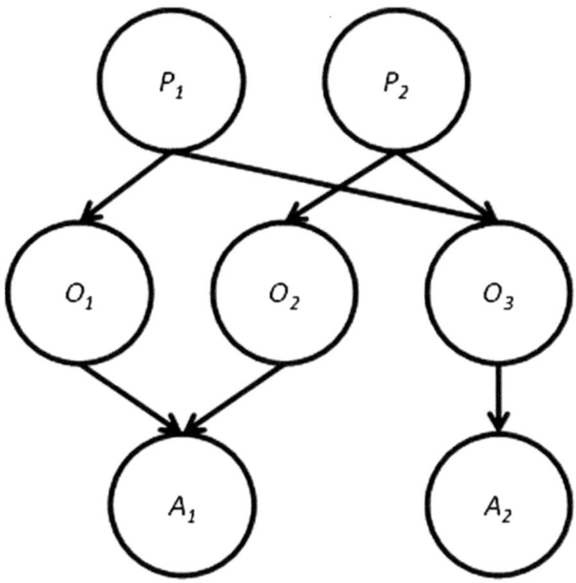

Fig. 7.1 Illustration of Multiple Problems (P_i), Objectives (O_i), and Attributes (A_i)

maximize profit and maximize savings, two distinct objectives, may both be measured using the same attribute of money (e.g., dollars). Measurement of objective achievement using attributes will be revisited during the Observation stage of the text (Part III).

While there are multiple problems within a single mess, we will concentrate on a singular problem of interest (from the potentially many generated per the guidance in Chap. 2) in this discussion in order to gain increased understanding. We understand that we wish to resolve our problems, but, in the absence of objectives, resolution is impossible. This chapter will help the reader to further decompose a problem in a manner that supports decision making. This process centers on objective identification and organization. But first, to gain some perspective, we address the importance of objectives.

7.2 The Importance of Objectives

An objective, simply put, is "a statement of something that one desires to achieve" (Keeney, 1992, p. 34). Hammond et al. (2002) get at the heart of objective selection by suggesting we ask ourselves "What do you really want? What do you really need?" (p. 29). They provide a number of reasons for selecting objectives:

- They help you determine what information to seek. Once objectives are selected, we can determine what information we may need to increase our understanding or gain insight about a problem.
- They can help you explain your choice. Armed with a justifiable set of objectives, an individual can explain the rationale of a particular choice to someone unfamiliar with the problem or to stakeholders of the problem, if necessary, to garner support.
- They determine a decision's importance, and thus, how much time or effort to spend on it. Effort spent to achieve purposes that are not identified as problem objectives is ill-spent and should be avoided.

Objectives are refinements of *values*—"the things that matter to us" (Clemen & Reilly, 2014, p. 24). Taken together, our objectives make up our values. Keeney (1992) adds, "The achievement of objectives is the sole reason for being interested in any decision. And yet, unfortunately, objectives are not adequately articulated for many important decisions." (p. 55). Ultimately, objectives help us understand what we are trying to achieve and give us a basis on which to make decisions; they give us something to care about. Clemen and Reilly (2014) elaborate:

> If we did not care about anything, there would be no reason to make decisions at all, because we would not care how things turned out. Moreover, we would not be able to choose from among different alternatives. Without objectives, it would not be possible to tell which alternative would be the best choice. (p. 24)

Clearly objectives are important to our problem. But how do identify them and how do we organize them? These two questions form the focus of the remainder of this chapter.

7.3 Objective Identification

The first step in understanding our problem further is to identify its objectives. Keeney (1992) describes objectives as "characterized by three features: a decision context, an objective, and a direction of preference" (p. 34). Hammond et al. (2002) suggest our objectives take the form of a succinct statement consisting of a verb and an objective such as *Minimize expenses* or *maximize revenue*. Clemen and Reilly (2014) provide an example of a new graduate that values compensation and whose objectives follow this guidance; his or her objectives might include *maximize salary*, *maximize medical benefits*, and *maximize retirement savings*, all of which further define compensation. The context of this decision is that of a new (presumably single for simplicity's sake) graduate whose objectives would be different, for example, from that of an individual with a young family (a distinct problem context) who may also wish to *maximize childcare benefits* or *maximize educational support* if he or she was interested in continuing his or her educational pursuits.

So how do we identify objectives? MacCrimmon (1969) identifies the following strategies:

1. Examine the relevant literature of similar problems,
2. Perform an analytical study by modeling the system under consideration, and
3. Observe individuals making decisions with the current system.

Keeney and Raiffa (1976) add the following fourth strategy:

4. Consult a set of knowledgeable subject matter experts.

Additionally, Keeney (1992, p. 57) offers the following strategies for objective identification:

- A wish list,
- Alternatives,
- Problems and shortcomings,
- Consequences,
- Goals, constraints, and guidelines,
- Different perspectives,
- Strategic objectives,
- Generic objectives,
- Structuring objectives, and
- Quantifying objectives.

Each of these strategies is ultimately a brainstorming exercise. We are decomposing our problem into its individual objectives in an effort to better understand its fundamental elements. Ultimately, objective identification is a delicate balance:

> The process of specifying the objectives is not done in a vacuum. At the same time, we may have relevant information about what data are accessible, the quality and quantity of other available resources (e.g., computers), various types of constraints that are in force (e.g., time, politics), the range of alternative courses of action, and so on. All of thesefactors might significantly affect the objectives hierarchy… (Keeney & Raiffa, 1976, pp. 64, 65)

Once we have identified what we believe to be an exhaustive list of our objectives, we can begin to organize them.

7.4 Objective Organization

After we have identified our objectives, we need to organize them in a logical manner that facilitates further understanding. It is important first to understand the type of objectives that we may be dealing with. Objectives may be *strategic*, *fundamental (ends)*, or *means*. Strategic objectives are beyond the control of the current decision frame, but important for long-term success. They do not factor directly into our problem assessment and, thus, are not considered any further. If they represent a direct concern for our problem, then they rise to the level of a fundamental objective. A fundamental (ends) objective "characterizes an essential reason for interest in the decision situation" (Keeney, 1992, p. 34). A means objective is "of interest in the decision context because of its implications for the degree to which another (more fundamental) objective can be achieved" (Keeney, 1992, p. 34). Both fundamental and means objectives are important to our increased understanding and further elaboration on their distinction is necessary.

In order to distinguish between fundamental and means objectives, we must ask, for each identified objective:

> "Why is this objective important in the decision context?" Two types of answers seem possible. One answer is that the objective is one of the essential reasons for interest in the situation. Such an objective is a candidate for a fundamental objective. The other response is that the objective is important because of its implications for some other objective. In this case, it is a means objective, and the response to the question identifies another objective. The "Why is it important?" test must be given to this objective in turn to ascertain whether it is a means objective or a candidate for a fundamental objective. (Keeney, 1992, p. 66)

Keeney (1992) continues:

> The fundamental objectives hierarchy specifies in detail the reasons for being interested in a given problem. For each of the fundamental objectives, the answer to the question "Why is it important?" is simply "It is important." With a means objective, the answer to the question "Why is it important?" is always an end that follows from that means. For instance, regarding the transport of nuclear fuel, the answer to the question "Why is the objective of minimizing radiation dosage important?" is that radiation doses can cause cancer and that cancer is important. (p. 78)

We must take care to ensure we have generated a sufficient set of fundamental objectives to fully characterize our problem. Keeney (1992) says that a set of fundamental objectives should be:

1. "*Essential*, to indicate consequences in terms of the fundamental reasons for interest in the decision situation.
2. *Controllable*, to address consequences that are influenced only by the choice of alternatives in the decision context.
3. *Complete*, to include all fundamental aspects of the consequences of the decision alternatives.
4. *Measurable*, to define objectives precisely and to specify the degrees to which objectives may be achieved.

5. *Operational*, to render the collection of information required for an analysis reasonable considering the time and effort available.
6. *Decomposable*, to allow the separate treatment of different objectives in the analysis.
7. *Nonredundant*, to avoid double-counting of possible consequences.
8. *Concise*, to reduce the number of objectives needed for the analysis of a decision.
9. *Understandable*, to facilitate generation and communication of insights for guiding the decision making process" (p. 92).

Properties "1–3 pertain to framing the decision situation, properties 4–5 pertain to the quality of thinking and analysis, properties 6–8 pertain to the difficulty of such thinking and analysis, and property 9 pertains to the quality of insights from the thinking and analysis" (Keeney, 1992, pp. 82, 83).

On the characteristics of essential and controllable, Keeney (1992) remarks:

> A set of objectives is essential if each of the alternatives in the decision context can influence the degree to which the objectives are achieved. A set of objectives is controllable is all of the alternatives that can influence the consequences are included in the decision context. (p. 82)

To be *complete*, we should aim to separate objectives that are uniquely important in addressing; for example, height and weight in choosing a mate or speed and maneuverability when designing an aircraft. Thus, by specifying the height, weight, and other objectives of a potential mate, a neutral third party can determine the extent to which someone has identified an ideal mate. The requirement for completeness is reinforced by the *principle of minimum critical specification* which states that we must identify what is essential, but strive to specify no more than is absolutely necessary (Cherns, 1976, 1987). This guiding principle provides bounds on our objective set.

To be *measurable*, it must be possible to obtain the information necessary to determine whether or not the objective has been achieved. This may occur directly via a mechanism such as maximize speed (measured in miles per hour) or indirectly. In the case of intangible quantities such as comfort or maneuverability, use a *proxy* measure that can serve as a substitute for the original measure (e.g., flexibility of work schedule can be measured as the percent of time that can be teleworked) or a constructed scale that directly measures the objective (e.g., 1–5 stars are used to rate products on consumer sites such as Amazon.com).

In order to be *operational*, a set of objectives:

> …must be meaningful to the decision maker, so that he can understand the implications of the alternatives. They should also facilitate explanations to others, especially in cases where the main purpose of the study is to make and advocate a particular position. (Keeney & Raiffa, 1976, p. 51)

A synonym for operational is usable. They must be able to be used by the individual or individuals trying to solve a problem. This connotes the difficult nature of complex problems. Inclusion of the human element in the analysis of a problem introduces considerations which must be accounted for but which nonetheless provide no improvement in objective attainment. For example, management decisions regarding layoffs may need to be couched in terms of jobs saved in order to maintain organizational morale.

In order to be *decomposable*, a set of objectives must be able to be broken down into smaller subsets. This can be useful, for example, in decomposing fundamental objectives across lower-level objectives. This also reinforces the *principle of hierarchy* (Pattee, 1973). Further, this requirement speaks to the complex nature of objectives. The objective of profit, for example, may be composed of income and expenditures. Income can be further broken down into direct sales, indirect sales, tax revenue, etc. Direct sales can be broken down by item, region, etc. The appropriate level of abstraction must be chosen in a manner which is tractable and meaningful for the problem owner.

Nonredundancy is achieved by ensuring that objectives "should be defined to avoid double counting of consequences" (Keeney & Raiffa, 1976, pp. 51, 52). A practical lower limit to redundancy is provided by the *principle of information redundancy* (Shannon & Weaver, 1949), which measures the amount of wasted information used in transmitting a message, thereby providing a lower bound for us to aim for (no redundant information), while also considering the principle of minimal critical specification. Adhering to these principles ensures that we do not avoid information that is necessary in order to fully capture our problem, while avoiding extraneous information.

On the criteria of *concise* and the number of objectives, "it is desirable to keep the set as small as possible" (Keeney & Raiffa, 1976, p. 52). This is a limiting factor which ensures we address the other characteristics in the limit. For example, while our objectives are decomposable, we should only decompose them to the point where it is meaningful and not beyond, to avoid a level of granularity that is neither discernible nor meaningful to relevant stakeholders. George Miller's seminal work *The Magical Number Seven, Plus or Minus Two: Some Limits on Our Capacity for Processing Information* (1956) provides practical limits for human information capacity. The accompanying *principle of requisite parsimony* (Miller, 1956; Simon, 1974) states that humans can only deal simultaneously with between five and nine items at one time. Thus, creating a set of objectives of greater than nine would not only violate the criteria of minimal, but also it would be unusable as well.

Finally, objectives should be *understandable* "so that they can be adequately communicated to and understood by individuals in positions to make or influence decisions" (Keeney, 1992, p. 85). Recall the *principle of redundancy of potential command* which states that power resides where information resides (Adams, 2011, p. 151). Equipping those in power with appropriate information affords them the agency to make effective decisions. In the absence of this information (or in the presence of ambiguous information), the ability to make decisions becomes a power grab among interested parties.

Additional guidance on objective development is found in the acronym SMART, developed by Doran (1981) [1939–2011], who provided the following five criteria that can be used to develop appropriate fundamental objectives:

1. Specific—target a specific area for improvement.
2. Measurable—quantify or at least suggest an indicator or progress.
3. Assignable—specify who will do it.
4. Realistic—state what results can realistically be achieved, given available resources.
5. Time-related—specify when the result(s) can be achieved (p. 36).

Armed with a number of guiding principles by which to develop a set of objectives, the next step stakeholders are faced with is to organize these objectives. Once we have separated our objectives into fundamental objectives and means objectives, we can organize them into two structures: (1) a fundamental objectives hierarchy; and (2) a means-ends network. Both share the same overall objective. "For a given decision situation, the overall objective is the same for both the fundamental and the means-ends objective structures. It characterizes the reason for interest in the decision situation and defines the breadth of concern" (Keeney, 1992, p. 77). For the purposes of our analysis, our problem, stated as an objective, is the most fundamental of our objectives. This objective can then be broken down into further fundamental objectives, as well as means objectives, using the guidance provided in the following sections.

7.5 Fundamental Objectives Hierarchy

In a fundamental objectives hierarchy,

> The lower-level objective is a part of the higher-level objective. The higher-level objective is defined by the set of lower-level objectives directly under it in the hierarchy. These lower-level objectives should be mutually exclusive and collectively should provide an exhaustive characterization of the higher-level objective. There should be at least two lower-level objectives connected to any higher-level objective. (Keeney, 1992, p. 78)

But how do we discern our fundamental objectives from our means objectives? Hammond et al. (2002) suggest using the Japanese technique of *Five why's*. That is, for a given objective, ask *why?* five times to get to the bottom of our concern. This helps to identify truly fundamental objectives. Hammond et al. (2002) elaborate on the importance of asking why:

> Asking "Why?" will lead you to what you really care about–your fundamental objectives, as opposed to your means objectives. Means objectives represent way stations in the progress toward a fundamental objective, the point at which you can say "I want this for its own sake. It is a fundamental reason for my interest in this decision." Fundamental objectives constitute the broadest objectives directly influenced by your decision alternatives. (p. 37)

To illustrate the identification of fundamental objectives, imagine you have just graduated from college and are planning your move to a new city in preparation for starting your new job. This is a mess, no doubt. You have concerns regarding fitting in at your new job, finding a place to live, finding your way around the city, etc. However, we will simply focus on a problem related to your residence.

Let us say you have an objective of the use of public transportation. Why is this important? It is important to both minimize cost and as an indicator of commute time (i.e., a choice to ride the bus is cheap, but time-consuming). Why is minimizing cost important? It helps to maximize housing satisfaction (with money saved on transportation, we can afford a nicer residence). Why is commute time important? For the same reason. Minimizing commute time (thereby leading to more time at home) maximizes our housing satisfaction. Why is housing satisfaction important? It just is. That is, it is the end we want to arrive at. So, maximizing housing satisfaction can be said to be a *fundamental objective* for us. Asking why helped us to arrive at our fundamental objective. This objective does not preclude a particular solution (apartment/house, rental/purchase, etc.) and it gets at the core of what we are trying to achieve (satisfactory residence to support our new job in this new city).

Within a fundamental objectives hierarchy, Clemen and Reilly (2014) suggest we ask *What do you mean by that?* to move down in the hierarchy, whereas, to move up in the hierarchy, you should ask *Of what more general objective is that an aspect?* Because fundamental objectives illustrate that which we value most directly, they are long-lasting for similar problems. Hammond et al. (2002) remark, "well-thought-out fundamental objectives for similar problems should remain relatively stable over time" (p. 40). Thus, maximizing housing satisfaction should be remain important to us any time we would be moving to a new city. "…However, the objectives hierarchy for a particular problem is not unique" (Keeney & Raiffa, 1976, p. 47). That is to say, while our fundamental objectives should remain relatively stable, our sub-objectives (and their organization) may not. This is in part due to the context element of objective identification addressed earlier in the chapter.

Returning to our new grad example, let us assume we have identified two initial objectives related to maximizing housing satisfaction, namely (1) minimize cost and (2) minimize commute time. We could imagine a scenario in which we could find housing near our work that is affordable. But what if our options in this neighborhood are unsafe and/or not close to any entertainment? We have just discovered another fundamental objective, maximize neighborhood quality. Now, we can organize our objectives by decomposing each of our fundamental objectives, (1) minimize cost, (2) minimize commute time, and (3) maximize neighborhood quality. The result is shown in Fig. 7.2.

Our first objective, cost, is broken down into initial cost, monthly housing cost, monthly utility cost, and monthly maintenance cost. These help to give us a richer understanding of what is intended by the idea of cost. Commute time is broken down into average commute time and variability of traffic, allowing us to account for normal and abnormal commute conditions. Finally, neighborhood quality is broken down into safety and proximity to entertainment.

We can now explore means-ends networks to gain a different perspective on our problem's objectives organization.

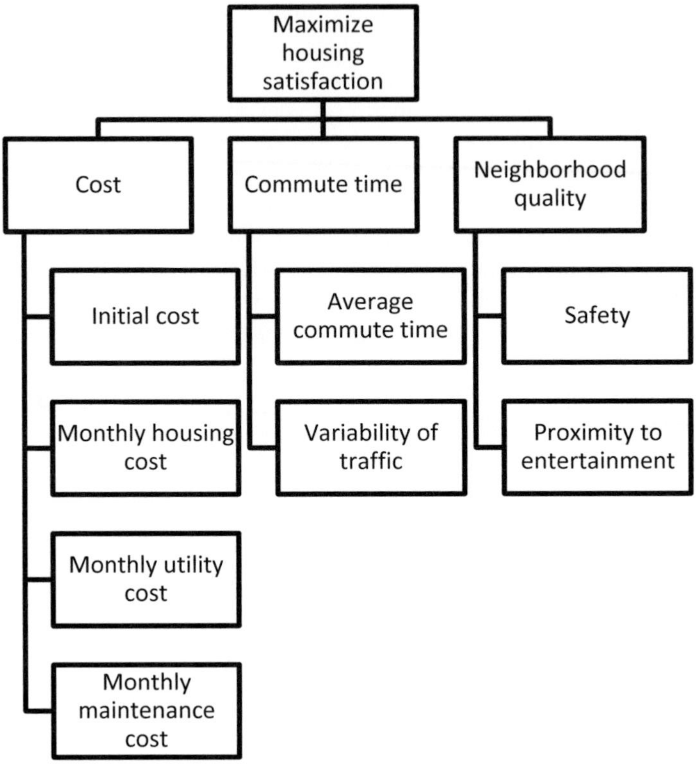

Fig. 7.2 Example fundamental objectives hierarchy

7.6 Means-Ends Network

A means-ends network fulfills a different purpose than a fundamental objectives hierarchy. Whereas a fundamental objectives hierarchy provides necessary decomposition of our objectives, a means-ends network expresses causal relationships between components:

> In a means-ends objectives network…the relationship between adjacent levels is causal. The lower-level objective is a means (that is, a causal factor) to the higher-level objective. Not all of the causal factors to a higher-level objective are included as lower-level objectives.…Thus, the means objectives are not in any sense a collectively exhaustive representation of the means to the higher-level ends. It may therefore be the case that a higher-level objective has only one lower-level means objective. (Keeney, 1992, p. 78)

Starting with our fundamental objectives (the same as those identified in the fundamental objectives' hierarchy), we must ask "How could you achieve this?" as opposed to "Why is this important?" to move toward fundamental objectives

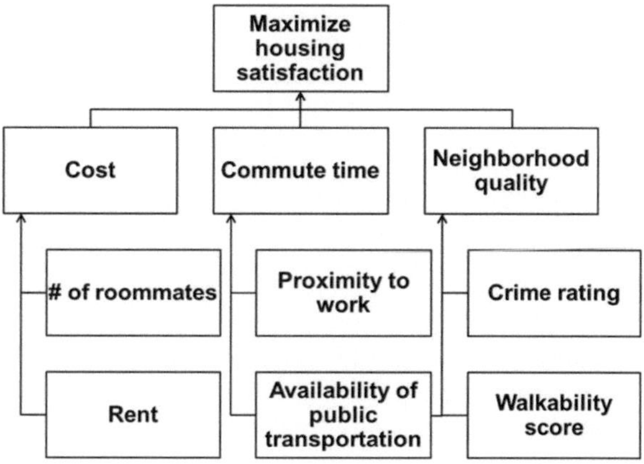

Fig. 7.3 Example means-ends network

(Clemen & Reilly, 2014, p. 52). Thus, beginning our new job problem with the same fundamental objectives of (1) minimize cost, (2) minimize commute time, and (3) maximize neighborhood quality, we can begin to build a means-ends network. Asking *how could we minimize cost?* We might identify having roommates and renting as two means. Similarly, searching for means to minimize our commute time, we might identify proximity to work and availability of public transportation as means to minimize commute times. Finally, identifying means for maximizing neighborhood quality, we might identify crime rating, walkability score, and availability of public transportation (recalling that a means objective can contribute to more than one fundamental objective) as means to achieve it. Graphically, these relationships are shown in Fig. 7.3.

Now that we know how to create a fundamental objectives hierarchy and a means-ends network, we can address the *what* perspective in our problem.

7.7 Framework for Addressing *What* in Messes and Problems

Addressing the *what* in our messes and problems requires that we complete the following steps for an identified problem:

1. Articulate our objectives.
2. Organize these objectives into a fundamental objectives hierarchy.
3. Using the same fundamental objectives, create a means-ends network.
4. Link the means-ends network to our existing FCM.

Each of these four steps is demonstrated on a simple example that follows.

7.7.1 Articulate Objectives

We are continuing with our real estate example from Chap. 6. Recall that the developer's goal is, simply put, to *maximize return on investment (ROI)*. Their problem is that they have a plot of land which needs to be rezoned in order to maximize their profit on it. In the absence of such a rezoning, they will be saddled with a vacant parcel of land. Using this as a starting point, we can begin to articulate our objectives.

In terms of the development company, it aims to balance both short- and long-term company performance. So, in terms of the project itself, we can establish one fundamental objective as *maximize project ROI*. However, we can conceive of numerous mechanisms for maximizing project ROI that may be problematic in the long run. For example, they could engage in illegal behavior by hiring undocumented workers and paying them a wage below minimum wage. They could also engage in unethical behavior such as delaying payment for services rendered, which, although not strictly illegal, might be considered unethical by some. Both of these actions would be detrimental to their long-term company health. So, we should also add a fundamental objective of *maximize long-term company viability*. Armed with our two fundamental objectives (*maximize project ROI* and *maximize long-term company viability*), we can begin to organize them.

7.7.2 Fundamental Objectives Hierarchy

Organizing our two fundamental objectives into a hierarchy yields Fig. 7.4, showing a further decomposition of our objectives. Project ROI is broken down into carrying costs (what it costs to keep the land until we sell it), capital investments (the cost of building out the property), and profit (what the company makes once it sells the land as usable real estate, i.e., condos). Long-term company viability is broken down into a balanced portfolio (a mix of short- and long-term investments) and strategic vision (to include, for example, making ethical business choices, which guide our daily decision making).

7.7.3 Means-Ends Network

The means-ends network shows our understanding of the means necessary to produce our desired ends (i.e., our fundamental objectives). Using the same two fundamental objectives as before, we can create the network shown in Fig. 7.5. Both project ROI and long-term company viability are means to achieve the end of

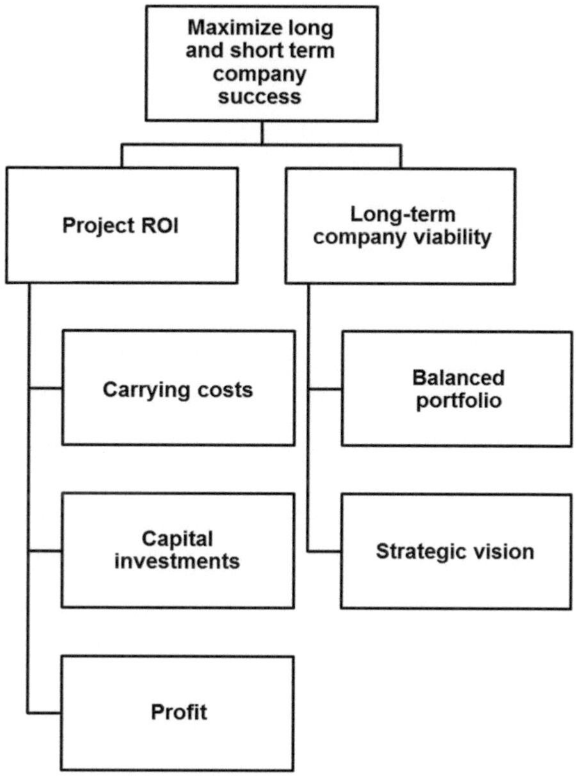

Fig. 7.4 Real estate fundamental objectives hierarchy

maximize long- and short-term company success. Additionally, maximizing project ROI is a means to the end of long-term company viability. Labor costs (i.e., hourly wages, overtime), material costs (e.g., lumber, equipment), sales volume (i.e., how many units have we sold), and profit per unit (based, in part on the type of product we are selling and the market) all contribute to project ROI. Compliance with regulations (i.e., is the company following its zoning constraints), quality products (short-time cost savings by using cheap materials will tarnish the company's reputation), and diversified projects (in the event of another economic downturn, it pays to diversify what projects the company is working on) are all means to achieve the end of long-term company viability.

7.7.4 FCM Update

Armed with our means-ends network, we can now integrate it into our existing FCM. In this case, we have revised our problem statement to reflect our short- and long-term focus, so this new objective replaces the old one in our FCM. This revised scenario depiction is shown in Fig. 7.6. It is worth noting that not only has

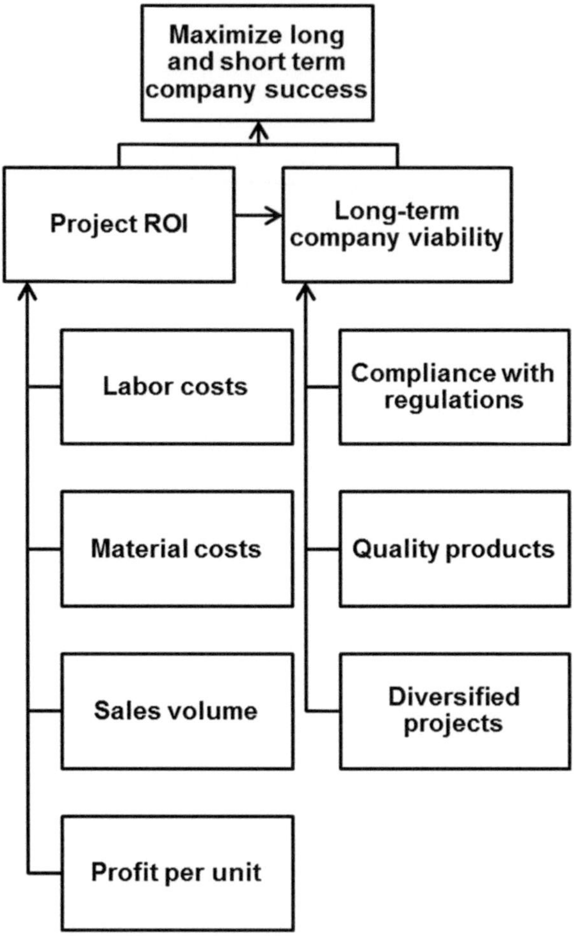

Fig. 7.5 Real estate means-ends network

the means-ends network been incorporated, but also a new causal linkage was introduced external to the means-ends network. A positive relationship was identified between sales volume and property values. Additionally, since the concept of compliance with regulations already existed as the city planning commission's objective, it was retained and causal connections shown in the means-ends network were duplicated in the FCM.

Now, after having investigated both the *who* and *what* perspectives, we have a more holistic perspective regarding our real estate problem.

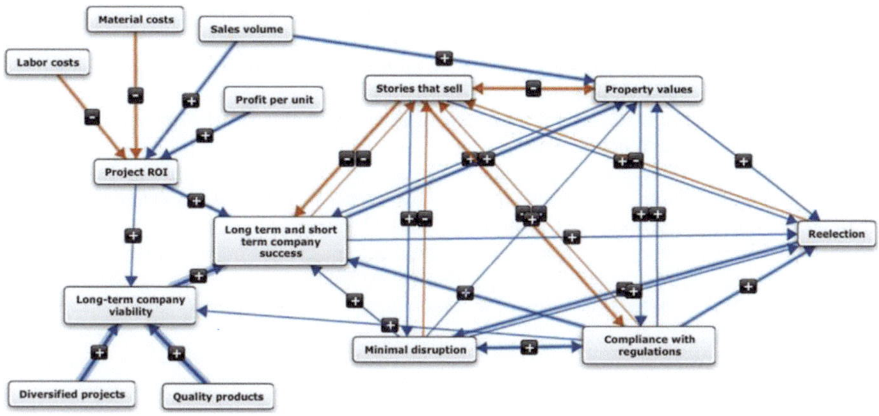

Fig. 7.6 Real estate FCM with means-ends network

7.8 Summary

This chapter began with some introductory information regarding the anatomy of a problem. Then, it discussed the importance of objectives. Objective articulation and organization were addressed. Finally, organization of objectives into both fundamental objectives hierarchy and means-ends network were presented. Consideration of these elements together allows us to answer the *what* question of systemic thinking.

After reading this chapter, the reader should:

1. Understand the role of objectives in systemic thinking;
2. Be able to differentiate between fundamental and means objectives;
3. Be able to create a fundamental objectives hierarchy; and
4. Be able to create a means-ends objective network.

References

Adams, K. M. (2011). Systems principles: Foundation for the SoSE methodology. *International Journal of System of Systems Engineering, 2*(2/3), 120–155. doi:10.1504/IJSSE.2011.040550.

Cherns, A. (1976). The principles of sociotechnical design. *Human Relations, 29*(8), 783–792. doi:10.1177/001872677602900806.

Cherns, A. (1987). The principles of sociotechnical design revisited. *Human Relations, 40*(3), 153–161.

Clemen, R. T., & Reilly, T. (2014). *Making hard decisions with decision tools* (3rd ed.). Mason, OH: South-Western.

Doran, G. T. (1981). There's a S.M.A.R.T. way to write management's goals and objectives. *Management Review, 70*(11), 35–36.

Hammond, J. S., Keeney, R. L., & Raiffa, H. (2002). *Smart choices: A practical guide to making better life decisions*. New York: Broadway Books.
Keeney, R. L. (1992). *Value-focused thinking: A path to creative decision making*. Cambridge, MA: Harvard University Press.
Keeney, R. L., & Raiffa, H. G. (1976). *Decisions with multiple objectives: Preferences and value tradeoffs*. New York: Wiley & Sons.
MacCrimmon, K. R. (1969). *Improving the system design and evaluation process by the use of trade-off information: An application to northest corridor transportation planning (RM-5877-DOT)*, CA. Retrieved 1969, from Santa Monica.
Miller, G. A. (1956). The magical number seven, plus or minus two: some limits on our capability for processing information. *Psychological Review, 63*(2), 81–97.
Pattee, H. H. (1973). *Hierarchy theory: The challenge of complex systems*. New York: George Braziller.
Shannon, C. E., & Weaver, W. (1949). *The mathematical theory of communication*. Champaign, IL: University of Illinois Press.
Simon, H. A. (1974). How big is a chunk? *Science, 183*(4124), 482–488.

Chapter 8
The *Why* of Systemic Thinking

Abstract The previous chapters in this section have addressed: (1) the who question through a discussion of problem stakeholders, their analysis, and management; and (2) the what question by decomposing our mess and constituent problems into its objectives and organizing them. In this chapter, we will address the why question through an analysis of motivation and how each problem has a unique model of motivation and feedback between and among the stakeholders. This chapter discusses motivation, its 20 major theories, and how we can incorporate motivation into our systemic thinking. A framework for assessing motivation is provided, and this framework is demonstrated on our example real estate problem.

8.1 Overview

The main focus of the *why* question of systemic thinking attempts to determine either (1) a premise, reason, or purpose for why something is the way it is, or (2) what the causal relationship is between an event and the actions that caused the event to occur. As a result, *why* can be treated as either a noun or an adverb:

adverb—for what reason or purpose,

noun—a reason or explanation.

Reason, purpose, and some explanation of causality are central elements expected in any answer to the question *Why?* The underlying premise for the *why* question is most often based upon the following assumption:

> "Why" questions presuppose that things happen for a reason and that those reasons are knowable. "Why" questions presume cause-effect relationships, an ordered world, and rationality. "Why" questions move beyond what has happened, what one has experienced, how one feels, what one opines, and what one knows to the making of analytical and deductive inferences. (Patton, 2002, p. 363)

The answer to the *why* question relates reason through explanation.

> Often such reasons are causes, but even when 'cause' is not the natural description, 'Because - - -' is the natural formula for answering why questions. 'Because - - -' answers,

P.T. Hester and K.M. Adams, *Systemic Decision Making*, Topics in Safety, Risk, Reliability and Quality 33, DOI 10.1007/978-3-319-54672-8_8

> usually becoming more informative in the process (the expansion will often indicate that the thing to be explained does some good, or—differently—aims at some good, these being two kinds of teleological explanation. (Honderich, 2005, p. 957)

The notion of a teleological explanation is important. A teleological explanation is one in which there is a belief in or the perception of purposeful development toward an end. This is contained within the *principle of purposive behavior* (Adams, Hester, Bradley, Meyers, & Keating, 2014) from the goal axiom of systems theory in Chap. 4 that states:

> Purposeful behavior is meant to denote that the act or behavior may be interpreted as directed to the attainment of a goal - i.e., to a final condition in which the behaving object reaches a definite correlation in time or in space with respect to another object or event. (Rosenblueth, Wiener, & Bigelow, 1943, p. 18)

In *systemic thinking*, the attainment of specific, purposeful goals is the most desirable answer to *why*. The reason for attaining the goals has some underlying rationale which includes:

1. The basis or motive for the goals and supporting objectives.
2. A declaration made to explain or justify the goals and supporting objectives.
3. An underlying fact or cause that provides logical sense for achieving goals and objectives.

Items 1 and 2 were addressed in Chap. 7*, The What of Systemic Thinking*. The sections that follow will address item 3—the underlying fact or cause that provides logical sense for achieving goals and objectives as part of solving messes and their constituent problems.

8.2 Motivation

The underlying fact or cause that provides logical sense for achieving goals and objectives can be labeled motivation. Motivation is defined as (Runes, 1983, p. 218):

> **Motivation**: Designation of the totality of motives operative in any given act of volition or of the mechanism of the operation of such motives. See *Motive*.
>
> **Motive**: (Lat. *motus*, from *movere*, to move) An animal drive or desire which consciously or unconsciously operates as a determinant of an act of volition.

As defined, motivation is the property central in the explanation of intentional conduct. Specifically, a motivational explanation is "a type of explanation of goal-directed behavior where the explanans appeals to the motives of the agent" (Audi, 1999, p. 592). Understanding the motives for the behaviors associated with *why* is the central tenet of theories associated with motivation.

8.3 Categorizing Theories of Motivation

There are a number of implicit theories for motivation in the literature. However, before we discuss the elements of these theories, it is important to understand how the scientific community has categorized theories of motivation. There are also two accepted methods for categorizing these theories.

The first method for grouping motivation theories has three categories: (1) content-based theories of motivation; (2) process-based theories of motivation; and (3) environmentally based theories of motivation (Bowditch, Buono, & Stewart, 2008). Figure 8.1 is a depiction of this categorization.

The second method for grouping motivation theories also has three categories: (1) hedonic/pleasure-based theories of motivation; (2) cognitive/need-to-know-based theories of motivation; and (3) growth/actualization-based theories of motivation (Roeckelein, 2006). Figure 8.2 is a depiction of this categorization.

The two categorization schemas for motivation theories present twenty principal motivation theories, which are listed in Table 8.1. The theories are arranged and presented in chronological order to provide a contextual setting for how the theories were revealed over the last hundred or so years of research in this field. The sections that follow will review each of these principal theories of motivation.

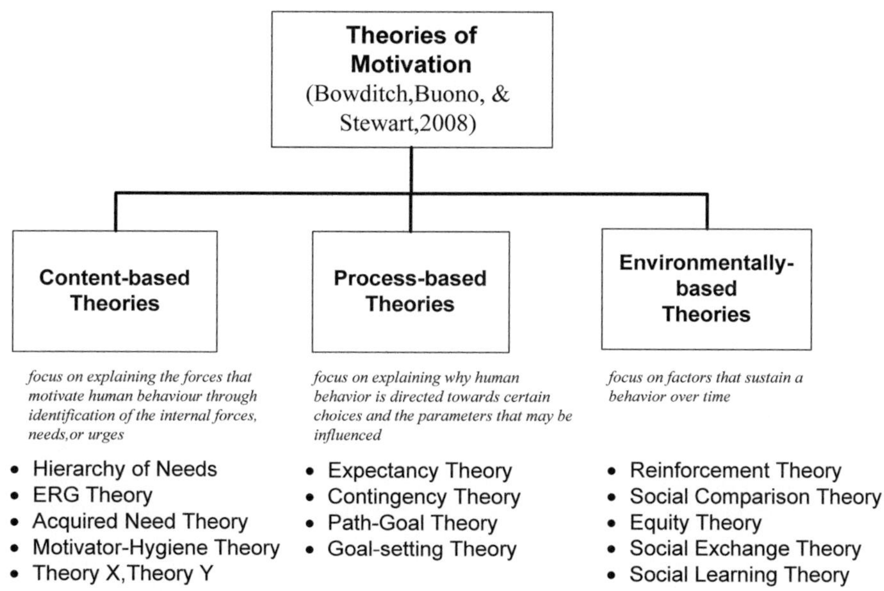

Fig. 8.1 Categorization of theories of motivation (Bowditch et al., 2008)

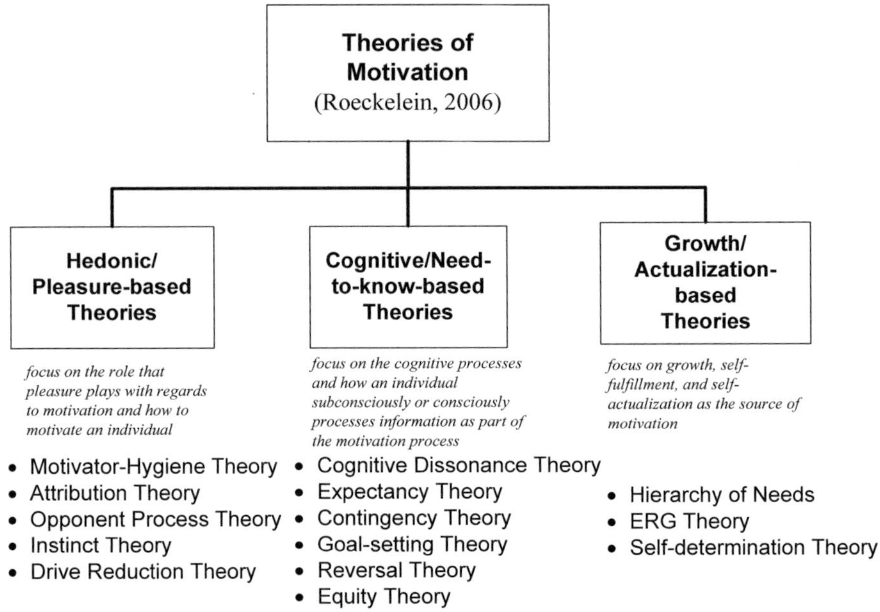

Fig. 8.2 Categorization of theories of motivation (Roeckelein, 2006)

8.4 Theories of Motivation

The sections that follow will present each of the major theories of motivation in a very broad fashion. The reader is encouraged to consult the cited references for more in-depth explanations of each of these theories. Note that the theories are presented chronologically in the same order they appear in Table 8.1.

8.4.1 Instinct Theory of Motivation

The instinct theory of motivation suggests that all living beings are supplied with innate tendencies that enable them to remain viable. The theory suggests that motivational behaviors are driven by instincts, where instincts are goal-directed and which have intrinsic tendencies that are not the result of learning or prior experience.

Wilhelm Wundt [1832–1920], the father of experimental psychology, coined the term *instinct* as a psychological term in the 1870s. Fellow psychologist William James [1842–1910] defined an instinct as an action which will "produce certain ends, without foresight of the ends, and without previous education in the performance" (James, 1887c, p. 355). James believed that motivation through instinct was

Table 8.1 Motivation theories and categorization schemas

Motivation theory and principal proponent (in chronological order)	Bowditch, Buono, and Stewart (2008)			Roeckelein (2006)		
	C	P	E	H	CO	G
1. Instinct theory of motivation (James, 1887a, b, c; McDougall, 1901)				✓		
2. Drive reduction theory (Hull, 1943, 1950)				✓		
3. Hierarchy of needs (Maslow, 1943, 1967, 1987)	✓					✓
4. Attribution theory (Heider, 1944; Kelley, 1973; Weiner, 1972, 1985)				✓		
5. Reinforcement theory (Skinner, 1953, 1956)			✓			
6. Social comparison theory (Festinger, 1954)			✓			
7. Path–goal theory (Georgopoulos, Mahoney, & Jones, 1957; House, 1971)		✓				
8. Social exchange theory (Blau, 1964; Homans, 1958)			✓			
9. Theory X and Theory Y (McGregor, 2006 (1960))	✓					
10. Cognitive dissonance theory (Festinger, 1962)					✓	
11. Equity theory (Adams, 1963)			✓		✓	
12. Social learning theory (Bandura, 1971; Bandura & Walters, 1963)			✓			
13. Expectancy theory (Vroom, 1964) and contingency theory (Porter & Lawler, 1965, 1968)		✓			✓	
14. Motivator-hygiene theory (Herzberg, 1964)	✓			✓		
15. Acquired needs theory (McClelland, 1961, 1965, 1978)	✓					
16. ERG theory (Alderfer, 1969, 1972)	✓					✓
17. Self-determination theory (Deci, 1971, 1972a, b; Gagné & Deci, 2005)						✓
18. Opponent process theory (Solomon & Corbit, 1973, 1974)				✓		
19. Goal-setting theory (Latham & Locke, 1979; Locke & Latham, 2002)		✓			✓	
20. Reversal theory of motivation (Apter, 1984)					✓	

Note *C* Content, *P* Process, *E* Environmental, *H* Hedonic, *CO* Cognitive, *G* Growth

important for human behavior and expounded upon 22 of these instincts in the monthly journal *Popular Science* (James, 1887a, b).

This theory of motivation remained popular or generally accepted into the early twentieth century. William McDougall [1871–1938] subscribed to the theory and felt that individuals are motivated by a significant number of inherited instincts, many of which they may not consciously comprehend and which may lead to misunderstood and misinterpreted goals (McDougall, 1901).

The main problem with this theory is that it did not really explain behavior; it just described it. The theory then led to the search for additional theories of motivation.

8.4.2 Drive Reduction Theory of Motivation

The drive reduction theory of motivation (Hull, 1943, 1950) became popular during the 1940s and 1950s as a way to explain behavior, learning, and motivation. The theory was created by behaviorist Clark Hull [1884–1952] and was based upon the systems *principle of homeostasis* from the operational axiom of systems theory in Chap. 4. Hull extended Cannon's (1929, 1967, 1932) ideas on physiological homeostasis to human behavior, proposing that behavior was one of the ways that an organism maintains equilibrium.

Hull's drive reduction theory uses the term *drive* to explain the state of tension that is caused by physiological needs. For instance, thirst and hunger are examples of specific drives caused by a physiological condition. In order to maintain equilibrium (i.e., homeostasis), the tension created by the drive must be balanced by an equal and opposite action—*reduction*, which will act to reduce the tension and return the human to a state of equilibrium. In the examples of thirst and hunger presented here, the human will act to reduce thirst by drinking and will act to reduce hunger by eating.

Hull and his partner Kenneth Spence [1907–1967] believed that drive reduction was a major factor in learning and behavior (Spence, 1936, 1937). They classified primary drives as innate drives (e.g., thirst, hunger, and sex) and secondary drives as learned drives (e.g., wanting money). Hull understood that human beings are routinely subjected to multiple drives and must balance these drives in an effort to maintain equilibrium. He developed a mathematical formula to express how a human balances these behaviors. The formula accounts for this using a stimulus–response relationship where a stimulus (i.e., drive) is followed by a corresponding response (i.e., reduction), in an effort to maintain equilibrium. Hull theorized that satisfactory stimulus–response patterns would lead to learning. Hull's *Mathematico Deductive Theory of Behavior* (Hull et al., 1940) is presented in Eq. 7.1:

$$sEr = (sHr \times D \times K \times V) - (sIr + Ir) \pm sOr \tag{7.1}$$

where sEr = Excitatory potential, or the likelihood that an organism will produce a response (r) to a stimulus (s),
sHr = Habit strength, established by the number of previous conditioning,
D = Drive strength, determined by the hours of deprivation of a need,
K = Incentive motivation, or value of a stimulus,
V = The measure of connectiveness,
sIr = Inhibitory strength or number of nonreinforcers,

Ir = Reactive inhibition, or fatigue based on work for a reward,
sOr = Random error.

The main problem with this theory is that it did not account for secondary or learned drives (i.e., wanting money) and how it reduces drives. An additional problem was that the theory does not account for why humans routinely increase tension by conducting exploratory ventures whether or not they were in a state of equilibrium. These shortcomings led researchers to search for more complete theories of motivation.

8.4.3 Hierarchy of Needs

The hierarchy of needs theory of motivation was proposed by Abraham Maslow [1908–1970] in the paper *A Theory of Human Motivation* (Maslow, 1943). In this paper, Maslow proposed that human needs are satisfied in an ordered hierarchy where critical lower-level needs would need to be satisfied before less critical higher-level needs. The five levels in the hierarchy, from bottom to top, are as follows: (1) physiological; (2) safety; (3) love; (4) self-esteem; and (5) self-actualization. In the 1943 paper, Maslow addresses the fixed order or *fixity* of the hierarchy and that “it is not nearly as rigid as we may have implied” (p. 386), and he goes on to list seven (7) exceptions to the general theory.

It is important to note that although this theory is often presented as a pyramid, none of Maslow’s published works (1943, 1967, 1987) on the hierarchy of needs include a visual representation of the hierarchy. This section will avoid using the pyramid to support Maslow’s notions that the hierarchy of needs is neither a fixed nor rigid sequence of progression, that human needs are relatively fluid, and that many needs are simultaneously present.

Finally, Maslow also coined the term *meta-motivation* to describe the motivation of people who go beyond the scope of the basic needs and strive for constant betterment (Maslow, 1967).

While Maslow’s hierarchy of needs remains a very popular framework, it has largely been surpassed or replaced by newer theories of motivation.

8.4.4 Attribution Theory of Motivation

Psychological research into attribution theory as a source of motivation began with the work of Fritz Heider [1896–1988], who is often described as the *father* of attribution theory. Heider was interested in how people explain their behaviors. He found that people explain themselves by *attributing* a particular behavior as being caused by either internal or external forces. Internal forces are labeled *dispositions* and include personality, motives, attitudes, and feelings. External forces are labeled *situations* and include societal norms, acts of nature, and random chance.

Heider's concepts were advanced by Kelley (1973, 1978) who published a *co-variation model* that includes three main types of information from which to make attribution decisions about individual behavior: (1) *Consensus information* includes data about how other people, faced with the same situation, behave. (2) *Distinctive information* includes data about how an individual will respond based upon different stimuli. (3) *Consistency information* includes data related to the frequency of the individual's behavior in a variety of situations. An observer may use this information when assessing the individual's behavior as either internally or externally attributable.

Weiner (1972, 1985) expanded upon the work of both Heider and Kelley by proposing that individuals search for attributions and analyze casual relations based on the behaviors they experience. This is the *achievement attribution model*. When the attributions they assign to causes are positive (i.e., lead to successful outcomes), these attributions should lead to additional attempts in this area. However, when the attributions they assign to causes are negative (i.e., lead to unsuccessful outcomes), these attributions result in a reluctance toward future attempts.

In summary, attribution theory attempts to explain the motivation of individuals by evaluating the processes in which individuals explain the causes of behavior. The term attribution theory is an umbrella term for a variety of models in which individuals look for explanations or causes that can be attributed to their own success or failure.

8.4.5 Reinforcement Theory of Motivation

The reinforcement theory of motivation was first proposed by B.F. Skinner [1904–1990] during the 1950s. The theory links behavior and consequence. It is based upon Edward Thorndike's [1874–1949] *law of effect* that was the result of his work on animal intelligence. Thorndike's *law of effect* proposed that responses that produce a satisfying effect in a particular situation are more likely to be repeated than responses that produce an uncomfortable effect in the same situation (Thorndike, 1898, 1911).

Skinner applied the concept of *reinforcement* to the law of effect by rewarding desired behaviors in an effort to motivate individuals (Skinner, 1953, 1956). This was a notable departure from theories of motivation which were concerned with the internal state of the individual (i.e., feelings, desires, and instincts) and focused on the outcomes of the individual's actions. Reinforcement theory includes four aspects.

1. *Positive reinforcement*: When desired behaviors occur, a reward is provided as motivation for continued behavior.

2. *Negative reinforcement*: When desired behaviors are problematic, assistance is provided in order to modify the behavior.
3. *Punishment*: When desired behaviors are not achieved and harm arises, a punishment is given.
4. *Extinction*: When desired behaviors are not achieved on a continual basis and harm is present, the individual will be disregarded and extinct.

Reinforcement theory also includes schedules for reinforcement that included both fixed and variable time intervals and fixed and variable ratios (based on the ratio of responses to reinforcements).

Reinforcement theory is important because it was relatively easy to understand and implement because the goal was to provide control through the manipulation of the consequences of behavior.

8.4.6 Social Comparison Theory of Motivation

The social comparison theory of motivation was first proposed by Leon Festinger [1919–1989]. Festinger's theory of social comparison is centered on the belief that "there exists, in the human organism, a drive to evaluate his opinions and abilities" (Festinger, 1954, p. 117). The theory also posits that "to the extent that objective, non-social means are not available, people evaluate their opinions and abilities by comparison respectively with the opinions and abilities of others" (Festinger, 1954, p. 118).

Festinger's initial 1954 framework has been advanced to include:

1. Understanding of the motivations that underlie social comparisons and the particular types of social comparisons that are made (Gruder, 1971).
2. The concept of downward comparison. Downward social comparison is a defensive tendency where the social comparison will be made with individuals who are considered to be worse off in order to make themselves feel better (Wills, 1981).
3. The concept of upward comparison. Research has suggested that comparisons with individuals that are considered to be better off can lower self-regard, whereas downward comparisons can elevate self-regard (Tesser, Millar, & Moore, 1988).

Social comparison theory is important because it introduced the notion that an individual is capable of self-evaluation and that the drive to understand strengths and weaknesses exists in order to provide a more accurate view of the self.

8.4.7 Path-Goal Theory of Motivation

The path-goal theory of motivation was first proposed by House (1971) and was based upon pioneering work conducted by Georgopoulos, Mahoney, and Jones (1957) and Evans (1970).

House's original theory proposed that behavior in leaders is contingent upon the satisfaction, motivation, and performance of subordinates in the organizational hierarchy (House, 1971). His revised version of the theory proposes that leaders exhibit behaviors that complement the abilities of subordinates and often compensate for skill deficiencies in the organizational hierarchy (House, 1996).

> The essence of the theory is the meta proposition that leaders, to be effective, engage in behaviors that complement subordinates' environments and abilities in a manner that compensates for deficiencies and is instrumental to subordinate satisfaction and individual and work unit performance. (House, 1996, p. 323)

The theory maintains that leaders are required to modify their behavior by implementing leadership behaviors dictated by the situation they face. The leader is required to adjust the leadership style to support the unique needs presented by the dynamic nature of the mission, goals, and objectives of the organization. As such, leader behaviors are the independent variables in the theory and consist of the following:

- **Directive path-goal clarifying leader behavior** is behavior directed toward providing psychological structure for subordinates: letting subordinates know what they are expected to do, scheduling and coordinating work, giving specific guidance, and clarifying policies, rules, and procedures.
- **Supportive leader behavior** is behavior directed toward the satisfaction of subordinates' needs and preferences, such as displaying concern for subordinates' welfare and creating a friendly and psychologically supportive work environment. Supportive leader behavior was asserted to be a source of self-confidence and social satisfaction and a source of stress reduction and alleviation of frustration for subordinates. (House & Mitchell, 1974)
- **Participative leader behavior** is behavior directed toward encouragement of subordinate influence on decision making and work unit operations: consulting with subordinates and taking their opinions and suggestions into account when making decisions. (House, 1996, pp. 326, 327)

In summary, the independent variable in the path-goal theory is the leaders' behavior. As such, the theory relies heavily upon the notion that individuals in leadership positions are flexible enough and have the cognizant ability to modify their behavior based upon the situation they face.

8.4.8 Social Exchange Theory of Motivation

Social exchange theory was first proposed by sociologist George Homans [1910–1989] and was codified by sociologist Peter Blau [1918–2002]. Emerson (1976) explains:

> …social exchange theory… is not a theory at all. It is a frame of reference within which many theories-some micro and some more macro-can speak to one another, whether in argument or in mutual support. (p. 336)

Blau (1964) explained that the frame of reference was "Social exchange as here conceived is limited to actions that are contingent on rewarding reactions from others" (p. 6). Social exchange proposes that as individuals interact over time, they develop the need to reciprocate favors. This need is termed the norm of reciprocity (Gouldner, 1960).

Homan's concept of social exchange theory relies upon three basic propositions of social behavior:

1. *The Success Proposition.* "For all actions taken by persons, the more often a particular action of a person is rewarded, the more likely the person is to perform that action" (Homans, 1974, p. 16).
2. *The Stimulus Proposition.* "If in the past the occurrence of a particular stimulus, or set of stimuli, has been the occasion on which a person's action has been rewarded, then the more similar the present stimuli are to the past ones, the more likely the person is to perform the action, or some similar action, now" (Homans, 1974, pp. 22, 23).
3. *The Deprivation-Satiation Proposition.* "The more often in the recent past a person has received a particular reward, the less valuable any further unit of that reward becomes for him" (Homans, 1974, p. 29).

Despite the apparently clear nature of the theory, there are a number of complications that can arise and compromise the exchange relationships. Equivalent reciprocity requires that each returned favor has some value at least equal to the initial favor. Failure to ensure the favor is equivalent or of comparable benefit is subjective and can be the source of conflict and resentment. Placing value on favors is difficult and often involves qualities that are hard to measure (i.e., convenience, time, and scarce resources).

8.4.9 Theory X and Theory Y

Theory X and Theory Y are contrasting theories of motivation proposed by Douglas McGregor [1906–1964] in the 1960s. Theory X and Theory Y describe two models of workforce motivation from the view of management. Management feels that employees are motivated by either (1) authoritative direction and control or (2) integration and self-control.

In Theory X, management assumes that employees are inherently lazy and dislike work. As a result, employees require close supervision and a system of controls must be developed to ensure compliance with work goals. In addition, a hierarchical structure of management and supervision is required.

In Theory Y, management assumes that employees are ambitious and self-motivated and enjoy work. As a result, employees will seek out and accept responsibility. Due to these conditions, employees are able to meet goals and objectives based on self-direction and their personal commitment to work.

> At the heart of McGregor's argument is the notion that managers' assumptions/attitudes represent, potentially, self-fulfilling prophecies. The manager who believes that people are inherently lazy and untrustworthy will treat employees in a manner that reflects these attitudes. Employees, sensing that there is little in the job to spur their involvement, will exhibit little interest and motivation. Consequently, and ironically, the manager with low expectations will lament that 'you can't get good help nowadays', oblivious as to the actual nature of cause and effect. Closing the self-reinforcing cycle, the manager feels vindicated; that is, his/her low expectations were warranted. Conversely, the manager who believes that employees are generally trustworthy and desirous of growth will facilitate their achievement. (Kopelman, Prottas, & Davis, 2008, pp. 256, 257)

The contrasting characteristics of Theory X and Theory Y are presented in Table 8.2.

Although McGregor's theories of motivation are seldom used explicitly, they have strongly influenced several generations of managers. A 2003 review of 73 established organizational behavior theories found that Theory X and Theory Y were tied for second in terms of recognition and in 33rd place with respect to importance (Miner, 2003).

8.4.10 Cognitive Dissonance Theory of Motivation

The cognitive dissonance theory of motivation was first proposed by Leon Festinger [1919–1989]. Festinger's theory of cognitive dissonance focuses on how individuals strive for internal consistency. When an inconsistent behavior (i.e., a dissonance) is experienced, individuals largely become psychologically distressed

Table 8.2 Characteristics of Theory X and Theory Y

Characteristic	Theory X	Theory Y
Attitude	Dislike work, find it boring, to be avoided	Want to work, find it interesting, can be enjoyed
Direction	Must be coerced into effort	Self-directed toward effort
Responsibility	Avoid responsibility	Seek and accept responsibility
Motivation	Money and fear	Desire to realize personal potential

and have a desire to return to a state of equilibrium (i.e., homeostasis). Festinger (1957) stated two basic hypotheses:

1. The existence of dissonance, being psychologically uncomfortable, will motivate the person to try to reduce the dissonance and achieve consonance.
2. When dissonance is present, in addition to trying to reduce it, the person will actively avoid situations and information which would likely increase the dissonance (p. 3).

In the presence of dissonance, an individual may return to equilibrium by adjusting their cognitions or actions. Adjustment results in one of three relationships between cognition and action:

- *Consonant relationship* This occurs when two cognitions or actions are consistent with one another (e.g., not wanting to go swimming while at the beach and then going for a walk in the sand instead of swimming).
- *Irrelevant relationship* This occurs when two cognitions or actions are unrelated to one another (e.g., not wanting to go swimming while hiking in the Mojave Desert).
- *Dissonant relationship* This occurs when two cognitions or actions are inconsistent with one another (e.g., not wanting to go swimming while surfing).

Cognitive dissonance theory posits that individuals desire consistency between expectations and the real world. As a result, individuals invoke *dissonance reduction* to balance their cognitions and actions. Dissonance reduction provides a means for homeostasis, where there is a reduction in psychological tension and a return to equilibrium. Festinger (1957, 1962) stated that dissonance reduction can be achieved in one of three ways: (1) changing the behavior or cognition; (2) justifying the behavior or cognition by changing the conflict; or (3) justifying the behavior or cognition by adding a new cognition.

Early experiments showed that

> 1. If a person is induced to do or say something which is contrary to his private opinion, there will be a tendency for him to change his opinion so as to bring it into correspondence with what he has done or said. 2. The larger the pressure used to elicit the overt behavior (beyond the minimum needed to elicit it) the weaker will be the abovementioned tendency. (Festinger & Carlsmith, 1959, pp. 209, 210)

In later experiments, researchers demonstrated cognitive dissonance in a learning environment. For instance, school children who completed activities with the promise of a reward were less interested in the activity later than those children who were offered no reward in the first place (Lepper & Greene, 1975).

In summary:

> Since it was presented by Festinger over 40 years ago, cognitive dissonance theory has continued to generate research, revision, and controversy. Part of the reason it has been so generative is that the theory was stated in very general, highly abstract terms. As a consequence, it can be applied to a wide variety of psychological topics involving the interplay of cognition, motivation, and emotion. (Harmon-Jones & Mills, 1999, p. 5)

8.4.11 Equity Theory of Motivation

The equity theory of motivation was first proposed by Adams (1963). In this theory of motivation, Adams proposed satisfaction and motivation in terms of an individuals' perception of the distribution of resources within an organizational or interpersonal setting. Adams (1965) asserted that individuals maintain equity by comparing the inputs that they provide against the outcomes they receive against the perceived inputs and outcomes of others. The theory proposed that individuals highly value equitable treatment which in turn causes them to remain motivated in order to maintain the equitable conditions established between individuals or within an organization.

Equity theory posits that when individuals perceive themselves in an inequitable relationship, they will experience stress, placing them in a state where equilibrium is disturbed. In order to restore the equilibrium state, the individual must restore the equity in the relationship (either personal or organizational). True equality is not required by the theory. That is, equity is determined by analysis of fairness in the distribution of resources. Two parties do not have to have equality; however, the perceived ratio of contributions and benefits to each individual is what matters. Adams (1965) proposed that anger is an outcome caused by underpayment inequity and guilt is caused by overpayment equity.

Criticism of equity theory has been focused on both the assumptions of the theory and application in the real world. The simplicity of the elements of the theory has been questioned, with arguments that additional variables are important to an individual's perceptions of equity. One such argument calls for a new construct that includes equity sensitivity, stating:

> The equity sensitivity construct suggests that individuals do not conform consistently to the norm of equity. Instead, individuals react consistently to specific, but different, preferences they have for the balance between their outcome/input ratios and that of a comparison other. Benevolents prefer that their outcome/input ratios be less than the comparison other's; Equity Sensitives, who adhere to the norm of equity, prefer balanced outcome/input ratios; and Entitleds prefer that their outcome/input ratios exceed the comparison other's. Furthermore, these general preferences for equity can be traced to internal standards that characterize the Benevolent as emphasizing own inputs exceeding own outcomes; the Entitled, own outcomes exceeding own inputs; and the Equity Sensitive, own outcomes equaling own inputs. (Huseman, Hatfield, & Miles, 1987, p. 231)

In summary, a generalized equity theory supports the notion that individuals value fair treatment, which causes them to remain motivated to maintain an equilibrium of fairness in the individual and organizational relationships. The structure of generalized equity is based on the ratio of contributions to benefits.

8.4.12 *Social Learning Theory of Motivation*

The social learning theory of motivation was proposed by Albert Bandura in the early 1960s. In social learning theory, Bandura proposes that behavior is learned from the environment through the process of observational learning.

> In the social learning view, man is neither driven by internal forces nor buffeted helplessly by environmental influences. Rather, psychological functioning is best understood in terms of a continuous reciprocal interaction between behavior and its controlling conditions. (Bandura, 1971, p. 2)

Bandura's theory postulates that new behavioral patterns can be learned either (1) through direct experience or (2) by observing the behavior of others. The theory supports the notion of reinforcement and that individual learning is largely governed by the reward-punishment consequences that follow the actions. Reinforcement is proposed as having the following incentive functions:

- *Informative function*. Individuals observe the range of consequences that accompany their actions.
- *Motivational function*. Individuals use the results of prior experience to expect that certain actions will result in outcomes that either: (1) have outcomes they value; (2) have no appreciable effect; or (3) have outcomes that are undesirable.
- *Cognitive function*. The onset of awareness in an individual is a function of the reward value of the actions' consequence.
- *Reinforcing function*. Individual responses can be strengthened through selective reinforcement imposed below the level of awareness.

Bandura summarizes reinforcement as:

> The overall evidence reveals that response consequences can be informative, motivating, and reinforcing. Therefore, in any given instance, contingent reinforcement may produce changes in behavior through any one or more of the three processes. People can learn some patterns of behavior by experiencing rewarding and punishing consequences, but if they know what they are supposed to do to secure desired outcomes they profit much more form such experiences. (Bandura, 1971, p. 5)

Most importantly, Bandura challenged the notion that behavior (B) was a function of (1) internal personal incentive (I) and (2) external or environmental pressure (E), where all behavior was a function of the joint effects of personal incentives and environmental pressures such that $B = f(I, E)$. Bandura noted that external, environmental pressure is not a fixed entity. In fact, it is only a *potentiality* and can itself be subject to behavior and vice versa, in a two-way causal process. In social learning theory, internal personal incentives (e.g., pride, satisfaction, and a sense of accomplishment) reinforce the cognitive element of the theory to cognitive developmental theories.

8.4.13 Expectancy Theory of Motivation

The Expectancy theory of motivation was first proposed in the 1960s by Vroom (1964) and expanded upon in the work of Porter and Lawler (1965, 1968).

The theory proposes that an individual will decide to behave or act in a certain way because they are motivated to select a specific behavior over other behaviors due to what they expect the result of that selected behavior will be. The motivation for how they will act is determined by the desirability of the outcome of the behavior or *expectancy*. Individual motivation is a product of the individual's expectancy that a certain effort will lead to the desired outcome. The theory has three variables that affect motivation:

- *Valence (V)* the attractiveness or desirability of various rewards or outcomes.
- *Expectancy (E)* the desirability of the result for the individual which the perceived relationship between effort and performance.
- *Instrumentality (I)* is the perceived relationship between performance and rewards.

Motivation in expectancy theory is labeled motivation force (M_f) and is the product of these three components, as shown in Eq. 7.2.

$$M_f = V \times E \times I \quad (7.2)$$

Each of the variables in the expectancy theory of motivation requires additional explanation.

Valence (V). Vroom defines valence as "… the affective orientation toward particular outcomes" (Vroom, 1964, p. 15). It is the attractiveness or desirability of various rewards or outcomes based on the value individuals place on the rewards of an outcome. The value is based on the unique needs, goals, values, and preferences of each unique individual. As such, valance is characterized by the extent to which a person values a given outcome or reward and is not an objective measure of satisfaction, but a subjective measure of the expected satisfaction of a particular outcome, for a particular individual.

> Outcomes desired by an individual are considered positively valent and those he wishes to avoid negatively valent; therefore valences are scaled over a virtually unbounded range of positive and negative values. Vroom emphasizes, as do most other expectancy theorists, the idea that the objective utilities associated with outcomes of working at a particular level are not of primary concern; rather, the crucial factor is the individual's perception of the satisfaction or dissatisfaction to be derived from working at a particular level. (Behling & Starke, 1973, p. 374)

Expectancy (E). "Expectancy is defined as a momentary belief on the part of an individual that acting in a particular way will actually be followed by a given outcome. The expectancy value associated with any action-outcome link may range from 0.0 (no relationship perceived) to 1.0 (complete certainty that acting in a particular way will result in the outcome)" (Behling & Starke, 1973, p. 374). There are three components associated with the individual's expectancy perception:

1. *Self-efficacy* the individual's belief about their ability to successfully perform a particular behavior.
2. *Goal difficulty* the individual's belief about the ability to achieve the goal or performance expectation.
3. *Perceived control* the individual's belief in their ability to control their performance.

Instrumentality (I). "Instrumentality theory hypothesizes that a person's attitude toward an outcome (state of nature) depends on his perceptions of relationships (instrumentalities) between that outcome and the attainment of other consequences toward which he feels differing degrees of liking or disliking (preferences)" (Graen, 1969, p. 1). In the perceived relationship between performance and rewards, rewards in organizational settings may be an increase in pay or responsibility, special recognition or award, or a personal sense of accomplishment.

Factors associated with the individual's instrumentality for outcomes are trust, control, and policies. If individuals trust their superiors, they are more likely to believe their leaders' promises. When there is a lack of trust in leadership, people often attempt to control the reward system. When individuals believe they have some kind of control over how, when, and why rewards are distributed, instrumentality tends to increase. Formalized written policies impact the individuals' instrumentality perceptions. Instrumentality is increased when formalized policies associate rewards with performance.

8.4.14 Motivator-Hygiene Theory of Motivation

The motivator-hygiene theory of motivation was first proposed in the 1960s by Frederick Herzberg [1923–2000]. The theory, which is also referred to as the *two-factor theory* and *dual-factor theory*, proposes that there are two sets of factors in the workplace that affect workers' satisfaction.

The motivator-hygiene theory (Herzberg, 1968; Herzberg, Mausner, & Snyderman, 1959) has built upon Maslow's hierarchy of needs theory by proposing the presence of one set of factors or incentives that lead to satisfaction and a separate and unique set of factors or detractors that leads to dissatisfaction. Herzberg abandons the idea of a continuum of satisfaction (ranging from highly satisfied to high dissatisfied) and proposes two independent phenomena. The motivator-hygiene theory requires management to consider each factor when addressing worker motivation. Herzberg's (1964) original list of motivators (lead to satisfaction) and hygiene (lead to dissatisfaction) factors was as follows:

- *Motivators*: "achievement, recognition for achievement, intrinsic interest in the work, responsibility, and advancement" (p. 487).
- *Hygiene factors*: "company policy and administrative practices, supervision, interpersonal relationships, working conditions, and salary" (p. 487).

In summary, motivating factors are needed to shift an employee to higher performance and hygiene factors are needed to ensure an employee is not dissatisfied.

8.4.15 Acquired Needs Theory of Motivation

The acquired needs theory of motivation was first proposed by David McClelland [1917–1998] in 1965. In this theory, which is also referred to as the *three needs theory* and the *learned needs theory*, McClelland (1961, 1965, 1978) proposed that individuals have three needs: (1) achievement; (2) affiliation; and (3) power. These motivations exist independent of age, sex, race, or culture. Furthermore, the dominant type of motivation that drives an individual is a function of the life experiences and the opinions of the culture in which the individual was immersed. The three needs are classified as:

1. *Achievement*: Individuals with this need desire to excel and seek timely recognition for their efforts. Their efforts do not involve risks and require some gain for themselves. The possibility of failure is strictly avoided.
2. *Affiliation*: Individuals with this need seek peaceful relationships and refrain from actions which would attract attention to themselves. They seek sufficient recognition and do not require overjustification for their work.
3. *Power*: Individuals with this need require power in order to exercise control over other individuals. The power is acquired to serve their needs and to achieve objectives. These individuals do not seek recognition or approval, consider themselves superior, require direct compliance, and expect agreement with their decisions.

In summary, McClelland believed that every individual has one of three main driving motivators and that these motivators are not inherent, but developed based upon life experiences and the culture in which the individual was immersed.

8.4.16 ERG Theory of Motivation

The existence, relatedness, growth (ERG) theory of motivation was first proposed by Clayton Alderfer [1940–2015] in 1969. In this theory, Alderfer redefines Maslow's hierarchy of needs theory in new terms. Alderfer (1969) does this by recategorizing Maslow's hierarchy of needs into three simpler and broader classes of needs.

1. *Existence needs*: "include all of the basic forms of material and physiological desires" (p. 145).
2. *Relatedness needs:* "include all of the needs which involves relationships with significant other people" (p. 146).
3. *Growth needs*: "include all of the needs which involves a person making creating or productive effects on himself and the environment" (p. 146).

The ERG theory of motivation differs significantly from Maslow's hierarchy of needs. Unlike Maslow's theory, Alderfer's ERG theory does not require the fulfillment of a lower level of need prior to moving to a higher level. In ERG theory, if a higher-level need causes aggravation and cannot be fulfilled, then an individual may revert to increase the satisfaction of a lower-level need. This is labeled the *frustration-regression* aspect of ERG theory. In this manner, ERG theory (Alderfer, 1972) explicitly states that any given point in time, more than one need may be operational.

8.4.17 Self-determination Theory of Motivation

The self-determination theory of motivation (SDT) was first proposed by Edward Deci and Richard Ryan in 1971 (Deci & Ryan, 1985). SDT proposes that individuals tend to be motivated by a need to grow and gain fulfillment. The first assumption of SDT is that individuals are activity-directed toward growth. While many theories propose that individuals are most often motivated extrinsically (i.e., external rewards such as money, prizes, and acclaim), SDT is focused on intrinsic motivation (i.e., need to gain knowledge or independence).

SDT proposes that in order to become self-determined, individuals need to feel the following:

- *Competence*: Individuals need to gain mastery of tasks and control outcomes.
- *Relatedness*: Individuals need to experience a sense of belonging and attachment to other people.
- *Autonomy*: Individuals need to feel in control of their own behaviors and goals.

Once individuals achieve self-determination, they are able to be intrinsically motivated. Deci's (1972b) findings show that:

> The general findings of this study and the Deci (1971) studies suggest that one who is interested in developing and enhancing intrinsic motivation in children, employees, students, etc., should not concentrate on external control systems such as monetary rewards, which are linked directly to performance, but, rather, he should concentrate on structuring situations that are intrinsically interesting and then be interpersonally supportive and rewarding toward the persons in the situation. While large payments can lead to increased performance due to feelings of inequity, these payments will, however, be making the people dependent on the money, thereby decreasing their intrinsic motivation. (pp. 119, 120)

In summary, Deci's and Ryan's SDT (2002) proposes that three basic psychological needs motivate individuals. SDT states that these needs are said to be universal, innate, and psychological and include the need for (1) competence; (2) autonomy; and (3) psychological relatedness.

8.4.18 Opponent Process Theory of Motivation

The opponent process theory of motivation was first proposed by Richard Solomon [1918–1995] in 1965. In this theory, Solomon proposed that every process has a primary element called an *affective valence* (i.e., is it pleasant or unpleasant) and is followed by a secondary or *opponent process*. The secondary opponent process begins to take effect after the primary affective valence is quieted. As this sequence is repeated, the primary process tends to become weaker, while the opponent process becomes stronger.

> The theory assumes that for some reason the brains of all mammals are organized to oppose or suppress many types of emotional arousals or hedonic processes, whether they are pleasurable or aversive, whether they have been generated by positive or by negative reinforcers. (Solomon, 1980, p. 698)

Solomon and his collaborator Corbit (1973, 1974) conducted experiments on work motivation and addictive behavior, showing (1) how the opponent process theory applies to drug addiction and is the result of a pairing of pleasure (affective) and the symptoms associated with withdrawal (opponent) and (2) how, over time, the level of pleasure from using addictive substances decreases, while the levels of withdrawal symptoms increase, providing motivation to continue using the addictive substance despite a decreasing lack of pleasure.

In summary, the opponent process theory of motivation may be generalized beyond addictions to understand why situations that are distasteful or unpleasant may still be treated as rewarding.

8.4.19 Goal-Setting Theory of Motivation

The goal-setting theory of motivation was first proposed in the late 1970s by Latham and Locke (1979). The theory proposes that individuals will be motivated to the extent that they accept specific, challenging goals and receive feedback that indicates their progress toward goal achievement. Their goal-setting theory is fully consistent with social cognitive theory in that both acknowledge the importance of conscious goals and self-efficacy. The goal-setting theory focuses primarily on motivation in work settings. The core components of goal-setting theory include the following:

- *Goal specificity* the extent to which goals are detailed, exact, and unambiguous.
- *Goal difficulty* the extent to which a goal is hard or challenging to accomplish.
- *Goal acceptance* the extent to which people consciously understand and agree to goals.

The theory includes four mechanisms that directly affect performance:

1. Goals serve a directive function where they direct attention and effort toward goal-relevant activities and away from goal-irrelevant activities.
2. Goals have an energizing function such that high goals lead to greater effort than low goals.
3. Goals affect persistence when participants are allowed to control the time they spend on a task, hard goals prolong effort.
4. Goals affect action indirectly by leading to the arousal, discovery, or use of task-relevant knowledge and strategies.

The theory states that goal moderators are factors that facilitate goal effects and include the following: (1) *Commitment*, whereby public recognition of the goal is enhanced by leaders communicating an inspiring vision and behaving supportively; (2) *Importance*, where leadership commits resources based upon the goals relative importance; (3) *Self-efficacy* or the extent or strength of leadership's belief in its ability to complete tasks and reach goals by providing adequate training, positive role models, and persuasive communication; (4) *Feedback*, as an element stating that "for goals to be effective, people need summary feedback that reveals progress in relation to their goals" (Locke & Latham, 2002, p. 708); and (5) *Task complexity*, as the complexity of tasks increases and higher-level skills and strategies are required, goal effects are dependent on the ability to provide proper resources and strategies for accomplishment.

Figure 8.3 depicts the integration of the essential elements of goal-setting theory.

8.4.20 Reversal Theory of Motivation

The reversal theory of motivation was first proposed in 1975 by Michael Apter and Ken Smith and fully detailed in Apter's book *The Experience of Motivation* (1982). Reversal theory describes how individuals regularly reverse between psychological states, reflecting their motivational style and the meaning they attach to a specific situation at a unique point in time.

Because the theory is focused on mental life, it is termed phenomenological (Apter, 1981), where the behavior of an individual can only be fully understood within the subjective meaning assigned to it by the individual. An example of a reversal is shown by the response to a simple cat's meow. Sometimes, the meow evokes a warm memory and a smile; other times, the meow can evoke a frown and a sense of annoyance.

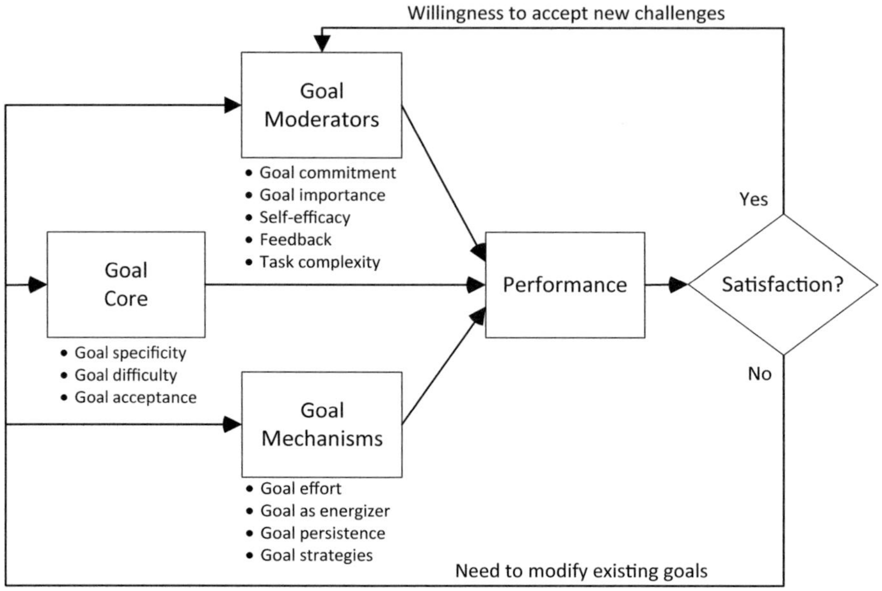

Fig. 8.3 Essential elements of goal-setting theory and the high-performance cycle (Locke and Latham, 2002, p. 714)

The theory proposes that individual experience is structurally organized into meta-motivational states which are opposing pairs labeled as *domains,* where only one of each pair can be active or experienced at a time. The states include the following:

- *Means-Ends State*: *Telic* (serious) and *Paratelic* (playful) which refer to whether an individual is motivated by achievement and future goals or the enjoyment of the moment.
- *Rules State*: Conforming and Rebellious which refer to whether an individual enjoys operating within rules and expectations or whether the individual desires to be free and rebels against rigid rules and structure.
- *Interaction State*: Mastery and Sympathy relate to whether an individual is motivated by transacting power and control or by sympathetic reaction demonstrated by care and compassion.
- *Orientation State*: *Autic* (self) and *Alloic* (other) which refer to whether an individual is motivated by self-interests or by the interests of others.

In summary, reversal theory proposes that individuals are changeable and move between different motivational states in the course of daily life. The theory serves as a means to understand why individual seems to contradict themselves in pursuit of satisfaction and provides a framework for improved understanding.

8.5 Applying Theories of Motivation

The twenty principal theories of motivation provide a variety of theoretical explanations for what motivates both individuals and groups. Many of the theories have similar notions and often augment one another. Because both of the authors primarily think systemically, the idea that a single, holistic, meta-theory that could synthesize the ideas presented in the twenty theories has great merit with each of us (and as it turns out, with others as well).

The idea of a meta-theory, or framework, for linking existing theories of motivation has been proposed by both Landy and Becker (1987) and Klein (1989). Klein's approach is to use control theory as an integrating framework for the theories of motivation. This has a great deal of appeal for systemic thinkers and will be presented as a functional framework for implementing a variety of useful aspects from the wide array of motivation theories.

8.5.1 Cybernetics and Control Theory

Tamotsu Shibutani [1920–2004] argues that two University of Chicago professors were responsible for introducing cybernetic features as important in explaining individual action long before Wiener (1965) coined the term *cybernetics*.

Philosopher-educator John Dewey [1859–1952] and psychologist George Mead [1863–1931], close colleagues at the University of Chicago, were heavily invested in the evolutionary view of individual action and interaction. Both Dewey and Mead felt that the individual and the environment were intimately interdependent, in disagreement with the prevailing stimulus–response theory of psychology in vogue at the time (Buckley, 1967). During the late nineteenth century, Dewey (1896) commented that the existing stimulus–response model was inadequate to explain human behavior or action, stating:

> It is the motor response of attention which constitutes that, which finally becomes the stimulus to another act. (p. 363)

Dewey also introduced the notions of communication and control (Dewey, 1916), which are two of the central principles of cybernetics and our notion of systems theory presented in Chap. 4. Similarly, Mead (1967) commented about both the individual:

> An act is an impulse that maintains the life-process by the selection of certain sorts of stimuli it needs. Thus, the organism creates its environment ... Stimuli are means, tendency is the real thing. Intelligence is the selection of stimuli that will set free and maintain life and aid in rebuilding it. (p. 6)

and the social act:

> The social act is not explained by building it up out of stimulus plus response; it must be taken as a dynamic whole – as something going on – no part of which can be considered or understood by itself – a complex organic process implied by each individual stimulus and response involved in it. (p. 7)

Both Dewey (1896, 1916) and Mead (1967; Morris, 1972) were pioneers in applying some fundamental features of a cybernetics to models of individual action. Cybernetics is the precursor to control theory and as such contains the foundation principles used to explain purposive action required in self-governing (i.e., cybernetic) models of human motivation.

8.5.2 Klein's Integrated Control Theory Model of Work Motivation

Howard Klein, of the Fisher College of Business at the Ohio State University, has constructed a framework, which is based on control theory that houses the salient features of a number of motivation theories (Klein, 1989, 1991, 1996). The control theory model integrates the works of a number of researchers who have developed control theory approaches in human behavior (Campion & Lord, 1982; Carver, 1979; Carver, Blaney, & Scheier, 1979; Carver & Scheier, 1981, 1982; Hollenbeck, 1989; Hollenbeck & Brief, 1988; Hollenbeck & Williams, 1987; Lord & Hanges, 1987; Taylor, Fisher, & Ilgen, 1984). The special features of Klein's model are as follows:

- *Parsimony*: The proposed model contains definitive elements of a limited number of motivation theories. As new theories are proposed, and older ones are supplanted, they can be incorporated into the model with relative ease. This is because the model is a framework, and even as other theories are included, "it can remain a simple heuristic" (Klein, 1989, pp. 150–151). This feature is noteworthy because it is invoking the goal axiom's *principle of requisite parsimony* (Miller, 1956).
- *Goal-setting*: The framework includes the ability to establish specific goals and objectives. The feature is invoking the goal axiom's *principle of purposive behavior* where the behavior is directed toward the attainment of a specific goal (Rosenblueth et al., 1943).
- *Feedback*: The framework contains feedback loops where sensors and comparators are used to provide signals based on an established standard or benchmark. This feature is invoking the viability axiom's *principle of feedback*. "Feedback control shows how a systems can work toward goals and adapt to a changing environment, thereby removing the mystery from teleology" (Simon, 1996, p. 172).
- *Motivation Theories*: The framework includes expectancy and attribution theories and can be extended to include social learning theory.

Klein's model is based upon the simple feedback model from cybernetics, which includes the following: (1) a *reference standard or benchmark;* (2) a *comparator* that differentiates between the signal and the standard or benchmark; (3) *feedback* which is the actual performance signal detected by the sensors and its transmission signal; and (4) an *effector* that implements corrective action based on the values generated in the comparator. The unique element in Klein's model is the inclusion of formal processes between the comparator and the effector that are based on four motivation theories included in the model. Figure 8.4 is a generic control theory model of work motivation based upon Klein's model that may be used as a process model for motivation in understanding the underlying *why* question when determining either (1) a premise, reason, or purpose for why something is the way it is, or (2) what the causal relationship is between the event and the actions that caused the event to occur.

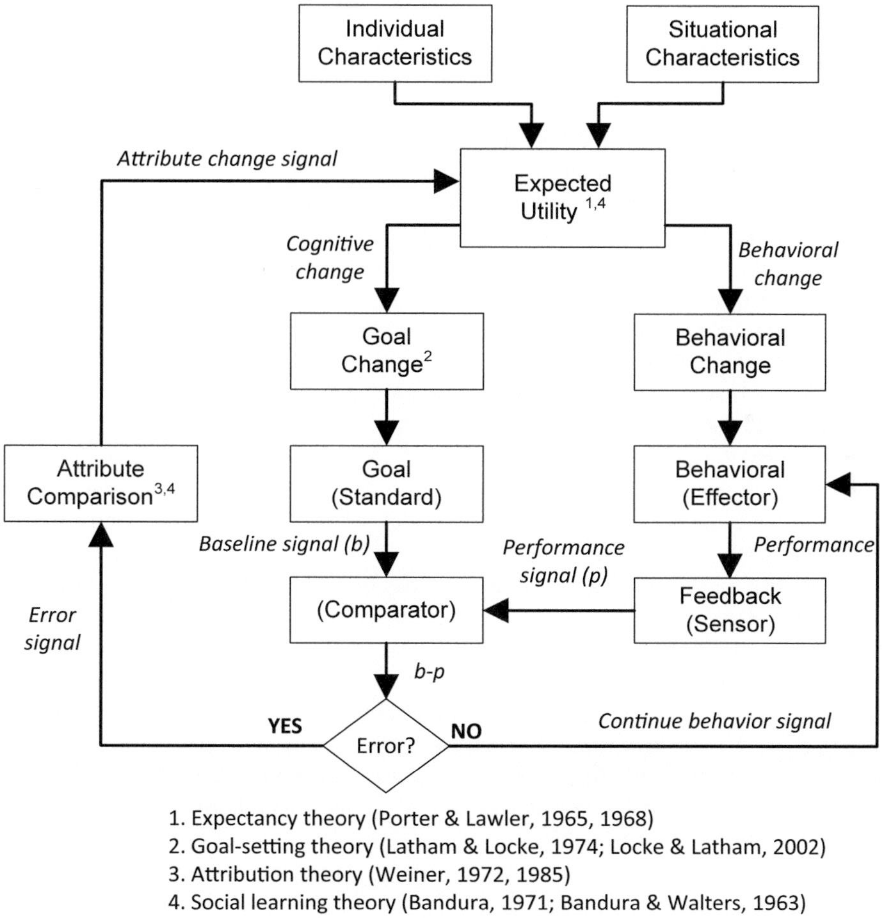

Fig. 8.4 Generic Control Theory of Motivation (based on Figure 2 in Klein, 1989, p. 153)

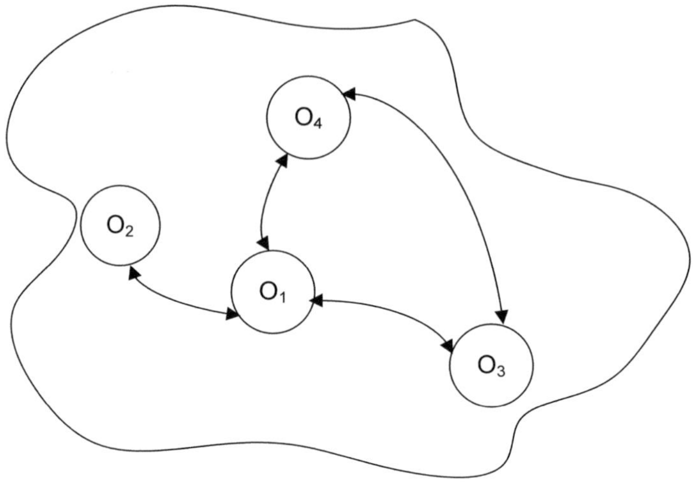

Fig. 8.5 Stakeholder objective relationships

The generalized control theory model of motivation depicted in Fig. 8.4 can be used to understand the unique relationship between stakeholders and their objectives in a problem. Each stakeholder has its own motivations for involvement in a mess. This motivation is formalized through its objectives. These objectives often involve relationships that extend to other stakeholders. If executed correctly, these motivational relationships are two-way in order to create a feedback loop. Figure 8.5 shows a set of relationships between four stakeholder objectives (O_1, O_2, O_3, and O_4).

Each of the two-way lines in Fig. 8.5 is unique and based on the control theory model in Fig. 8.4. As a result, there are both motivational goal signals ($M_{i,j}$) driving achievement of stakeholder objectives and feedback response signals ($F_{i,j}$) occurring between stakeholder objectives. Figure 8.6 shows how each stakeholder objective relationship contains a mini-model of motivation and feedback that influences each relationship.

Models of motivation based on stakeholder objective relationships need not be quantified or formalized, but the fact that each objective pair has unique motivating factors (and affiliated feedback) is the important point for practitioners invoking a systemic thinking perspective. When creating feedback mechanisms, care should be taken to avoid vicious circles and promote virtuous circles, as described by the *principle of circular causality* (Korzybski, 1994). As such, it may be necessary to use a hierarchy of regulation as described by the *principle of requisite hierarchy* (Aulin-Ahmavaara, 1979), in order to achieve ample regulatory control and motivational feedback.

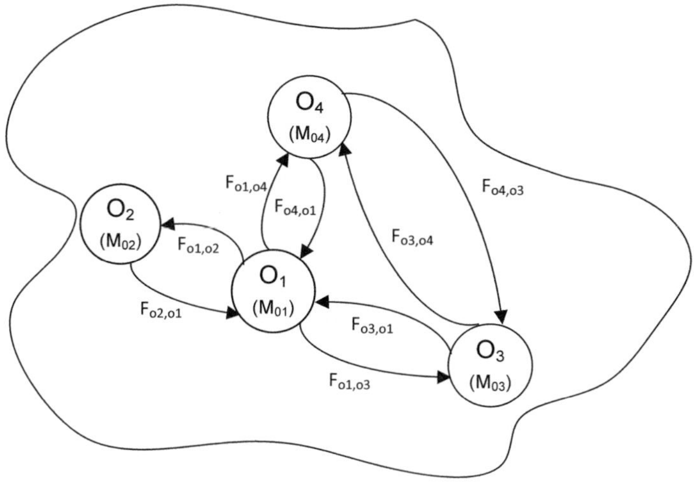

Fig. 8.6 Stakeholder objective relationship with motivation (*M*) and feedback (*F*) signals

8.6 Framework for Addressing *Why* in Messes and Problems

Addressing the *why* perspective requires that we complete the following steps for an identified problem:

1. Examine our FCM and its stakeholder objective relationships. Pay specific attention to the motivation/feedback cycles exhibited by each stakeholder's objective(s).
2. If necessary, modify the FCM by adding additional concepts and/or connections as appropriate to ensure feedback is provided for stakeholder objectives.
3. If these connections do not exist in the real system but they should, make note of them as changes to propose during the *Act* stage.

The following section demonstrates each step on our real estate problem.

8.7 Example Problem

We return to our real estate example pictured in Fig. 8.7.

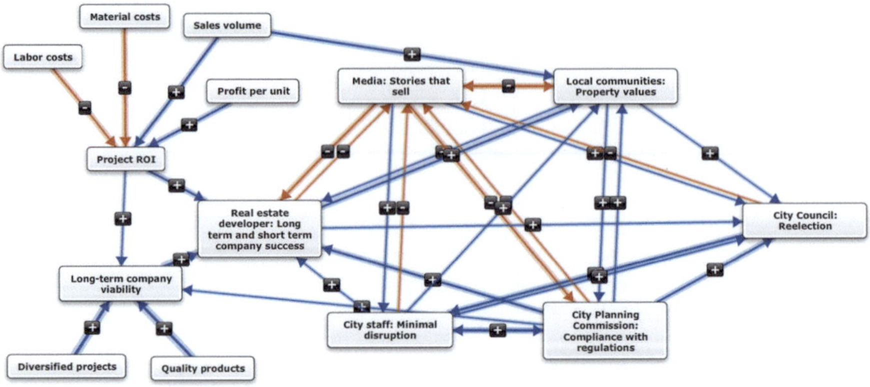

Fig. 8.7 Real estate example FCM

8.7.1 Motivation/Feedback Analysis

If we are interested primarily in *long-term and short-term company success (abbreviated company success)*, then we can make a few observations regarding its feedback loops:

- *Stories that sell* and *company success* are involved in a virtuous circle (assuming we believe that less media is a positive). As *company success* goes up, *stories that sell* go down. As *stories that sell* go down, the *company success* goes up, and so on. Motivationally, this causes the media (whose objective is maximizing stories that sell) to become less interested in the story. As the developer, this is a positive for us.
- The only other direct feedback loop exists between *company success* and *property values*. As *company success* goes up, *property values* go up. As *property values* goes up, *company success* goes up, and so on. This is another virtuous circle. If property owners see their property values go up as a result of the development, they will be supportive and maintain interest in the project.
- No direct feedback loops exist between *company success* and *minimal disruption*, *compliance with regulations*, and *reelection*. However, each provides feedback indirectly:
  - An increase in *company success* causes an increase in *reelection*, which causes *minimal disruption* to increase. This in turn causes a rise in *company success*.
  - An increase in *company success* also causes a decrease in *stories that sell*, which causes a decrease in *compliance with regulations*. This does not make sense. So, we have found an inconsistency. The FCM can be modified by removing this connection. Under the revised scenario, an increase in *company success* causes an increase in *property values*, which causes an increase in *compliance with regulations*. This in turn causes an increase in *company success*.

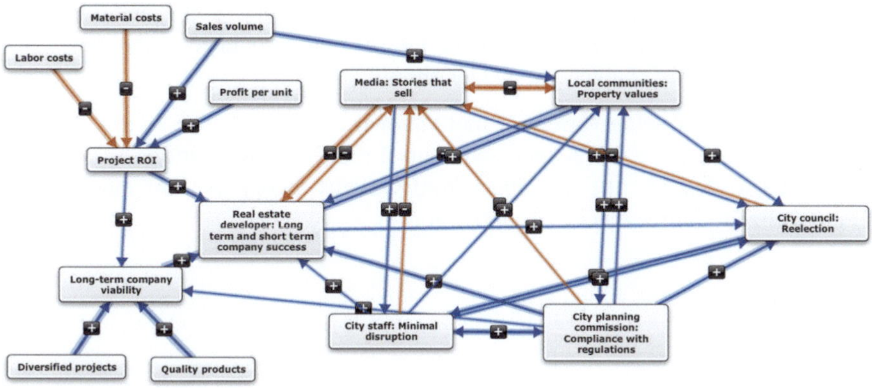

Fig. 8.8 Updated FCM with feedback analysis incorporated

8.7.2 FCM Update

Our feedback analysis in the previous subsection indicates the need to remove the causal link from *stories that sell* to *compliance with regulations*. This change is reflected in Fig. 8.8.

8.7.3 Proposed Changes During Act Stage

No proposed changes to the FCM are required during the Act stage; however, if we feel it is necessary, we could investigate the creation of a direct feedback loop between *company success* and *minimal disruption*, *compliance with regulations*, and *reelection*. In terms of feedback, this would expedite communication between the relevant stakeholders. However, it may also unnecessarily complicate the interactions between these entities.

8.8 Summary

Utilization of a formal model for motivation, based on the generic processes depicted in Fig. 8.4, may prove useful when attempting to understanding messes and their constituent problems. The initial and continued motivation serves as the incentive, the stimulus, and the inspiration for continued involvement. Using a cybernetic model with clear feedback loops ensures continued performance by ensuring goals remain synchronized with the individual and situational characteristics that form the context of the messes and constituent problems. This provides a congruent, current, and logical framework for achieving goals and objectives developed to address the elements of the messes and associated problems.

After reading this chapter, the reader should

1. Be aware of the wide variety of motivational theories.
2. Be able to describe the generalized control theory of motivation.
3. Be able to describe the principles from systems theory that are satisfied by a generalized control theory of motivation.

References

Adams, J. S. (1963). Towards an understanding of inequity. *The Journal of Abnormal and Social Psychology, 67*(5), 422–436. doi:10.1037/h0040968.

Adams, J. S. (1965). Inequity in social exchange. In L. Berkowitz (Ed.), *Advances in experimental social psychology* (Vol. 2, pp. 267–299). New York: Academic Press.

Adams, K. M., Hester, P. T., Bradley, J. M., Meyers, T. J., & Keating, C. B. (2014). Systems theory: The foundation for understanding systems. *Systems Engineering, 17*(1), 112–123. doi:10.1002/sys.21255.

Alderfer, C. P. (1969). An empirical test of a new theory of human needs. *Organizational Behavior and Human Performance, 4*(2), 142–175.

Alderfer, C. P. (1972). *Existence, relatedness, and growth; human needs in organizational settings*. New York: The Free Press.

Apter, M. J. (1981). The possibility of a structural phenomenology: The case of reversal theory. *Journal of Phenomenological Psychology, 12*(2), 173–187.

Apter, M. J. (1982). *The experience of motivation: The theory of psychological reversals*. New York: Academic Press.

Apter, M. J. (1984). Reversal theory and personality: A review. *Journal of Research in Personality, 18*(3), 265–288. doi:10.1016/0092-6566(84)90013-8.

Audi, R. (Ed.). (1999). *Cambridge dictionary of philosophy*. New York: Cambridge University Press.

Aulin-Ahmavaara, A. (1979). The law of requisite hierarchy. *Kybernetes, 8*(4), 259–266.

Bandura, A. (1971). *Social learning theory*. New York: General Learning Corporation.

Bandura, A., & Walters, R. H. (1963). *Social learning and personality development*. New York: Holt, Rinehart & Winston.

Behling, O., & Starke, F. A. (1973). The postulates of expectancy theory. *Academy of Management Journal, 16*(3), 373–388. doi:10.2307/254999.

Blau, P. M. (1964). *Exchange and power in social life*. New York: Wiley.

Bowditch, J. L., Buono, A. F., & Stewart, M. M. (2008). *A primer on organizational behavior* (7th ed.). Hoboken, NJ: Wiley.

Buckley, W. (1967). *Sociology and modern systems theory*. Englewood Cliffs, NJ: Prentice-Hall.

Campion, M. A., & Lord, R. G. (1982). A control systems conceptualization of the goal setting and changing process. *Organizational Behavior and Human Performance, 30*(2), 265–287.

Cannon, W. B. (1929). Organization for physiological homeostasis. *Physiological Reviews, 9*(3), 399–431.

Cannon, W. B. (1967/1932). *The wisdom of the body* (Revised ed.). New York: W.W. Norton & Company.

Carver, C. S. (1979). A cybernetic model of self-attention. *Journal of Personality and Social Psychology, 37*(8), 1251–1281. doi:10.1037/0022-3514.37.8.1251.

Carver, C. S., Blaney, P. H., & Scheier, M. F. (1979). Reassertion and giving up: The interactive role of self-directed attention and outcome expectancy. *Journal of Personality and Social Psychology, 37*(10), 1859–1870. doi:10.1037//0022-3514.37.10.1859.

Carver, C. S., & Scheier, M. F. (1981). *Attention and self regulation: A control theory approach to human behavior*. New York: Springer.

Carver, C. S., & Scheier, M. F. (1982). Control theory: A useful conceptual framework for personality—Social, clinical, and health psychology. *Psychological Bulletin, 92*(1), 111–135. doi:10.1037/0033-2909.92.1.111.

Deci, E. L. (1971). Effects of externally mediated rewards on intrinsic motivation. *Journal of Personality and Social Psychology, 18*(1), 105–115. doi:10.1037/h0030644.

Deci, E. L. (1972a). The effects of contingent and non-contingent rewards and controls on intrinsic motivation. *Organizational Behavior and Human Performance, 8,* 217–229. doi:10.1016/0030-5073(72)90047-5.

Deci, E. L. (1972b). Intrinsic motivation, extrinsic reinforcement, and inequity. *Journal of Personality and Social Psychology, 22*(1), 113–120. doi:10.1037/h0032355.

Deci, E. L., & Ryan, R. M. (1985). *Intrinsic motivation and self-determination in human behavior*. New York: Plenum Press.

Deci, E. L., & Ryan, R. M. (Eds.). (2002). *Handbook of self-determination research*. Rochester, NY: University of Rochester Press.

Dewey, J. (1896). The reflex arc concept in psychology. *Psychological Review, 3*(4), 357–370.

Dewey, J. (1916). *Democracy and education: An introduction to the philosophy*. New York: MacMillan.

Emerson, R. M. (1976). Social exchange theory. *Annual Review of Sociology, 2,* 335–362. doi:10.2307/2946096.

Evans, M. G. (1970). The effects of supervisory behavior on the path-goal relationship. *Organizational Behavior and Human Performance, 5*(3), 277–298.

Festinger, L. (1954). A theory of social comparison processes. *Human Relations, 7*(2), 117–140. doi:10.1177/001872675400700202.

Festinger, L. (1957). *A theory of cognitive dissonance*. Stanford, CA: Stanford University Press.

Festinger, L. (1962). Cognitive dissonance. *Scientific American, 207*(4), 93–107. doi:10.1038/scientificamerican1062-93.

Festinger, L., & Carlsmith, J. M. (1959). Cognitive consequences of forced compliance. *The Journal of Abnormal and Social Psychology, 58*(2), 203–210. doi:10.1037/h0041593.

Gagné, M., & Deci, E. L. (2005). Self-determination theory and work motivation. *Journal of Organizational Behavior, 26*(4), 331–362. doi:10.1002/job.322.

Georgopoulos, B. S., Mahoney, G. M., & Jones, N. W. (1957). A path-goal approach to productivity. *Journal of Applied Psychology, 41*(6), 345–353. doi:10.1037/h0048473.

Gouldner, A. W. (1960). The Norm of Reciprocity: A Preliminary Statement. *American Sociological Review, 25*(2), 161–178. doi:10.2307/2092623

Graen, G. (1969). Instrumentality theory of work motivation: Some experimental results and suggested modifications. *Journal of Applied Psychology, 53*(2, Pt.2), 1–25. doi:10.1037/h0027100.

Gruder, C. L. (1971). Determinants of social comparison choices. *Journal of Experimental Social Psychology, 7*(5), 473–489. doi:10.1016/0022-1031(71)90010-2.

Harmon-Jones, E., & Mills, J. (1999). An introduction to cognitive dissonance theory and an overview of current perspectives on the theory. In E. Harmon-Jones & J. Mills (Eds.), *Cognitive dissonance: Progress on a pivotal theory in social psychology* (pp. 3–21). Washington, DC: American Psychological Association.

Heider, F. (1944). Social perception and phenomenal causality. *Psychological Review, 51*(6), 358–374. doi:10.1037/h0055425.

Herzberg, F. I. (1964). The motivation-hygiene concept and problems of manpower. *Personnel Administration, 27*(1), 486–497.

Herzberg, F. I. (1968). One more time: How do you motivate employees? *Harvard Business Review, 46*(1), 53–62.

Herzberg, F. I., Mausner, B., & Snyderman, B. B. (1959). *The motivation to work*. New York: Wiley.
Hollenbeck, J. R. (1989). Control theory and the perception of work environments: The effects of focus of attention on affective and behavioral reactions to work. *Organizational Behavior and Human Decision Processes, 43*(3), 406–430.
Hollenbeck, J. R., & Brief, A. P. (1988). Self-regulation in the workplace: Towards a unified approach to understanding worker attitudes and behaviors. In R. Schuler (Ed.), *Readings in personnel and human resource management* (pp. 496–508). St. Paul, MN: West.
Hollenbeck, J. R., & Williams, C. R. (1987). Goal importance, self-focus, and the goal setting process. *Journal of Applied Psychology, 72*(2), 204–211.
Homans, G. C. (1958). Social behavior as exchange. *American Journal of Sociology, 63*(6), 597–606. doi:10.1086/222355.
Homans, G. C. (1974). *Social behavior*. New York: Harcourt Brace.
Honderich, T. (2005). *The oxford companion to philosophy* (2nd ed.). New York: Oxford University Press.
House, R. J. (1971). A path goal theory of leader effectiveness. *Administrative Science Quarterly, 16*(3), 321–339. doi:10.2307/2391905.
House, R. J. (1996). Path-goal theory of leadership: Lessons, legacy, and a reformulated theory. *The Leadership Quarterly, 7*(3), 323–352. doi:10.1016/S1048-9843(96)90024-7.
House, R. J., & Mitchell, T. R. (1974). Path-goal theory of leadership. *Journal of Contemporary Business, 3*(4), 81–97.
Hull, C. L. (1943). *Principles of behavior*. New York: D. Appleton-Century Co., Inc.
Hull, C. L. (1950). Behavior postulates and corollaries—1949. *Psychological Review, 57*(3), 173–180. doi:10.1037/h0062809.
Hull, C. L., Hovland, C. I., Ross, R. T., Hall, M., Perkins, D. T., & Fitch, F. B. (1940). *Mathematico-deductive theory of rote learning: A study in scientific methodology*. New Haven, CT: Yale University Press.
Huseman, R. C., Hatfield, J. D., & Miles, E. W. (1987). A new perspective on equity theory: The equity sensitivity construct. *The Academy of Management Review, 12*(2), 222–234. doi:10.2307/258531.
James, W. (1887a). Some human instincts (Part I). *Popular Science Monthly, 31*(6), 160–170.
James, W. (1887b). Some human instincts (Part II). *Popular Science Monthly, 31*(9), 666–681.
James, W. (1887c). What is an instinct. *Scribner's Magazine, 1,* 355–365.
Kelley, H. H. (1973). The processes of causal attribution. *American Psychologist, 28*(2), 107–128. doi:10.1037/h0034225.
Kelley, H. H., & Thibault, J. W. (1978). *Interpersonal relations: A theory of interdependence*. New York: Wiley.
Klein, H. J. (1989). An integrated control theory model of work motivation. *The Academy of Management Review, 14*(2), 150–172. doi:10.2307/258414.
Klein, H. J. (1991). Control theory and understanding motivated behavior: A different conclusion. *Motivation and Emotion, 15*(1), 29–44. doi:10.1007/bf00991474.
Klein, H. J. (1996). An integrated control theory model of work motivation. In R. Steers, L. Porter, & G. Bigley (Eds.), *Motivation and leadership at work* (6th ed., pp. 122–148). New York: McGraw-Hill/Irwin.
Kopelman, R. E., Prottas, D. J., & Davis, A. L. (2008). Douglas McGregor's Theory X and Y: Toward a construct-valid measure. *Journal of Managerial Issues, 20*(2), 255–271. doi:10.2307/40604607.
Korzybski, A. (1994). *Science and sanity: An introduction to non-aristotelian systems and general semantics* (5th ed.). Englewood, NJ: Institute of General Semantics.
Landy, F. J., & Becker, W. S. (1987). Motivation theory reconsidered. In L. L. Cummings & B. M. Staw (Eds.), *Research in organizational behavior* (Vol. 9, pp. 1–38). Greenwich, CT: JAI Press Inc.
Latham, G. P., & Locke, E. A. (1979). Goal setting—A motivational technique that works. *Organizational Dynamics, 8*(2), 68–80.

Lepper, M. R., & Greene, D. (1975). Turning play into work: Effects of adult surveillance and extrinsic rewards on children's intrinsic motivation. *Journal of Personality and Social Psychology, 31*(3), 479–486.

Locke, E. A., & Latham, G. P. (2002). Building a practically useful theory of goal setting and task motivation: A 35-year odyssey. *American Psychologist, 57*(9), 705–717. doi:10.1037/0003-066X.57.9.705.

Lord, R. G., & Hanges, P. J. (1987). A control systems model of organizational motivation: Theoretical development and applied implications. *Behavioral Science, 32*(3), 161–178.

Maslow, A. H. (1943). A theory of human motivation. *Psychological Review, 50*(4), 370–396.

Maslow, A. H. (1967). A theory of metamotivation: The biological rooting of the value-life. *Journal of Humanistic Psychology*, *7*(2), 93–127. doi:10.1177/002216786700700201.

Maslow, A. H. (1987). *Motivation and personality* (3rd ed.). New York: Harper & Row Publishers.

McClelland, D. C. (1961). *The achieving society*. New York: D. Van Nostrand Company Inc.

McClelland, D. C. (1965). Toward a theory of motive acquisition. *American Psychologist, 20*(5), 321–333. doi:10.1037/h0022225.

McClelland, D. C. (1978). Managing motivation to expand human freedom. *American Psychologist, 33*(3), 201–210.

McDougall, W. (1901). On the seat of the psycho-physical processes. *Brain, 24*(4), 530–579. doi:10.1093/brain/24.4.579.

McGregor, D. [2006 (1960)]. *The human side of enterprise* (Annotat ed.). New York: McGraw-Hill.

Mead, G. H. (1967). Mind, self, and society: From the standpoint of a social behaviorist. In C. W. Morris (Ed.). Chicago: University of Chicago Press.

Miller, G. A. (1956). The magical number seven, plus or minus two: Some limits on our capability for processing information. *Psychological Review, 63*(2), 81–97.

Miner, J. B. (2003). The rated importance, scientific validity, and practical usefulness of organizational behavior theories: A quantitative review. *Academy of Management Learning & Education, 2*(3), 250–268. doi:10.2307/40214194.

Morris, C. W. (Ed.). (1972). *Works of George Herbert Mead volume 3: The philosophy of the act*. Chicago: University of Chicago Press.

Patton, M. Q. (2002). *Qualitative research & evaluation methods* (3rd ed.). Thousand Oaks, CA: Sage Publications.

Porter, L. W., & Lawler, E. E. (1965). Properties of organization structure in relation to job attitudes and job behavior. *Psychological Bulletin, 64*(1), 23–51.

Porter, L. W., & Lawler, E. E. (1968). *Managerial attitudes and performance*. Homewood, IL: Irwin-Dorsey.

Roeckelein, J. F. (Ed.). (2006). *Elsevier's dictionary of psychological theories*. Amsterdam: Elsevier B.V.

Rosenblueth, A., Wiener, N., & Bigelow, J. (1943). Behavior, purpose and teleology. *Philosophy of Science, 10*(1), 18–24.

Runes, D. D. (Ed.). (1983). *The standard dictionary of philosophy*. New York: Philosophical Library.

Simon, H. A. (1996). *The sciences of the artificial* (3rd ed.). Cambridge, MA: MIT Press.

Skinner, B. F. (1953). *Science and human behavior*. New York: The MacMillan Company.

Skinner, B. F. (1956). A case history in scientific method. *American Psychologist, 11*(5), 221–233. doi:10.1037/h0047662.

Solomon, R. L. (1980). The opponent-process theory of acquired motivation: The costs of pleasure and the benefits of pain. *American Psychologist, 35*(8), 691–712. doi:10.1037/0003-066X.35.8.691.

Solomon, R. L., & Corbit, J. D. (1973). An opponent-process theory of motivation: II. Cigarette addiction. *Journal of Abnormal Psychology, 81*(2), 158–171. doi:10.1037/h0034534.

Solomon, R. L., & Corbit, J. D. (1974). An opponent-process theory of motivation: I. Temporal Dynamics of Affect. *Psychological Review, 81*(2), 119–145. doi:10.1037/h0036128.

Spence, K. W. (1936). The nature of discrimination learning in animals. *Psychological Review, 43*(5), 427–449. doi:10.1037/h0056975.

Spence, K. W. (1937). The differential response in animals to stimuli varying within a single dimension. *Psychological Review, 44*(5), 430–444. doi:10.1037/h0062885.

Taylor, M. S., Fisher, C. D., & Ilgen, D. R. (1984). Individuals' reactions to performance feedback in organizations: A control theory perspective. In K. Rowland & J. Ferris (Eds.), *Research in personnel and human resources management* (Vol. 2, pp. 81–124). Greenwich, CT: JAI Press.

Tesser, A., Millar, M., & Moore, J. (1988). Some affective consequences of social comparison and reflection processes: The pain and pleasure of being close. *Journal of Personality and Social Psychology, 54*(1), 49–61. doi:10.1037/0022-3514.54.1.49.

Thorndike, E. L. (1898). Animal intelligence: An experimental study of the associative processes in animals. *The Psychological Review: Monograph Supplements, 2*(4), i-109. doi:10.1037/h0092987.

Thorndike, E. L. (1911). *Animal intelligence: Experimental studies*. New York: The MacMillan Company.

Vroom, V. H. (1964). *Work and motivation*. New York: Wiley.

Weiner, B. (1972). Attribution theory, achievement motivation, and the educational process. *Review of Educational Research, 42*(2), 203–215. doi:10.2307/1170017.

Weiner, B. (1985). An attributional theory of achievement motivation and emotion. *Psychological Review, 92*(4), 548–573. doi:10.1037/0033-295X.92.4.548.

Wiener, N. (1965). *Cybernetics: Or control and communication in the animal and the machine* (2nd ed.). Cambridge, MA: MIT Press.

Wills, T. A. (1981). Downward comparison principles in social psychology. *Psychological Bulletin, 90*(2), 245–271. doi:10.1037/0033-2909.90.2.245.

Chapter 9
The *Where* of Systemic Thinking

Abstract The previous chapters in this section have addressed: (1) the *who* question through a discussion of problem stakeholders, their analysis and management; (2) the *what* question by deconstructing a mess and its constituent problems into relevant elements such as fundamental and means objectives; and (3) the *why* question through an analysis of motivation and how each problem has a unique model of motivation and feedback between and among its stakeholders. This chapter will answer the *where* question. This *where* we refer to is not associated with physical location and geographical coordinates, but with the circumstances, factors, conditions, values and patterns that surround the problem, and the boundaries that separate the problem from its environment.

9.1 Introduction

The sections that follow will focus on two elements of *where*. The first section will review *context*—the circumstances, factors, conditions, values, and patterns that surround messes and problems. The second section will review *boundaries*—the representations we use that provide lines of demarcation between messes and problems and the surrounding environment. Both of these elements can be used as a method for decreasing problem complexity, thereby improving understanding and analysis.

9.2 Context

As problems have evolved from simple systems to complex systems, the associated complexity surrounding each problem has also increased. Problems are no longer (1) isolated from the surrounding environment, or (2) responsive to detached technical solutions. Modern complex systems problems require approaches that include additional complementary perspectives that encompass viewpoints beyond

P.T. Hester and K.M. Adams, *Systemic Decision Making*, Topics in Safety, Risk, Reliability and Quality 33, DOI 10.1007/978-3-319-54672-8_9

a simplified technical perspective. The aperture of the problem lens must be widened to include multiple perspectives in order to permit a broader and more improved understanding of problem context.

The application of multiple perspectives offers a more inclusive framework through which complex systems problems may be viewed. The integration of technical, organizational, political, and human perspectives widens the aperture of the viewing lens by allowing contextual elements that surround the problem to be included as part of the solution domain. This section will discuss the following: (1) the development of perspectives using context; (2) provide a description and some definitions for context; (3) reveal the essential elements of context; (4) relate the temporal aspect of context; (5) define a relationship between data, information, knowledge, and context; (6) explain how to extract procedural context; and (7) present a framework that accounts for context in understanding problems and messes.

9.2.1 *Perspectives and Context*

In order to achieve a *holistic* understanding in systems age problems, problem solvers must formally account for known contextual elements by invoking as many unique perspectives as possible. According to Eq. 2.2, an observer will need to include as many perspectives as possible in order to understand a problem accurately. As part of this understanding, an observer will employ a number of lenses to focus these observations, much like a microscope or telescope would do. The lenses are *contextual lenses*. A contextual lens serves to focus our powers of understanding onto a particular element of context. Figure 9.1 is a depiction of two

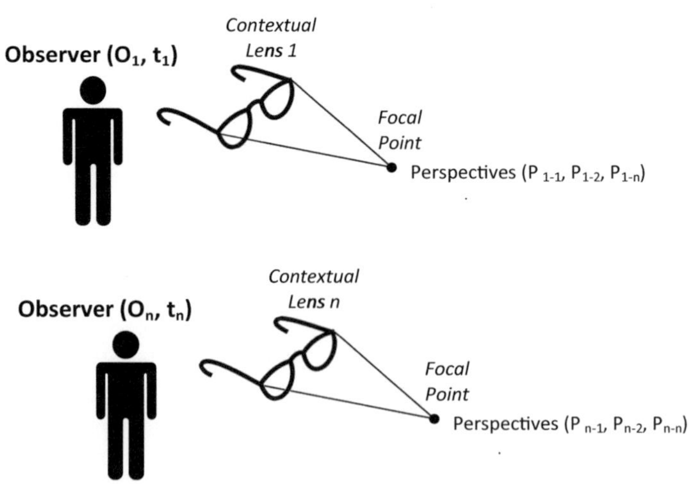

Fig. 9.1 Contextual lens and perspectives

Table 9.1 Contextual lenses used with complex systems

Contextual lens	Associated field of science	Associated perspectives
Individual	Psychology	Motivation, personality
Group	Sociology, management	Organizational behavior, control, finance, etc.
Political	Political science	Power, policies, etc.
Research	Philosophy of science	Ontological, epistemological, axiological, methodological, etc.
Engineering	Systems science, systems engineering	Systems-based methodologies, systems life cycle models, etc.
Science	Mathematics, physics, chemistry, biology	Logic, hierarchies, thermodynamics, etc.

observers; notice how there is an observer O_1 at time t_1 using contextual lens 1 and an observer O_n at time t_n using contextual lens n. Their observations result in a large number of different views, each providing a particular focus on understanding in a series of related perspectives P_{1-1}, P_{1-2}, … P_{n-n} and so on.

As an example, an observer beginning a research project will want to consciously select a research-related lens in order to develop a research view. The research lens will ensure that a particular number of research perspectives are included as part of the context (i.e., ontological, epistemological, axiological, and methodological perspectives) associated with the research project.

When a systems practitioner (who is always an observer) becomes involved with a complex systems mess and its constituent problems, there are a number of contextual lenses that must be invoked. Table 9.1 is a representative sample of contextual lenses with the associated science and resulting perspectives that may be used when developing context for complex systems.

The contrast between the various views focused by the contextual lenses leads to significantly different perspectives of the problem encountered by the problem solver or problem solving team. The *principle of holism* (Smuts, 1961 (1926)) posits that multiple perspectives be included as part of the development of problem context. The section that follows will define context and how it is integrated into part of the problem definition.

9.2.2 Description and Definitions for Context

A number of formal definitions for context exist in the extant literature and are provided in Table 9.2.

Table 9.2 Definitions for context

Definition	Source
1. (a) The parts of a discourse that surround a word or passage and can throw light on its meaning. (b) The interrelated **conditions** in which something exists or occurs	Mish (2009, p. 270)
2. The set of all **knowledge** that could be evoked by a human being facing a **situation**, assuming that he has an unlimited time to think about it	Brézillon & Pomerol (1999, p. 230)
3. Any information that can be used to characterize the situation of entities (i.e., whether a **person**, **place**, or **object**) that are considered relevant to the interaction between a user and an application, including the user and the application themselves. Context is typically the **location**, **identity**, and **state** of people, groups, and computational and physical objects	Dey, Abowd & Salber (2001, p. 106)
4. Context is any information that can be used to characterize the situation of an entity. An entity is a person, place, or object that is considered **relevant** to the interaction between a user and an application, including **location**, **time**, activities, and the **preferences** of each entity	Dey (2001, p. 5)
5. Context has been shown to be an emergent as well as a generative property of **knowledge**. Indeed, contexts are sets of **relations** and not self-evident things in themselves. We must therefore be alive to the possibility that there are two parallel processes of construing context: for us from within our own bodies of knowledge; and for them within theirs	Dilley (2002, p. 454)
6. Context may be defined loosely as the **setting** or **circumstances** in which an event or behavior of interest occurs. Context affects how we interpret the world around us	Keppie (2006, p. 242)
7. Context is a generalization of a collection of **assumptions** … may correspond to an infinite and only partially known collection of assumptions	McCarthy (1999, p. 557)
8. Context is mainly considered as a way to cluster **knowledge** for search efficiency, for representing counter-factual or hypothetical situations, for circumscribing the effects of particular actions to particular situations, and for directing an agent's focus of attention to **salient features** of a situation	Brézillon (1999, p. 61)
9. Context is a conceptual idea which can only be approximated by models, can be defined and used from different **perspectives**, is shared knowledge space	Maskery & Meads (1992, p. 13)

From these definitions, a number of key words and concepts can be extracted.

- Definition 1—*condition,*
- Definition 2—*situation* and *knowledge*
- Definition 3—*location, state, identity, person, place, and object*
- Definition 4—*relevant, location, time, and preferences*
- Definition 5—*knowledge and relations*
- Definition 6—*setting and circumstances*
- Definition 7—*assumptions,*
- Definition 8—*knowledge and salient features, and*
- Definition 9—*perspectives.*

Using these terms, we can develop a high-level definition for problem context which may be described as *the circumstances, factors, conditions, values, and patterns (i.e., salient features) that surround a particular situation, person, place, or object.*

9.2.3 Elements of Context

Based on our new definition of context, the five essential elements or salient features of context are as follows: (1) circumstances, (2) factors, (3) conditions, (4) values, and (5) patterns. These salient features can be grouped exclusively under the broad headings of abstraction and culture.

1. Abstraction: The removal, in thought, of some characteristics or features or properties of an object or a system that are not relevant to the aspects of its behavior under study. (Psillos, 2007, p. 6)
2. Culture: refer to systems of shared ideas, to the conceptual designs, the shared systems of meanings that underlie the ways people live. (Keesing, 1976, p. 139)

Based on these definitions of context, Table 9.3 is a simplified characterization of context that includes the five primary elements or salient features drawn from Adams and Meyers (2011).

Examples of how each contextual element contributes to problem context are as follows:

1. *Circumstances*: Certain elements of the problem context are bound by fixed parameters. An example could be the requirement to comply with an existing law or regulation.
2. *Factors*: A variable in the situation has a specific value that cannot be changed. For instance, a fixed budget.
3. *Conditions*: A state exists during a specific period of time and as a result may directly influence outputs or outcomes. An example is the threat of a terrorist attack.

Table 9.3 Contextual elements

	Abstraction	Culture
Elements	1. *Circumstances*: Particulars of the situation that define the state of affairs	1. *Values*: General beliefs for which the systems stakeholders have an emotional investment
	2. *Factors*: Specific characteristics or variables that affect the situation	2. *Patterns*: A perceived structure, operation, or behavior that is recurring
	3. *Conditions*: The prevailing state of the situation that influences outcomes	

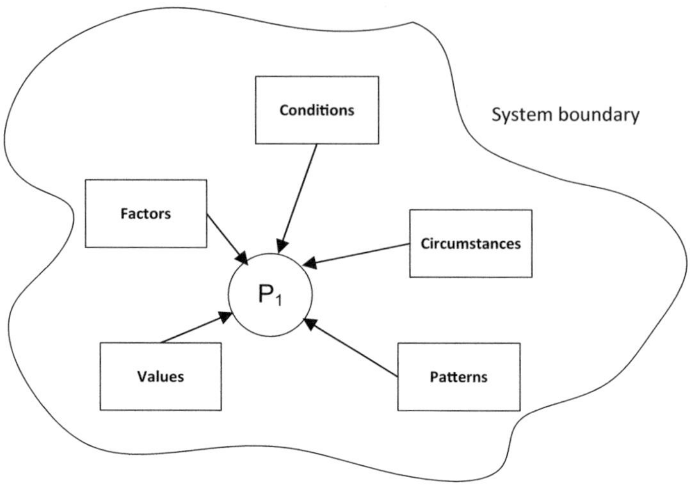

Fig. 9.2 Contextual elements acting on a problem

4. *Value*: A strongly held and guarded belief. For instance, workers will not be expected to work more than 40 h per week.
5. *Patterns*: A recurring behavior that is generally accepted. An example could be that artisans are allowed to make modifications to the established system design without having to consult the designers.

The five salient features of context can be viewed as nodes acting on our problem P_1 in Fig. 9.2.

It is important to note that the contextual elements depicted in Fig. 9.2 are within the system boundary. Because they reside within the system boundary, they can be controlled and manipulated. Conversely, elements outside of the system boundary merely act upon the problem in uncontrollable ways.

Because context can be a combination of tacit or explicit construct, and formal or informal construct, we believe that context must be meaningfully defined when addressing complex systems messes and their constituent problems. Ensuring that context is both explicit and formal facilitates the sharing of knowledge with others required to achieve a mutually agreeable perspective.

9.2.4 Temporal Aspects of Context

Because context is infinite (due to its being generated from an infinite number of perspectives), it is problematic to try to understand, know, or manage everything that surrounds a particular situation, person, place, or object. Therefore, we must

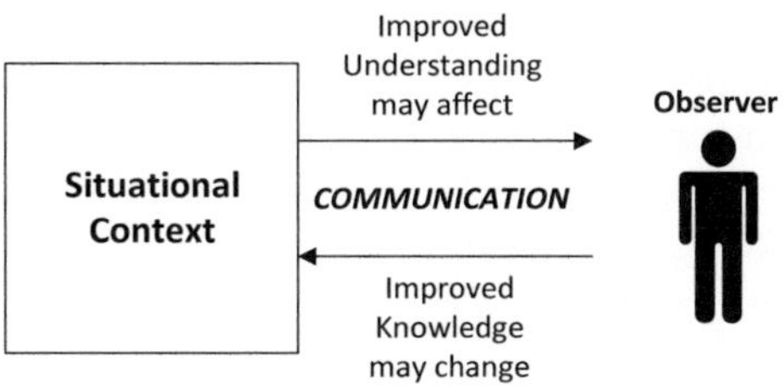

Fig. 9.3 Situational context and observer behavior

purposefully limit the information by creating a subset of the information in the context. By limiting the dimensions of the problem space, we are controlling what we choose to use to understand the problem and what we are to do next (Bainbridge, 1997).

Context has an important temporal element. Context is not static, but dynamic, changing over time. Change is a function of both external and internal interactions. Communication is a constant cycle of interaction because it includes ever-improving explanations with respect to the context and the subsequent behaviors by the observers who gain this improved understanding. Figure 9.3 depicts this interaction and supports Mittal's and Paris' (1995) notion that "in order to be able to build systems that take into account the context, it is necessary to identify how context constrains behavior and how context in turn is changed by action" (p. 492).

The main point being made in Fig. 9.3 is that communication serves as the link between the context of the situation and the observer, and that this is a continuous process.

The idea that both context and the degree of ambiguity change over time is an important one (Maskery & Meads, 1992). If problem perspectives remain rigid and fixed, then there is (1) the potential for a decreased perception of ambiguity in the process of improved understanding and (2) subsequent changes occurring in the context surrounding the problem may fail to be incorporated into plans associated with problem resolution.

9.2.5 Cultural Values and Their Impact on the Development of Context

In Table 9.3, we stated that cultural values *are general beliefs for which the systems stakeholders have an emotional investment*. This very general statement requires additional explanation with respect to how stakeholders acquire and treat the knowledge claims that underlie their cultural values. Knowledge has an enormous sociological element that we will briefly examine.

"Facts are not objectively given, but collectively created" (Fleck, 1979, p. 157). This statement by Ludwik Fleck [1896–1961], developer of the first system of the historical philosophy and sociology of science, proposes that all scientific knowledge is a three-way dialogue between the knowing subject, the known object, and a thought collective. Fleck (1979) defines the thought collective as:

> A community of persons mutually exchanging ideas or maintaining intellectual interaction, we will find by implication that it also provides the special "carrier" for the historical development of any field of thought, as well as for the given stock of knowledge and level of culture. This we have designated thought style. (p. 39)

Fleck's thought collective is representative of what we have termed cultural values. It is these cultural values of the thought collective that directly influence facts, which are the foundation for the data, information, knowledge, decisions, and metrics (DIKDM) framework we will present in the next section. Simply stated, the cognitive process employed by any individual is a function of his or her exposure to the thought collective, and is not a function, solely, of their own consciousness. It is a function of their interaction with the existing stock of information. The existing stock of information directly influences the individual's process for acquiring knowledge and understanding (i.e., cognition). "This cognition, in turn, enlarges, renews, and gives fresh meaning to what is already known" (Fleck, 1979, p. 38). This is a process of *cultural conditioning* that supports the notion that all facts are collectively created.

In the next section, we will present a framework for data, information, knowledge, decisions, and metrics (DIKDM) that relies upon facts which are a function of both empirical observations (see Chap. 15) and the cultural values discussed in this section.

9.2.6 Data, Information, and Knowledge

The term *data* refers to a set of observations already interpreted in some way. Because human observation involves personal interpretation, data may contain bias. Knowing this, we will further propose that data are *symbols* that *represent* properties of objects, events, and their environments. "Data is a set of discrete, objective facts about events" (Davenport & Prusak, 2000, p. 2).

Most pieces of *data* are of limited value until they are processed into a useable form. Processing data into a useable form requires human intervention, most often accomplished with the use of an automated system. The output of the processing of data is *information*. Information is contained in descriptions and in answers to questions that begin with such words as who, what, where, when, and how many. These functional operations are applied to data and transform it into information. It is important to note that the difference between data and information is functional, not structural (Ackoff, 1989). Data is transformed into information in the following ways (Davenport & Prusak, 2000):

- Contextualized: we know for what purpose the data was gathered
- Categorized: we know the units of analysis or key components of the data
- Calculated: the data may have been analyzed mathematically or statistically
- Corrected: errors may have been removed from the data
- Condensed: the data may have been summarized in a more concise form (p. 4).

Like data, *information* has little utility without additional processing. Processing information into useful elements is a higher-order process that requires a purposeful human intervention.

> Knowledge is a fluid mix of framed experience, values, contextual information, and expert insight that provides a framework for evaluating and incorporating new experiences and information. It originates and is applied in the minds of knowers. In organizations, it often becomes embedded not only in documents or repositories but also in organizational routines, processes, practices, and norms. (Davenport & Prusak, 2000, p. 5)

Because knowledge is know-how, a human cognitive process is required to transform information into knowledge and, subsequently, into a possible basis for decisions. Information is transformed into knowledge much like the process in which data is transformed into information (Davenport & Prusak, 2000). This process involves the following:

- Comparison: how does information about this situation compare to other situations we have known?
- Consequences: what implications does the information have for decisions and actions?
- Connections: how does this bit of knowledge relate to others?
- Conversation: what do other people think about this information? (p. 6)

Our previous descriptions and depictions of the basic relationship between data, information, knowledge, decisions, and metrics in our other text (Hester & Adams, 2014) are updated in this text to include the language utilized by Boisot (1998) "Knowledge is a capacity that is built upon information extracted from data" (p. xiv). These relationships are depicted in Fig. 9.4.

Although processed, the knowledge at this point is raw, and termed *viscous knowledge*. Viscous knowledge is "rich, qualitative, ambiguous" (Boisot, 1998, p. xiv). In order to use this knowledge properly, it must be codified and abstracted. Codified and abstracted knowledge is termed *fluid knowledge*; it is knowledge where all extraneous data and information has been sifted out and discarded. The two elements of this sifting process which can change viscous information to fluid knowledge are as follows:

1. *Codification*: This is the systematic formalization or representation of knowledge through the use of drawings, specifications, etc. Codifying permits transmission of the knowledge into perhaps a more direct and useful form.

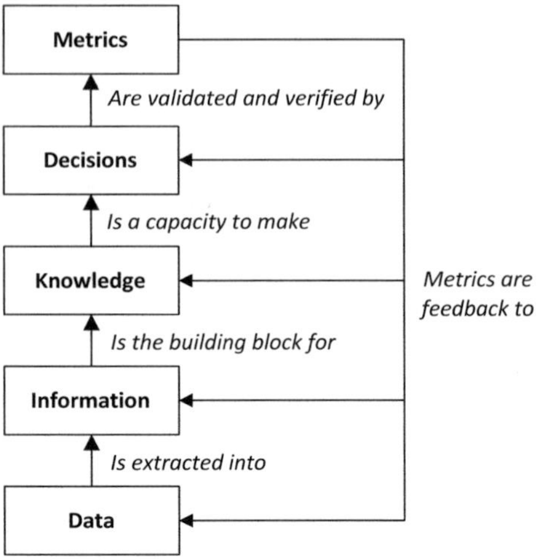

Fig. 9.4 Relationship between data, information, knowledge, decisions, and metrics

2. *Abstraction:* Abstraction was formally defined in Sect. 9.2.3. In this application, it is the process where cognitive simplification occurs. The level of complexity is decreased through the removal of extraneous details that continue to surround the knowledge.

"Codification and abstraction lower the cost of converting potentially useful knowledge into knowledge assets" (Boisot, 1998, p. 14).

9.2.7 Inclusion of Context

Context has a major role in the knowledge transformation process. Context is the wrapper that must be supplied and overlaid over the rest of the elements in a problem or mess in order to transform potentially useful knowledge into meaningful and useful knowledge assets. Once again, human intervention is essential in this process because "an explanation always takes place relative to a space of alternatives that require different explanations according to current context" (Brézillon, 1999, p. 57). Because context is dynamic (i.e., the situational information is changing over time), it can only be represented a posteriori.

So, faced with infinite context that is constantly changing, how does a systems practitioner establish the context for a mess and its constituent problems? The process through which relevant context is extracted is labeled *proceduralized context* and defined as:

> That part of the contextual knowledge that is invoked, structured and situated according to a given focus and which is common to the various people involved in decision making. (Brézillon & Pomerol, 1999, p. 233)

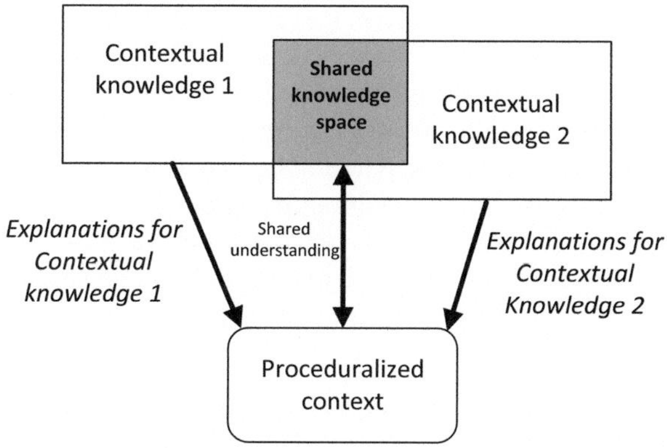

Fig. 9.5 Proceduralized context based on a figure in Brézillon and Pomerol (1999, p. 11)

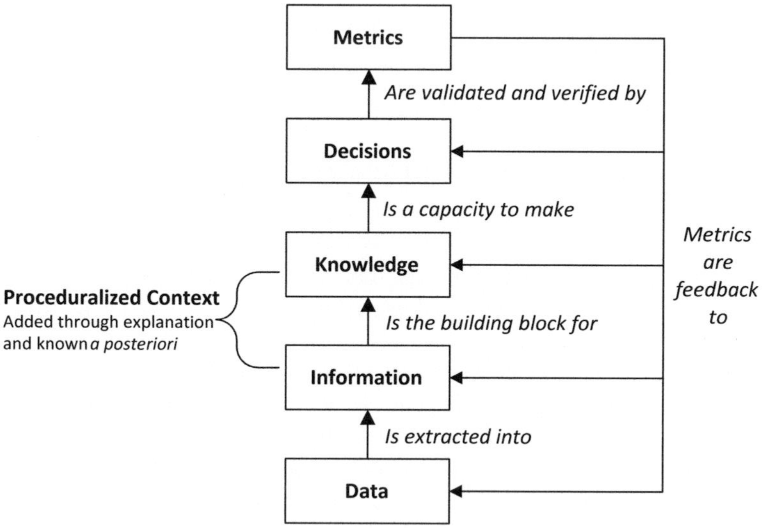

Fig. 9.6 Context as an element of the information transformation process

It is important to note that each of the contextual elements presented in Table 9.3 is influential in the development of the unique explanations present in each of the contextual knowledge situations. It is the sum of these elements that will both constrain and enable analysis of the problem. Proceduralized context accounts for the contextual knowledge present in each situation associated with the problem. Figure 9.5 is a depiction of proceduralized context.

Figure 9.6 shows how proceduralized context is an essential element in the knowledge transformation process.

As a final note on this process, the reader is encouraged to contemplate the explicit positioning of metrics in both Figs. 9.4 and 9.6. This incorporates the *feedback principle* of systems theory that was presented in Chap. 4, ensuring that decisions are validated (i.e., did we use the proper data, information, and knowledge for the decision?) and verified (i.e., are we using the data, information, knowledge for the decision correctly?) through appropriate metrics. The verification and validation process provides insights about the knowledge-based decision.

In summary, it is important to note that messes and problems cannot be separated from their context. Context is an inherent part of both a problem and its attendant solution.

> Neither problems nor solutions can be entertained free of context. A phenomenon that can be a problem in one context may not be one in another. Likewise, a solution that may prove effective in a given context may not work in another. (Gharajedaghi, 1999, p. 116)

The section that follows will address how a problem's boundaries, which set it apart from its surroundings, provide formal lines of demarcation between the defined mess and its associated problems with what we will define as the environment.

9.3 Boundaries and the Environment

While we have discussed the first element of the where question, context, the *where* question also has a second element, boundaries. This element, boundaries, addresses lines that separate or demark messes and problems from their environment.

As we improve our understanding of complex systems messes and their constituent problems, we recognize that the issues associated with boundaries—the representations we use to demark a problem from its environment—require additional understanding. It is no longer adequate to construct a demarcation line around a problem arbitrarily and to have confidence that the boundary is properly placed. Complex messes and problems are not easily bounded by the surrounding environment. Modern complex systems problems require finesse in the determination and establishment of boundaries. This section will provide definitions for both boundary and environment, discuss the significance of proper boundaries, provide a classification for boundaries, and propose a framework for boundary development.

9.3.1 Definitions for Boundary and Environment

Before we can discuss the characteristics associated with the establishment of a boundary between a problem system and its environment, we must create some relatively precise definitions for the primary terms we will use—boundary and environment.

Boundary is a term that has wide use in a variety of applications. However, when used by systems practitioners, it has a more precise definition that conveys specific meaning.

> Boundary: In the formal system model the area within which the decision-taking process of the system has power to make things happen, or prevent them from happening. More generally, a boundary is a distinction made by an observer which marks the difference between an entity he takes to be a system and its environment. (Checkland, 1999, p. 312)

Interestingly, the above definition uses the term environment as part of the definition. Environment is defined as:

> The environment for a system can be described as "a set of elements and their relevant properties, which elements are not part of the system, but a change in any of which can cause or produce a change in the state of the system. (Ackoff & Emery, 2006, p. 19)

Armed with relatively precise definitions for boundary and environment, the concepts related to the establishment of a boundary, which separates a problem system from its environment, may now be explored.[1]

9.3.2 The Significance of Boundary Establishment

The establishment of a boundary is a fundamental undertaking that has significant ramifications for all aspects of problem discovery and subsequent solution alternatives. Murray Gell-Mann, the 1967 Nobel Laureate in Physics, comments that "As always, determining the boundaries of the problem is a principal issue in problem formulation" (Gell-Mann, 1994, p. 270). Establishment of a problem boundary defines the limits of the knowledge that may be used during problem analysis. Not only does the establishment of the boundary set limits on knowledge, but also it defines those people and organizations (the *Who* stakeholders from Chap. 5) involved in the generation of the knowledge. By establishing a boundary, the systems practitioner has limited the perspectives and worldviews that can be brought to bear in support of problem analysis and resolution.

With acceptance of the premise that no system can be completely known (i.e., the contextual axiom's principle of darkness) and with acknowledgment that no single perspective or view of a system can provide complete knowledge of the system (i.e., the contextual axiom's principle of complementarity), the systems practitioner embraces the fact that setting boundaries is not only fundamental, but

[1]It is important to note that there are additional ideas about systems boundaries that exist beyond the scope of this book. Readers interested in advanced topics on system boundaries should consider both *structural coupling* [(Luhmann, 2013)] and *autopoiesis* [(Maturana & Varela, 1980)].

also critical to the successful understanding and subsequent dissolution, resolution, or solution (Ackoff & Emery, 2006) of a systems problem.

C. West Churchman [1913–2004], a leader in the systems movement, understood that boundaries are purely personal and group constructs that serve to define the limits for what is included during problem understanding and subsequent dissolution, resolution, or solution.

> The reason why the nature of the decision maker and the system boundaries are correlative problems is easy to see. We say that the environment of a system consists of those aspects of the natural world which influence the effectiveness of a system relative to the client, but which the decision maker cannot change. The resources, on the other hand, also influence the effectiveness, and can be changed by the decision maker. (Churchman, 1970, p. B44)

In summary, problem boundaries are artificial and also arbitrary representations (i.e., personal and group constructs), established to purposefully limit knowledge associated with the problem. We acknowledge that the delineation of boundaries can often have deleterious effects. The section that follows will discuss how boundaries may be classified.

9.3.3 Boundary Classification

Because problem boundaries are established based on a perceived personal or group reality, they should be established purposefully. Sufficiency requires that boundary conditions be classified using three characteristics: (1) temporal aspects; (2) the range of stakeholders included; and (3) ideas related to the scope of the dissolution, resolution, or solution endeavors (Midgley, 2000; Mingers, 2006).

- *Temporal characteristic*: In order to construct a sufficient boundary condition, the aspect of time must be included. Inclusion of a time boundary ensures that the problem is viewed from a similar perspective or worldview by all of the problem's stakeholders. This is often termed a *time horizon* and is an essential characteristic of any problem system's boundary definition.
- *Scope characteristic*: A sufficient boundary condition must include the notion of scope—what is to be included as part of the effort to understand the problem and any subsequent dissolution, resolution, or solution endeavors. This should include (a) what range of stakeholders are to be included, (b) the resources available for the effort, and (c) a generalized notion of what aspects of the problem are to be included as part of the endeavor.
- *Value characteristic*: A sufficient value characteristic should contain formal definitions that address the expected values of the participants within the defined boundary. These values should encompass the general beliefs for which the systems stakeholders have an emotional investment.

Armed with definitions for both boundary and environment, as well as an understanding of both the significance and the characteristics associated with the process of constructing a sufficient boundary, we have developed powerful concepts essential to problem bounding. The next section will provide a framework that may be used to ensure that the essential definitions are included as integral elements of a coherent approach to problem bounding.

9.3.4 Ulrich's Framework of Twelve Critically Heuristic Boundary Categories

Ulrich (1983) proposes that boundary judgments and value judgments are linked and that the boundaries adopted for the problem will be a direct function of the value judgments of participants. His answer is to construct practical guidelines that will permit participants to engage in meaningful dialogue where boundaries may be developed based on the critical reflection. Ulrich's boundary critique, which he labels *Critical Systems Heuristics* (CSH) , is a systemic effort of handling boundary judgments critically by viewing the boundary issues within specific categories.

> The boundary categories are arranged in four groups of three categories each. The first category of each group refers to a social role (rather than an individual person) who is or should be involved in defining the system of concern. For instance, in the first group, this is the "client" – the group of those who benefit or who ought to benefit. The second category addresses role specific concerns that are or should be included. Again taking the example of the first group, this is the client's "purpose" – the interests or concerns that are motivating a proposal. The third category relates to key problems that are crucial for understanding the previous two boundary judgements [sic]. (Ulrich, 2000, p. 258)

Table 9.4 presents Ulrich's framework. The first column has the social role category and its principal concerns. The second column addresses the underlying

Table 9.4 Ulrich's framework of twelve critically heuristic boundary categories (Ulrich, 2000)

Boundary category	Boundary issue	Participant category
1. Client	Sources of motivation	Those involved
Purpose		
Measure of improvement		
2. Decision maker	Sources of power	
Resources		
Decision environment		
3. Professional	Sources of knowledge	
Expertise		
Guarantee		
4. Witness	Sources of legitimation	Those affected
Emancipation		
Worldview		

boundary issues that are to be addressed. The third column categorizes the participants as either involved or affected.

A useful feature of CSH is a checklist of twelve boundary questions that are used to evaluate heuristically what the problem system *is* and what it *ought* to be.

> For systematic boundary critique, each question needs to be answered both in the "is" and in the "ought" mode. Differences between "is" and "ought" answers point to unresolved boundary issues. There are no definitive answers, in that boundary judgements [sic] may always be reconsidered. By means of systematic alteration of boundary judgements [sic], it is possible to unfold the partiality (selectivity) of an assumed system of concern from multiple perspectives, so that both its empirical content (assumptions of fact) and its normative content (value assumptions) can be identified and can be evaluated without any illusion of objectivity. (Ulrich, 2000, p. 259)

Table 9.5 presents Ulrich's checklist of critically heuristic boundary questions.

It is important to note that by using Ulrich's systemic guide for boundary critique, problem boundaries that were previously viewed as artificial and arbitrary representations are now exposed and challenged within a formal framework. Use of a formal framework for the critique permits participation and ownership by all problem stakeholders, and formulates the problem boundary using both *is* (i.e., descriptive) and *ought* (i.e., normative) modes. We propose that systems practitioners should adopt and utilize Ulrich's checklist of critically heuristic boundary questions in Table 9.5 as a means for operationalizing a formal process when identifying problem boundaries.

In summary, the establishment of problem boundaries requires a thorough understanding of both the significance of, and the characteristics associated with, the boundaries between a problem and its environment. This knowledge must be supported by a formal framework in which boundary knowledge is operationalized in a formal process where problem boundaries are proposed, critiqued, and accepted.

9.3.5 Force Field Diagrams

The boundary and contextual elements created using the techniques in this chapter can be represented together in a coherent articulation using a force field diagram (Lewin, 1938, 1939, 1943), a technique introduced in Chap. 4. The force field technique depicts opposing force types (driving and restraining) as vectors with both magnitude and direction. These forces act on either a current (as-is) or envisioned (ought-to-be) state. Thus, both boundary and context elements can be considered in their depiction as shown in Table 9.6. Those elements identified as context elements can be captured, as appropriate, as forces, whereas boundary investigation can lead to a description of the problem both in terms of its present state and its idealized state.

Table 9.5 Ulrich's checklist of critically heuristic boundary questions (Ulrich, 2000, p. 259)

Boundary issue	Boundary question
Sources of motivation	1. Who is (ought to be) the **client**? That is, whose interests are (should be) served?
	2. What is (ought to be) the **purpose**? That is, what are (should be) the consequences?
	3. What is (ought to be) the **measure of improvement**? That is, how can (should) we determine that the consequences, taken together, constitute an improvement?
Sources of power	4. Who is (ought to be) the **decision maker**? That is, who is (should be) in a position to change the measure of improvement?
	5. What **resources** are (ought to be) controlled by the decision maker? That is, what conditions of success can (should) those involved control?
	6. What conditions are (ought to be) part of the **decision environment**? That is, what conditions can (should) the decision-maker *not* control (e.g., from the viewpoint of those not involved)?
Sources of knowledge	7. Who is (ought to be) considered a **professional**? That is, who is (should be) involved as an expert, e.g., as a researcher, planner, or consultant?
	8. What **expertise** is (ought to be) consulted? That is, what counts (should count) as relevant knowledge?
	9. What or who is (ought to be) assumed to be the **guarantor of success**? That is, where do (should) those involved seek some guarantee that improvement will be achieved—for example, consensus among experts, the involvement of stakeholders, the experience and intuition of those involved, and political support?
Sources of legitimation	10. Who is (ought to be) **witness** to the interests of those affected but not involved? That is, who is (should be) treated as a legitimate stakeholder, and who argues (should argue) the case of those stakeholders who cannot speak for themselves, including future generations and nonhuman nature?
	11. What secures (ought to secure) the **emancipation** of those affected from the premises and promises of those involved? That is, where does (should) legitimacy lie?
	12. What **worldview** is (ought to be) determining? That is, what different visions of "improvement" are (should be) considered, and how are they (should they be) reconciled?

Table 9.6 Illustration of problem force field diagram

Driving force	Strength as-is	Strength ought-to-be	Problem	Strength ought-to-be	Strength as-is	Restraining force
Driving force 1	1	0.25	Present state (Idealized state)	−0.5	−1	Restraining force 1
Driving force 2	1	0.5		−0.25	−0.55	Restraining force 2
Driving force 3	0.25	1		−0.5	−0.5	Restraining force 3
Driving force 4	1	0.5		−0.25	−0.5	Restraining force 4
Driving force 5	0.5	0.25		−1	−1	Restraining force 5

We can now integrate the techniques developed in this chapter to create a framework for addressing the *where* perspective.

9.4 Framework for Addressing *Where* in Messes and Problems

Addressing the *where* perspective in our messes and problems requires that we complete the following steps for an identified problem:

1. Create an articulation of our boundary, drawing from Ulrich's boundary issues
2. Articulate relevant contextual elements, namely its circumstances, factors, conditions, values, and patterns
3. Generate a force field diagram using boundary and context elements
4. Revisit our FCM and adjust accordingly based on boundary and context concerns[2]
5. Capture proposed *what ought-to-be* changes for later analysis.[3]

The following section demonstrates each step on our real estate problem.

9.5 Example Problem

We return to our real estate example pictured (Fig. 9.7).

9.5.1 Boundary Articulation

Based on the analysis of critically heuristic boundary issues of our problem using guidelines found in (Ulrich, 2000) we can generate the boundary critique shown in Table 9.7.

Assessment of our boundary yields a few insights. There is a need for the real estate company to balance near-term and long-term company viability. Local residents have a great deal of power in this decision, more than the developer wishes for them to have. The developer would like to be involved in the decision making process. Finally, zoning regulations play a large part in the project's success.

---

[2]Considering boundary conditions, are any connections or concepts missing or do they require removal (i.e., are they outside the boundary of our decision context)? Do any require modification (i.e., should their weight be adjusted?).

[3]These serve as potential targets for mechanisms or for potential courses of action in the Act stage.

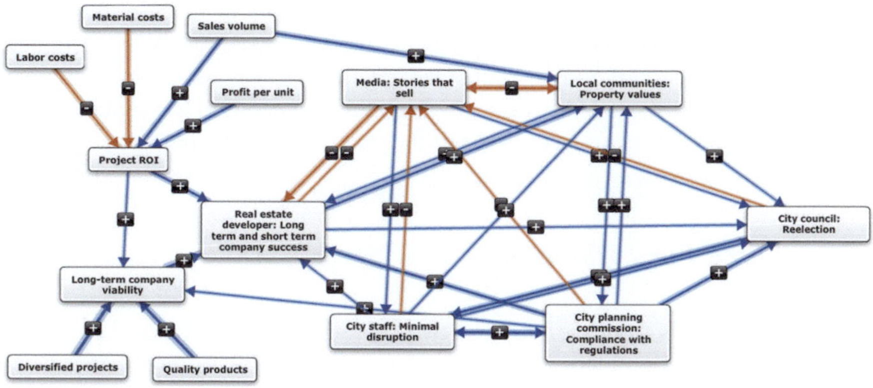

Fig. 9.7 Real estate problem

Table 9.7 Boundary critique

Boundary issue	What is	What ought to be
Sources of motivation	Profit	Profit and long-term viability
Sources of power	Developer, city planning commission, zoning regulations	Developer, city planning commission
Sources of knowledge	City planning commission	Developer, city planning commission
Sources of legitimation	City planning commission, local communities, developer, and customers	City planning commission, developer, and customers

9.5.2 Context

We can assess the contextual elements of our problem, including its abstraction (circumstances, factors, and conditions) and culture (values and patterns) as shown in Table 9.8.

This problem's context has several competing elements at play. Driving the problem toward resolution is the pattern of a business-friendly environment found in the city and society at large, as well as the factor of the developer's balance sheet,

Table 9.8 Context articulation

Category	Element
Circumstance	Current zoning status and regulations
Factor	Developer's balance sheet
Condition	Housing bubble status
Value	Homeowner perception of developer
Pattern	Business-friendly environment

which affords them some latitude regarding waiting to make a rash decision regarding the available land. Working against the real estate developer is the circumstance of the current zoning status and regulations, the condition of a recent housing bubble burst, and the value of declining homeowner perception of the developer (which is the primary source of the opposition to the developer's plans).

9.5.3 Force Field Diagram

Combining elements from the previous two subsections yields Table 9.9.

The nature of the problem, ultimately, stems from an "us-vs-them" perspective on the part of the local communities. This has caused problems for the developer and an inability to develop their land. Alternatively, they should seek to embrace an opportunity to work with the communities to find a mutually agreeable solution in an effort to move forward and finally achieve a return on their real estate investment.

9.5.4 Updated FCM

Many elements are missing from our FCM as a result of the boundary and context analysis. They include current zoning status and regulations, business-friendly environment, housing bubble status, and homeowner perception of developer. The developer's balance sheet, the only other element identified as a result of the boundary and context analysis, is already reflected in the FCM as the project ROI concept. The new concepts, as well as their connections, are reflected in the updated FCM shown in Fig. 9.8.

9.5.5 Proposed Ought-to-Be Changes

The real estate development company may wish to focus more on collaboration than they originally intended to. This is difficult as it may affect their bottom line in the short time, but it will lead to improved company viability, as well as short-term gains from a successful project that is not opposed by the local communities.

Table 9.9 Force field diagram

Driving force	Strength as-is	Strength ought-to-be	Problem	Strength ought-to-be	Strength as-is	Restraining force
Business-friendly environment	1	1	*The real estate developer sees their problem as an "us-vs-them" circumstance as it relates to the local communities. (They should embrace an opportunity to work with the communities to find a mutually agreeable solution, as the communities have significant power over the end result of the project)*	−0.5	−1	Current zoning status and regulations
Developer's balance sheet	1	0.5		−0.25	−0.5	Homeowner perception of developer
				−0.25	−0.5	Housing bubble status

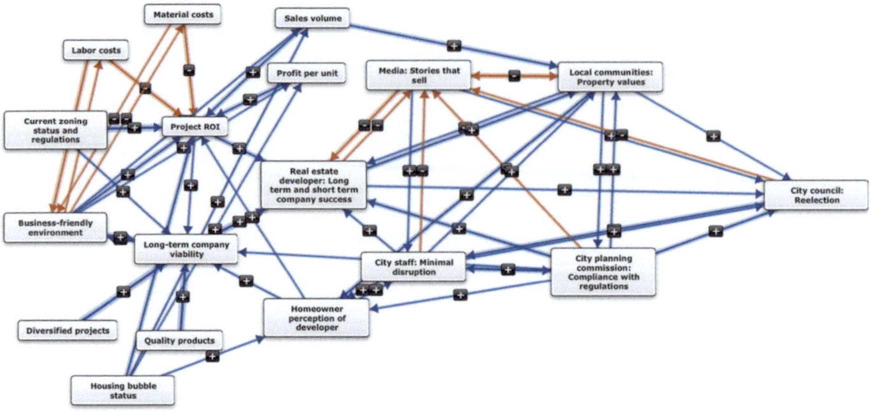

Fig. 9.8 Updated FCM with boundary and context incorporated

9.6 Summary

This chapter has provided a foundation for understanding the *where* in a mess or problem. The importance of *context*—the circumstances, factors, conditions, values and patterns that surround messes and problems—was presented. The importance of problem *boundaries*—the representations we use that provide lines of demarcation between messes and problems and the surrounding environment—was discussed, along with a framework for operationalizing the process of assessing problem boundary and context issues and updating our problem understanding based on these results.

After reading this chapter, the reader should:

1. Understand the five unique elements of context;
2. Be able to explain the relationship between data, information, and knowledge;
3. Recognize the importance of establishing a boundary between a system and its environment;
4. Be able to apply articulate how the environment and a boundary are essential elements of the problem; and
5. Be capable of applying a checklist of critical heuristic boundary questions as a framework for establishing a systems boundary.

References

Ackoff, R. L. (1989). From data to wisdom. *Journal of Applied Systems Analysis, 16*(1), 3–9.

Ackoff, R. L., & Emery, F. E. (2006). *On purposeful systems—An interdisciplinary analysis of individual and social behavior as a system of purposeful events*. Piscataway, NJ: Aldine.

Adams, K. M., & Meyers, T. J. (2011). Perspective 1 of the SoSE methodology: Framing the system under study. *International Journal of System of Systems Engineering, 2*(2/3), 163–192. doi:10.1504/IJSSE.2011.040552.

Bainbridge, L. (1997). The change in concepts needed to account for human behavior in complex dynamic tasks. *IEEE Transactions on Systems, Man, and Cybernetics—Part A: Systems and Humans, 27*(3), 351–359.

Boisot, M. H. (1998). *Knowledge assets: Securing competitive advantage in the information economy*. New York: Oxford University Press.

Brézillon, P. J. (1999). Context in problem solving: A survey. *The Knowledge Engineering Review, 14*(01), 47–80.

Brézillon, P. J., & Pomerol, J. C. (1999). Contextual knowledge sharing and cooperation in intelligent assistant systems. *Le Travail Humain, 62*(3), 223–246.

Checkland, P. B. (1999). *Systems thinking, systems practice*. New York John Wiley & Sons.

Churchman, C. W. (1970). Operations research as a profession. *Management Science, 17*(2), B37–B53. doi:10.2307/2629214.

Davenport, T. H., & Prusak, L. (2000). *Working knowledge: How organizations manage what they know*. Boston: Harvard Business School Press.

Dey, A. K., Abowd, G. D., & Salber, D. (2001). A conceptual framework and a toolkit for supporting the rapid prototyping of context-aware applications. *Human-Computer Interaction, 16*(2–4), 97–166.

Dilley, R. M. (2002). The problem of context in social and cultural anthropology. *Language & Communication, 22*(4), 437–456. doi:10.1016/s0271-5309(02)00019-8.

Fleck, L. (1979). *Genesis and development of a scientific fact* (T. J. Trenn & R. K. Merton, Eds., F. Bradley & T. J. Trenn, Trans.). Chicago: University of Chicago Press.

Gell-Mann, M. (1994). *The Quark and the Jaguar*. New York: Henry Holt and Company.

Gharajedaghi, J. (1999). *Systems thinking: Managing chaos and complexity: A platform for designing business architecture*. Boston: Butterworth-Heinemann.

Hester, P. T., & Adams, K. M. (2014). *Systemic thinking—Fundamentals for understanding problems and messes*. New York: Springer.

Keesing, R. (1976). *Cultural anthropology: A contemporary perspective*. New York: Holt, Rinehart and Winston.

Keppie, D. M. (2006). Context, emergence, and research design. *Wildlife Society Bulletin, 34*(1), 242–246.

Lewin, K. (1938). *The conceptual representation and the measurement of psychological forces*. Durham, N.C.: Duke University Press.

Lewin, K. (1939). Field theory and experiment in social psychology: Concepts and methods. *American Journal of Sociology, 44*(6), 868–897.

Lewin, K. (1943). Defining the 'field at a given time'. *Psychological Review, 50*(3), 292–310.

Luhmann, N. (2013). In *Introduction to systems theory*. P. Gilgen, Trans. D. Baecker (Eds.). Malden, MA: Polity Press.

Maskery, H., & Meads, J. (1992). Context: In the eyes of users and in computer systems. *SIGCHI Bulletin, 24*(2), 12–21. doi:10.1145/142386.142389.

Maturana, H. R., & Varela, F. J. (1980). *Autopoiesis and cognition: The realization of the living*. Hingham, MA: Kluwer.

McCarthy, J. (1999). Notes on formalizing context. *Proceedings of the 13th International Joint Conference on Artificial Intelligence* (Vol. 1, pp. 555–560). San Francisco: Morgan Kaufmann Publishers Inc.

Midgley, G. (2000). *Systemic intervention: Philosophy, methodology, and practice*. New York: Kluwer.

Mingers, J. (2006). *Realising systems thinking: Knowledge and action in management science*. Canterbury: Springer.

Mish, F. C. (Ed.). (2009). *Merriam-Webster's collegiate dictionary* (11th ed.). Springfield, MA: Merriam-Webster, Incorporated.

Mittal, V. O., & Paris, C. L. (1995). Generating explanations in context: The system perspective. *Expert Systems with Applications, 8*(4), 491–503. doi:10.1016/0957-4174(94)e0038-v.
Psillos, S. (2007). *Philosophy of science A-Z.* Edinburgh: Edinburgh University Press.
Smuts, J. C. (1961 (1926)). *Holism and evolution.* New York: Viking Press.
Ulrich, W. (1983). *Critical heuristics of social planning: A new approach to practical philosophy.* New York: John Wiley & Sons.
Ulrich, W. (2000). Reflective practice in the civil society: The contribution of critically systemic thinking. *Reflective Practice, 1*(2), 247–268.

Chapter 10
The *How* of Systemic Thinking

Abstract The previous chapters in this section have addressed the *who*, the *what*, the *why*, and the *where* questions associated with messes and their constituent problems. This chapter will address the *how* question. When we refer to *how*, we are interested in the specific means used in the attainment of specific, purposeful goals. The means are the mechanisms utilized in moving from the current problem state toward a new desired state where the goals and associated objectives have been satisfied. Mechanisms produce the effects that, when taken in concert, move a mess from the current state to the desired state.

10.1 Overview

The sections that follow will focus on nine elemental mechanisms that serve as the means of how. The first section will reveal the mechanisms of how, while the second section will examine the abstract mechanism of *method*. Finally, the third section will provide a framework that may be used when understanding messes and their constituent problems.

10.2 Mechanisms

When we speak of mechanism, we are using the term to describe *the means by which a desired effect or purpose is accomplished*. These mechanisms are the means which transform our problematic situation or existing states into more desirable state(s).

> Two central notions are involved in a problem-solving procedure: first, a problem state – a description of a problem situation including goals, available resources, and intermediate results; and second, a set of relevant moves that can be applied from a state to obtain new states. (Amarel, 1966, p. 112)

P.T. Hester and K.M. Adams, *Systemic Decision Making*, Topics in Safety, Risk, Reliability and Quality 33, DOI 10.1007/978-3-319-54672-8_10

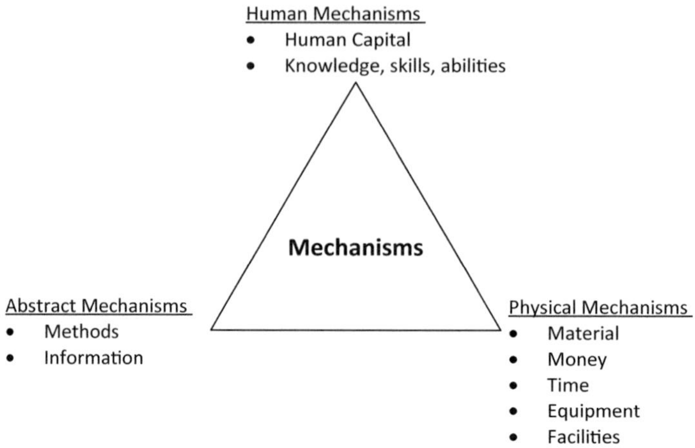

Fig. 10.1 Schema for mechanism classification

The mechanisms do not act alone, but in concert, to affect the movement from the problematic current state to the more desirable state(s). There are nine unique mechanisms which may be classified into three further categories: (1) abstract mechanisms; (2) physical mechanisms; and (3) human mechanisms. The schema for mechanism classification is depicted in Fig. 10.1.

The schema in Fig. 10.1 presents nine (9) unique mechanisms based on how these mechanisms are envisioned and utilized as the means for moving from a current state to a new, desired state. The sections that follow will briefly describe each of these mechanisms.

10.2.1 Physical Classification for Mechanisms

The mechanisms that are classified as physical are relatively easy to understand. Material, money, time, equipment, and facilities are measurable mechanisms (i.e., things) that may be used as means to accomplish objectives.

1. *Materials* are items that are consumed in the production of an output or outcome. Materials are often referred to as raw materials since they are the substances from which products are made. An example of a material is uranium dioxide which is the material from which nuclear pellets for fuel rods are manufactured.
2. *Money* is the medium of exchange used to pay for the exchange of products and services. Money is used as a means to reimburse the providers of products and services used to accomplish objectives. An example of money as a material is the paycheck offered to an employee for work.

3. *Time* is a measure of the duration of events and the intervals between them. Time is fundamental to understanding the movement between the current state and the desired state. All actions taking place between these states are evaluated using time, which serves as a universal measurement. An example of time is the specific period allocated for the accomplishment of a task. Finishing prior to the specified time period is an early finish, a coveted state. Finishing after the specified time period is a late finish, an undesirable state.
4. *Equipment* is movable objects used to create products and services in the production of an output or outcome. An example of equipment would be an ambulance which can be moved to a location where emergency medical services are required.
5. *Facilities* are immovable objects (i.e., they are not easily moveable) used to create products and services in the production of an output or outcome. An example of a facility is a hospital which exists to provide medical services, but cannot be relocated easily or quickly.

The application of physical mechanisms in support of outputs and outcomes is an easy concept to understand. Physical mechanisms are measureable and quantifiable means through which specific functions and processes produce desired outputs and outcomes.

10.2.2 Human Classification for Mechanisms

The mechanisms that are classified as human, much like physical mechanisms, are also relatively easy to understand. Manpower and knowledge are the attributes of human capital that are used to create outputs and outcomes.

1. *Human Capital* is the term used to specify the application of human beings (i.e., men and women) required to accomplish a specific process or function in support of a desired output or outcome. Human Capital is usually focused on the quantity rather than the quality of human capital required. It therefore does not address the knowledge, skills, or abilities of the human capital.
2. *Knowledge*, as a mechanism, consists of knowledge (with a small k), skills, and abilities (KSA), which are the unique list of qualifications and attributes required to successfully accomplish a process or function in support of a desired output or outcome. Table 10.1 provides formal definitions for these terms.

Table 10.1 defines knowledge, skills, and abilities. It also demonstrates how these three mechanisms can be assessed, measured, or observed as part of the process of ensuring that qualified human capital is provided to perform functions in support of desired outputs and outcomes.

We will propose a new term, *knowledge worker*, which is:

Table 10.1 Descriptions of knowledge, skills, and abilities/competence

Term	Description
Knowledge	• Knowledge refers to organized factual assertions and procedures that, if applied, makes adequate performance of a task possible (Cheney, Hale, & Kasper, 1990; Vitalari, 1985) • Knowledge can be assessed through formal examination (Gunnells, Hale, & Hale, 2006) • Knowledge is sometimes viewed as if it was a concrete manifestation of abstract intelligence, but it is actually the result of an interaction between intelligence (capacity to learn) and situation (opportunity to learn), so is more socially constructed than intelligence (Winterton, Delamare-Le Deist, & Stringfellow, 2006, p. 25)
Skills	• Skill refers to the proficient manual, verbal, or mental manipulation of tools, techniques, and methods (Cheney et al., 1990) • Skills can be readily measured by a performance test where quantity and quality of performance are tested, usually within an established time limit (Gunnells et al., 2006) • A combination of factors resulting in "competent, expert, rapid and accurate performance" (Welford, 1968, pp. 11–12)
Abilities	• Ability refers to the power to perform an observable activity at the present time (Cheney et al., 1990; Renck, Kahn, & Gardner, 1969) • Abilities can be observed and measured through behaviors that are similar to those required in a given role. Abilities are realized aptitudes. Aptitudes are only the potential for performing a behavior (Gunnells et al., 2006)

> A human being that possesses the required knowledge, skills, and abilities to perform a specific function in support of a desired output or outcome in the achievement of a purposeful objective or goal.

Once again, the application of human mechanisms is an easy concept to understand. Knowledge workers, in sufficient quantities, are required to successfully accomplish functions and processes required to deliver desired outputs and outcomes. It is the knowledge workers that add value to the process of delivering services or producing products, by transforming information into knowledge. At this point, it is important to note that knowledge, skills, and abilities (KSAs) are not static mechanisms. The environment within which the knowledge workers exist should be one where KSAs are acquired, refreshed, expanded, and validated. Because KSAs are such important mechanisms of *how*, the process in which capital K **K**nowledge (which, for this discussion, we will say encompasses knowledge, skills, and abilities) is created, acquired, articulated, and applied, the next two sections will review knowledge from both personal and organizational perspectives.

10.2.2.1 Personal Knowledge

Michael Polanyi [1891–1976] was a medical doctor, renowned physical chemist, and philosopher who proposed an individual theory of knowledge which he labeled *Personal Knowledge* (Polanyi, 1962, 2009). Polyani's theory was revolutionary. He sought to refute the established view that knowledge is discovered through the separation of the observer from the subject being studied and that the process of knowledge is a neutral one in which empirical data are collected and conclusions are drawn. Instead, Polyani's notion of Personal Knowledge posits that true discovery is guided by the passionate dedication and intellectual stimulation of the inquiring mind of an individual investigator. Polyani's (1962) theory of knowledge claims that humans experience the world in what he terms *tacit knowledge* by integrating *subsidiary awareness* into a *focal awaren*ess. For example:

> Subsidiary and focal awareness are mutually exclusive. If a pianist shifts his attention from the piece he is playing to the observation of what he is doing with his fingers while playing it, he gets confused and may have to stop. This happens generally if we switch our focal attention to particulars of which we had previously been aware only in in their subsidiary role. (p. 56)

Polyani believed that both tacit knowledge and explicit knowledge coexist along a continuum and that language was a relevant component of the explicit, as depicted in Fig. 10.2.

Tacit knowledge and explicit knowledge are at opposite ends of the continuum. Tacit knowledge is inherent physical experiences, intuition or implicit rules of thumb, while explicit knowledge is that which can be spoken, formulated through language, or presented graphically. Polyani's Personal Knowledge is categorized in current thinking as a first-generation model of knowledge generation.

10.2.2.2 Organizational Knowledge

The paradigm shift started by Polyani was expanded from the concept of individual or Personal Knowledge to the realm of organizational knowledge. Organizational knowledge, as a process, is defined as:

> Organizational knowledge creation is the process of making available and amplifying knowledge created by individuals as well as crystallizing and connecting it to an organization's knowledge system. In other words, what individuals come to know in their (work-) life benefits their colleagues and, eventually, the larger organization. (Nonaka, von Krogh, & Voelpel, 2006, p. 1179)

The dominant theory of organizational knowledge (Grant, 2007) is a knowledge conversion process of Socialization, Externalization, Combination, and Internalization or SECI (Nonaka, 1991, 1994, 2007; Nonaka & Toyama, 2003).

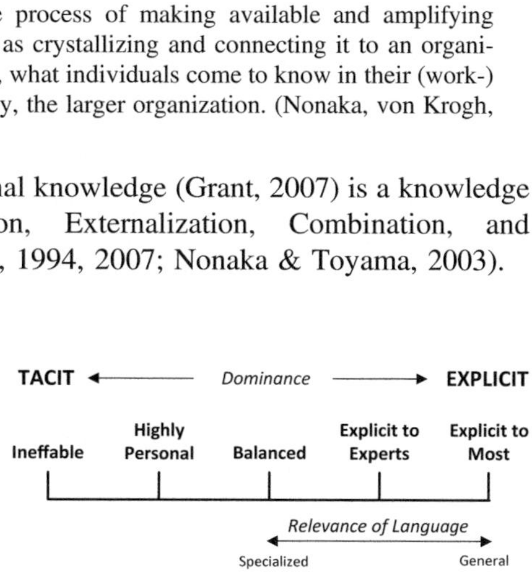

Fig. 10.2 The tacit–explicit continuum of knowledge, adapted from a figure in (Grant, 2007, p. 177)

The SECI model is a depiction of four processes. The first process, *socialization*, is where tacit knowledge gained through direct experience is passed on through practice, imitation, and observation. The tacit knowledge of the first process is passed to the second process, externalization. In *externalization,* the tacit knowledge is extracted from the practitioners through interview, observations, and similar activities. This is the most difficult process because tacit knowledge is most often impossible to fully codify. As such, the extent of the tacit knowledge conversion requires tremendous resources. From externalization, the process moves to the third process, combination. During *combination,* tacit knowledge is integrated into the organizations' instructions, documents, manuals, etc., so that it may be used within the larger organization. The final process is internalization. In *internalization*, the newly codified knowledge is utilized by the organization, and as it is used and learned, the knowledge becomes internalized, modifying and updating the organizational user's existing tacit knowledge. The four-node SECI model is depicted in Fig. 10.3.

A more thorough description of each of the processes in the four nodes of the SECI model of knowledge creation is presented in Table 10.2.

Another view of the SECI model has been constructed by French (2013) and is depicted in Fig. 10.4. In this view, there is a very clear delineation between (1) explicit knowledge (i.e., knowledge that can be easily encoded) and (2) tacit knowledge (i.e., learned skills and expertise that are not easily encoded). As such, it provides another perspective for how to view the SECI knowledge creation processes that are described in Table 10.3.

In summary, knowledge has both personal and organizational dimensions. However, when we are thinking systemically about situations of messes and

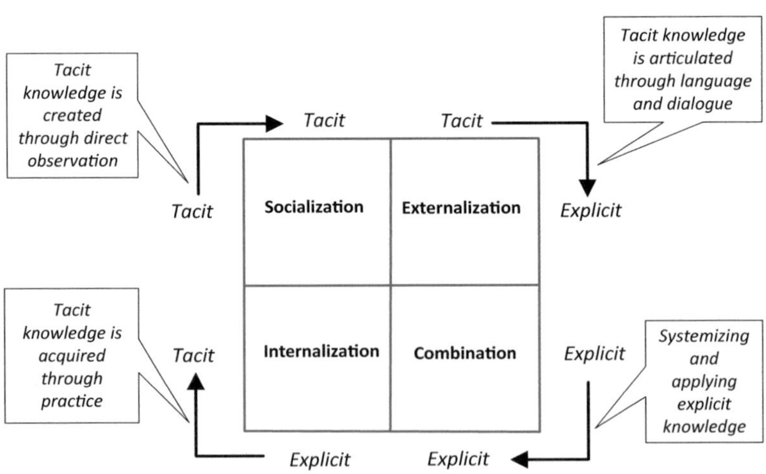

Fig. 10.3 Four-node SECI model of knowledge creation

Table 10.2 Process description for the SECI model of knowledge creation

SECI node	Description of process
Socialization	"The process of converting new tacit knowledge through shared experiences in day-to-day social interaction. Since tacit knowledge is difficult to formalize and often time and space-specific, tacit knowledge can be acquired only through shared direct experience, such as spending time together or living in the same environment, typically a traditional apprenticeship where apprentices learn the tacit knowledge needed in their craft through hands-on experiences" (Nonaka & Toyama, 2003, p. 4)
Externalization	"Tacit knowledge is made explicit so that it can be shared by others to become the basis of new knowledge such as concepts, images, and written documents. During the externalization stage, individuals use their discursive consciousness and try to rationalize and articulate the world that surrounds them. Here, dialogue is an effective method to articulate one's tacit knowledge and share the articulated knowledge with others" (Nonaka & Toyama, 2003, p. 5)
Combination	"The new explicit knowledge is then disseminated among the members of the organization. Creative use of computerized communication networks and large-scale databases can facilitate this mode of knowledge conversion. The combination mode of knowledge conversion can also include the "breakdown" of concepts. Breaking down a concept, such as a corporate vision, into operationalized business or product concepts also creates systemic, explicit knowledge" (Nonaka & Toyama, 2003, p. 5)
Internalization	"This stage can be understood as praxis, where knowledge is applied and used in practical situations and becomes the base for new routines. Thus, explicit knowledge, such as product concepts or manufacturing procedures, has to be actualized through action, practice, and reflection so that it can really become knowledge of one's own" (Nonaka & Toyama, 2003, p. 5)

Fig. 10.4 SECI model by French (2013)

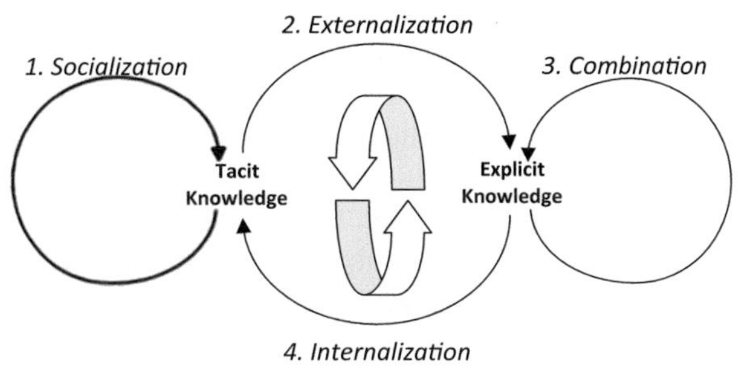

problems, we are most interested in organizational knowledge. Organizations must ensure that they have methods and processes in place that ensure knowledge workers are able to create, acquire, articulate, and apply knowledge in support of the organization, its mission, goals, and objectives. The SECI model is categorized as a second-generation model of knowledge generation.

Table 10.3 Description of SECI knowledge creation in Fig. 10.4

Process	Description
1. Socialization	Sharing tacit knowledge with users in communities of interest. This includes mentoring, workshops, and demonstrations
2. Externalization	Articulating tacit knowledge in words, diagrams, and models
3. Combination	Gathering and formalizing explicit knowledge into simple forms
4. Internalization	Understanding explicit knowledge and deploying it with tacit understanding as newly acquired (learned) knowledge in decision making

10.2.3 Abstract Classification of Mechanisms

The mechanisms that are classified as abstract, unlike the physical and human classifications previously discussed, are less easy to understand. Nevertheless, the abstract characteristics of information and methods are mechanisms that are just as essential in creating outputs and outcomes.

1. *Information.* As we described in Chap. 9, most data are of limited value until they are processed into a useable form. Once processed into a useable form, data become information. Information is the building block of knowledge (refer to Figs. 9.4 and 9.6 for a depiction). Without information, the knowledge workers described in the previous section would have nothing with which to work. Relevant and timely information, combined with a contingent of knowledge workers, is what provides one organization a competitive advantage over another, separating a viable organization from one which faces extinction. The ability of a corporation's knowledge worker to acquire knowledge is an essential mechanism. Without information, the development of proper perspectives in support of both relevant context and sufficient boundary conditions cannot occur.
2. *Methods.* Methods include the "systematic procedure, technique, or mode of inquiry employed by or proper to a particular discipline or art" (Mish, 2009, p. 781). Methods, like the access to relevant and timely information, also contribute to an organization's competitive advantage. Methods ensure replicable processes, act as a precursor to quality, and serve as the basis for evaluation of performance.

Because methods are such an important mechanism in systemic thinking, the invocation and implementation of adequate methods for problem solving will serve as the focus for the rest of this chapter.

10.3 Methods as Mechanisms for Messes and Constituent Problems

Conducting a review of this one important abstract mechanism of *how* is a non-trivial task. In general, there are as many unique methods for addressing situations that involve messes and problems as there are messes and problems. Our task is not to select one all-encompassing method for approaching problems and messes, but to provide an approach for matching the mess-problem system with an approach that is capable of shifting the mess-problem system from a problem state to a new, more desirable state.

Movement from an undesirable state to a new, desirable state requires us to make sense of the situation with which we are faced. *Sensemaking* is the formal process by which humans give meaning to experience when attempting to understand real-world situations and any associated data and information.

10.3.1 Sensemaking

Sensemaking has been defined by a number of practitioners. Some relevant definitions, arranged chronologically, are presented in Table 10.4.

Table 10.4 Definitions for sensemaking

Definition	Sources
"A label for a coherent set of concepts and methods used in a now 8-year programmatic effort to study how people construct sense of their worlds and, in particular, how they construct information needs and uses for information in the process of sense-making"	(Dervin, 1983, p. 3)
"The basic idea of sensemaking is that reality is an ongoing accomplishment that emerges from efforts to create order and make retrospective sense of what occurs"	(Weick, 1993, p. 635)
"The making of sense"	(Weick, 1995, p. 4)
"Sensemaking involves turning circumstances into a situation that is comprehended explicitly in words and that serves as a springboard into action"	(Weick, Sutcliffe, & Obstfeld, 2005, p. 409)
"A motivated, continuous effort to understand connections (which can be among people, places and events) in order to anticipate their trajectories and act effectively"	(Klein, Moon, & Hoffman, 2006, p. 71)
"Sensemaking, a term introduced by Karl Weick, refers to how we structure the unknown so as to be able to act in it. Sensemaking involves coming up with a plausible understanding—a map—of a shifting world; testing this map with others through data collection, action, and conversation; and then refining, or abandoning, the map depending on how credible it is"	(Ancona, 2012, p. 3)

From these definitions, we can clearly see that sensemaking has, at its core, a structured approach to understanding. Sensemaking has become an accepted practice in a number of programs, with practical applications in:

- The studies of organizations (Weick, 1995; Weick et al., 2005).
- The fields of communication and library and information science (Dervin, 1983, 1992, 1993).
- The design of interactive systems with the computer–human interaction (CHI) community (Russell, Stefik, Pirolli, & Card, 1993).
- Naturalistic decision making (Klein, Phillips, Rall, & Peluso, 2007).
- Military decision making process (Leedom, 2001).
- A generalized method for inquiry in complex systems (Kurtz & Snowden, 2003; Snowden, 2002; Snowden & Boone, 2007).

The next section will review how sensemaking may be applied as a generalized method for inquiry in situations like our messes and related problems.

10.3.2 Pragmatic Intersection of Knowledge and Information

Because sensemaking is a structured approach to understanding based on both knowledge and information, we will approach its application in a pragmatic manner. Donald Rumsfeld, a former naval aviator, member of the House of Representatives, the White House chief of staff, and Secretary of Defense, described his view of the various states of knowledge in response to a question during a post-Iraq War press conference, stating:

> Reports that say that something hasn't happened are always interesting to me, because as we know, there are known knowns; there are things we know we know. We also know there are known unknowns; that is to say we know there are some things we do not know. But there are also unknown unknowns – the ones we don't know we don't know. And if one looks throughout the history of our country and other free countries, it is the latter category that tend to be the difficult ones. (Rumsfeld, 2002)

Rumsfeld was quite right in his characterization of the four states of knowledge and the intersection between an observer's knowledge state and the state of information in the real world. These intersections, along with some associated phraseology, are depicted in Fig. 10.5.

Armed with this pragmatic view of sensemaking, we are prepared to review a third-generation framework for knowledge generation.

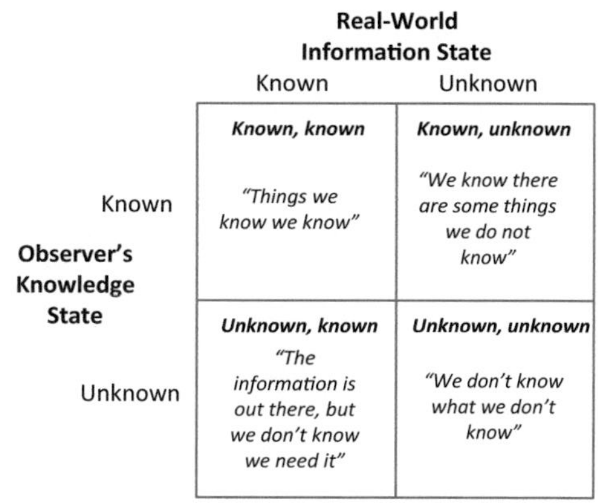

Fig. 10.5 Information and knowledge domain

10.3.3 Framework for Sensemaking

A particularly useful sensemaking framework has been developed to improve understanding based upon the degree of order present in systems. The framework is entitled Cynefin and is a third-generation model of knowledge generation, following Polanyi's first–generation concept of Personal Knowledge and Nonaka's second-generation SECI Model. "Cynefin (pronounced cun-ev'-vin) is a Welsh word with no direct equivalent in English. As a noun, it is translated as *habitat*, as an adjective *acquainted* or *familiar* ..." (Snowden, 2000, p. 236).

The Cynefin framework is organized around five domains that exist along a continuum from order to unorder. The area between the zones is purposefully *fuzzy* (Zadeh, 1965), as these are areas of instability where systems are transitioning between domain states—in what are termed transition zones. The Cynefin framework, with the four knowledge states using the combinations of the terms *known* and *unknown* (see Fig. 10.5), is depicted in Fig. 10.6.

Using the construct in Fig. 10.6, there are unique causal relationships and approaches to be used when dealing with situations in each of the five Cynefin framework domains.

1. *Simple*—in this domain, the relationship between cause and effect is obvious. The approach for dealing with this domain is to sense, categorize, and respond.
2. *Complicated*—in this domain the relationship between cause and effect requires analysis or some other form of investigation. The approach is to sense, analyze, and respond.
3. *Complex*—in this domain, the relationship between cause and effect can only be perceived in retrospect, rather than perceived in advance. The approach is to probe, sense, and respond.
4. *Chaotic*—in this domain, there is no relationship between cause and effect at the systems level. The approach is to act, sense, and respond.

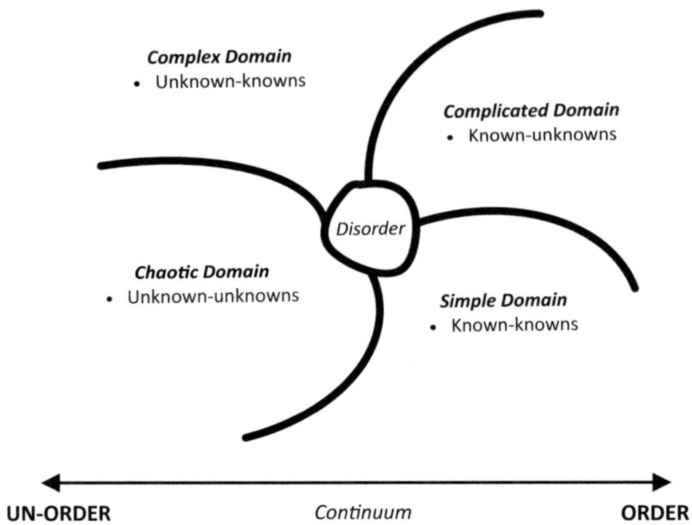

Fig. 10.6 Cynefin (cun-ev-vin) framework for complexity

5. *Disorder*—this is the domain where neither cause nor effect is known. This domain in which people will revert to their own comfort zones in making decisions. It exists in the gray space between the other four domains.

Kurtz and Snowden (2003) describe the totality of the framework's domains:

> The framework actually has two large domains, each with two smaller domains inside. In the right-side domain of order, the most important boundary for sense-making is that between what we can use immediately (what is known) and what we need to spend time and energy finding out about (what is knowable). In the left-side domain of unorder, distinctions of knowability are less important than distinctions of interaction; that is, distinctions between what we can pattern (what is complex) and what we need to stabilize in order for patterns to emerge (what is chaotic). (p. 470)

The principal approaches and examples of appropriate actions (i.e., sense—categorize—respond) that may be invoked in each of the domains are presented in Fig. 10.7. It is important to note that we use the terms *simple* and *complicated* where Kurtz and Snowden (2003) use the terms *known* and *knowable*. We do this in order to be consistent in applying already discussed terms for complexity (i.e., simple, complicated, complex, and chaotic) that we have adopted in this book.

The placement of the 5th domain, disorder, is purposeful. By being in the middle of the model, each of the other four states is on the boundary of disorder, which is representative of real-world state transition conditions.

> The very nature of the fifth context – disorder – makes it particularly difficult to recognize when one is in it. Here, multiple perspectives jostle for prominence, factional leaders argue with one another, and cacophony rules. The way out of this realm is to break down the

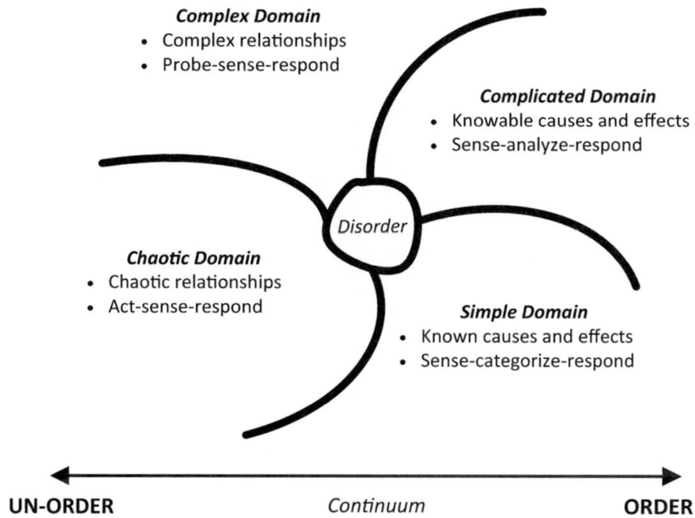

Fig. 10.7 Cynefin domains, relationships, and approaches

> situation into constituent parts and assign each to one of the other four realms. (Snowden & Boone, 2007, p. 72)

Movement between domains, which include term state transition and boundary shifting, is an important element in the Cynefin framework. As a problem's degree of order changes, there is a need to shift to a new domain and to new modes required for proper understanding and interpretation.

> Boundaries are possibly the most important elements, in sense-making, because they represent differences among or transitions between the patterns we create in the world that we perceive. (Kurtz & Snowden, 2003, p. 474)

Kurtz and Snowden (2003) recommend the use of metaphors when describing boundaries. For instance:

> - The shallow river can be crossed by anyone at any place, and thus control over crossing is difficult to achieve. However, it is easy to tell when one has crossed it (or when others have) because one's feet get wet.
> - The deep chasm can be crossed only at bridges, which can be built, demolished, and controlled at will. It is not easy to tell when one has crossed the boundary, but such a marker is not required because only some are allowed through.
> - The high plateau is the boundary with the most potential danger, because you may not be aware that you have crossed the boundary until it is too late and you drop off the other side. (p. 474)

There are ten recognizable patterns of movement across the boundaries of the Cynefin model which are described in Table 10.5.

The ten boundary shifts between domains are depicted in Fig. 10.8.

Table 10.5 Boundary shifts within the Cynefin model (Kurtz & Snowden, 2003)

Label	Domain movements	Description
1. Incremental improvement	• Complicated to simple • Simple to complicated	Movement from the knowable to the known and back, repeatedly
2. Exploration	• Complicated to complex	Movement from the knowable to the complex, selectively
3. Exploitation	• Complex to complicated	Movement from the complex to the knowable, selectively
4. Divergence–convergence	• Complex to chaotic to complex	Movement from the complex to the chaotic and back, repeatedly
5. Imposition	• Chaotic to simple	Movement from the chaotic to the known, forcefully
6. Asymmetric collapse	• Simple to chaotic	Movement from the known to the chaotic, disastrously
7. Swarming	• Chaotic to complex to complicated	Movement from the chaotic to the complex, to the knowable; first, in an emergent manner and then selectively
8. Liberation	• Simple to complex to complicated	Movement from the known to the complex to the knowable, periodically
9. Entrainment making	• Complicated to disorder to chaotic to complex	Movement from the knowable to the chaotic to the complex, periodically
10 Immunization	• Simple to chaotic	Movement from the known to the chaotic, temporarily

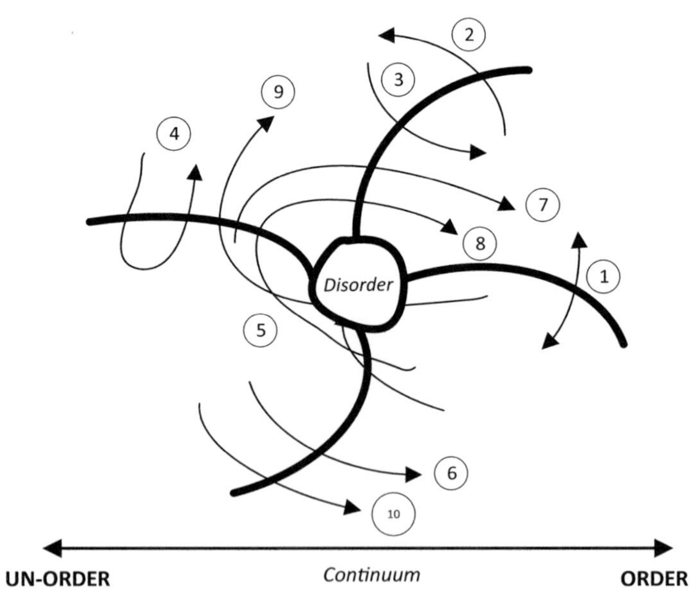

Fig. 10.8 Boundary shifting within the Cynefin framework

Cynefin provides a framework that the systems' practitioner may use to carefully view problematic situations and then to purposefully shift the problem to a more manageable situation. By doing this, the practitioner is shifting knowledge flows to where appropriate (i.e., available and accessible) models of decision making may be utilized.

> The purpose of the Cynefin model is to enable sense making by increasing the awareness of borders and triggering with a border transition a different model of decision making, leadership or community. Cynefin argues strongly against single or idealized models, instead focusing on diversity as the key to adaptability. The law of requisite variety is well understood in ecology; if the diversity of species falls below a certain level then the ecology stagnates and dies. Excessive focus on core competence, a single model of community of practice or a common investment appraisal process are all examples of ways in which organizations can destroy requisite variety. (Snowden, 2002, p. 107)

By increasing information flow and associated knowledge during the transition between and among domains, both connectivity and variety increase thus serving to break down existing patterns and to create the conditions where new patterns will emerge.

In summary, knowledge of the types of complexity and how to address them are essential when working with complex systems. The Cynefin framework provides a means for understanding how to approach and deal with complex systems based upon their level of complexity.

10.4 Cynefin Domain and Mechanism Types

Cynefin, introduced in the previous subsection, is a third-generation knowledge generation model for improved understanding of complex systems. The unique causal relationships and associated decision models (i.e., sense, categorize, probe, analyze, act, and respond) of the Cynefin framework will now be related to the mechanism types discussed earlier in the chapter. Table 10.6 shows how the five Cynefin domains, the causal relations, decision models, mechanism choice, and goal are related.

The characteristics in Table 10.6 are included on the Cynefin framework and shown in Fig. 10.9. This figure may be used by practitioners who are looking for mechanisms to invoke to address a problem in order to achieve objectives or to improve their understanding of a complex system or a situation such as a mess and its constituent problems.

10.4.1 Cynefin and the Strategic Decision Making Pyramid

The hierarchy of decision making in organizations is traditionally represented in a pyramid where decisions flow from the highest to the lowest level. Although there

Table 10.6 Relationship between Cynefin domains and mechanism types

Cynefin domain	Causal relations	Decision model	Mechanism choice	Goal
Simple	Cause and effect is obvious	• Sense • Categorize • Respond	Human (Manpower, KSAs), physical (material, money time, equipment, facilities), abstract (methods, information)	Achievement of objectives or increased understanding
Complicated	Cause and effect requires analysis	• Sense • Analyze • Respond		
Complex	Cause and effect can only be perceived in retrospect	• Probe • Sense • Respond	Abstract (methods, information)	Increased understanding
Chaotic	No relationship between cause and effect at the systems' level	• Act • Sense • Respond		
Disorder	Unknown	• Comfort zone decision making		

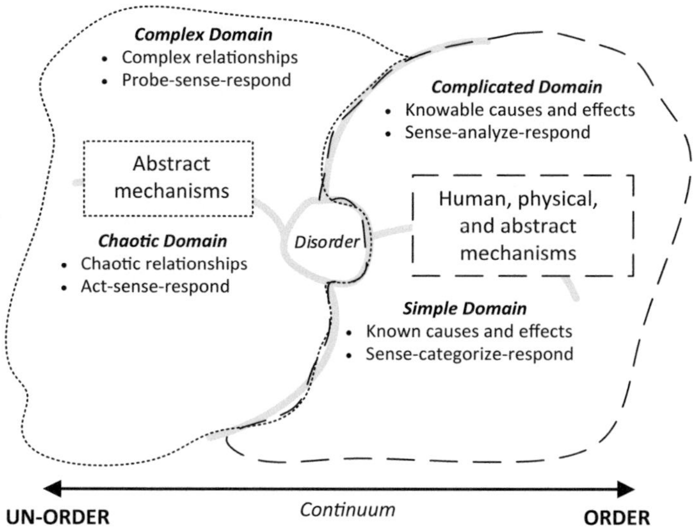

Fig. 10.9 Relationship between Cynefin domains and mechanism types

Fig. 10.10 Organizational decision pyramid

are many unique levels in a typical organization, most fall within four generic levels or strata characterized by the following identifying terms: (1) strategic, corporate; (2) tactical, integrative; (3) operational, middle; and (4) instinctive, first-line (Jaques, 1989).The pyramid is depicted in Fig. 10.10.

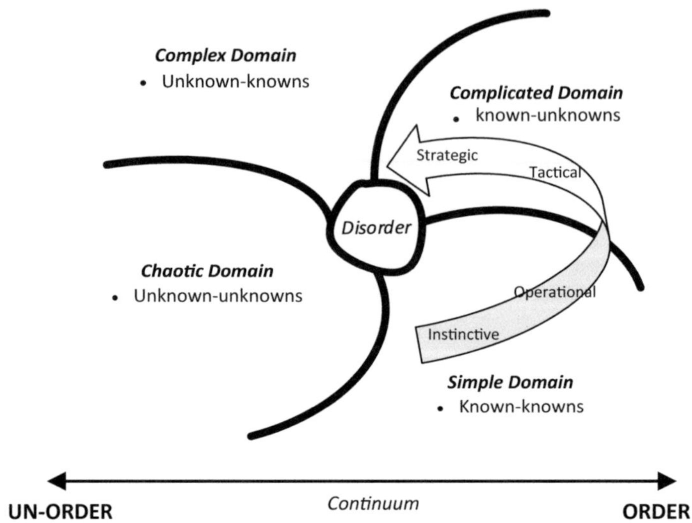

Fig. 10.11 Relationship between the perspectives offered by the strategy pyramid and Cynefin (French, 2013)

The four levels in the organizational decision pyramid can be related to domains in the Cynefin framework (French, Maule, & Papamichail, 2009). The relationships are shown by the curved arrow in Fig. 10.11.

It is important to note that the developers of this relationship do not claim to know the precise location of the four decision strata within the Cynefin framework's four domains. "While the appropriate domain for instinctive decision making may lie entirely within the known space, operational, tactical and strategic decision making do not align quite so neatly, overlapping adjacent spaces. Indeed, the boundaries between the four spaces in Cynefin should not be taken as hard" (French, 2013, p. 549). Despite this warning, the relationships provide a valuable guide for decision makers.

10.5 Framework for Addressing *How* in Messes and Problems

Addressing the *how* perspective in our messes and problems requires that we complete the following steps for an identified problem:

1. Identify the appropriate Cynefin domain for each problem;
2. Select an appropriate mechanism for use in each problem based on the Cynefin domain; and
3. Modify our FCM as appropriate to include the proposed use of a mechanism.

The following section demonstrates each on our real estate problem.

10.6 Example Problem

We return to our real estate example pictured in Fig. 10.12.

10.6.1 Cynefin Analysis

Our real estate problem seems relatively well ordered. It appears to be in the complicated domain, however, as there are some major uncertainties stemming from a lack of information regarding what the city planning commission will do. Thus, we should be able to achieve our objective (as the real estate developer) by using the appropriate mechanisms.

10.6.2 Mechanism Analysis

We should employ time and information to our problem. As appropriate, we should try to gain information about how the planning commission will rule. Given the status of this ruling, we may have no choice but to wait (invoking time). We should capture the importance of a ruling delay as a concept in our FCM. As the problem unfolds, the scenario may change.

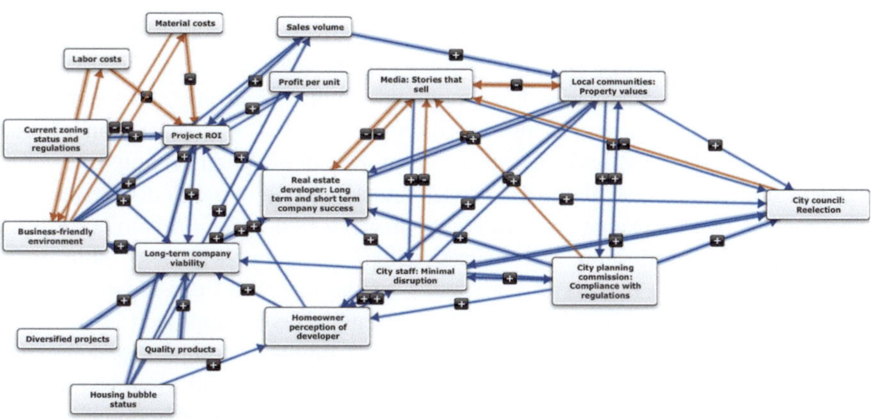

Fig. 10.12 Real estate example FCM

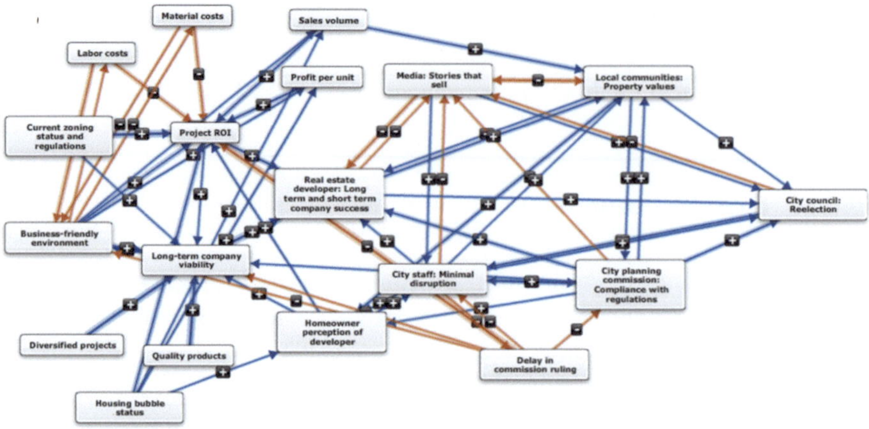

Fig. 10.13 Real estate example FCM with mechanisms

10.6.3 Updated FCM

Reflecting the inclusion of a concept representing the *delay in commission ruling*, Fig. 10.13 shows the updated FCM based on mechanism analysis.

10.7 Summary

This chapter addressed the *how* question as it relates to the attainment of specific, purposeful goals. Moving our mess from a current state toward a desired state is achieved through mechanisms. Nine physical, human, and abstract mechanisms were identified, and each was discussed. Specific focus was placed on abstract mechanisms, namely methods and information, because of their nonintuitive nature and their importance in achieving increased understanding in problem domains. The development of knowledge was addressed using a first-order (Polyani's Personal Knowledge), second-order (Nonaka's SECI framework), and third-order (Snowden's Cynefin) models of knowledge generation. The third-order model for sensemaking, the Cynefin framework, was discussed as an approach by which to achieve increased understanding in five domains found in complex systems. The Cynefin framework was addressed and analyzed as it pertained to eleven specific decision analysis techniques and their intersection with Cynefin's domains. This resulted in the development of a framework for analysis of the *how* question for our systemic thinking endeavor.

After reading this chapter, the reader should:

1. Understand the nine mechanism types and three broad categories into which these mechanisms fall;

2. Describe the five complexity domains within which messes and problems exist;
3. Be able to relate the complexity domains with appropriate mechanisms; and
4. Be capable of identifying an appropriate mechanism for a mess or problem based on its complexity domain.

References

Amarel, S. (1966). On the mechanization of creative processes. *IEEE Spectrum, 3*(4), 112–114. doi:10.1109/MSPEC.1966.5216589.

Ancona, D. (2012). Sensemaking: Framing and acting in the unknown. In S. Snook, N. Nohria, & R. Khurana (Eds.), *The handbook for teaching leadership: Knowing, doing, and being* (pp. 3–19). Thousand Oaks, CA: Sage Publications.

Cheney, P. H., Hale, D. P., & Kasper, G. M. (1990). Knowledge, skills and abilities of information systems professionals: Past, present, and future. *Information and Management, 19*(4), 237–247. doi:10.1016/0378-7206(90)90033-e.

Dervin, B. (1983). *An overview of sense-making research: Concepts, methods and results*. Dallas, TX, May: Paper presented at the Annual Meeting of the International Communication Association.

Dervin, B. (1992). From the mind's eye of the 'user': The sense-making qualitative-quantitative methodology. In J. D. Glazier & R. R. Powell (Eds.), *Qualitative research in information management* (pp. 61–84). Englewood, CO: Libraries Unlimited.

Dervin, B. (1993). Sense-making theory and practice: An overview of user interests in knowledge seeking and use. *Journal of Knowledge Management, 2*(2), 36–46.

French, S. (2013). Cynefin, statistics and decision analysis. *Journal of the Operational Research Society, 64*(4), 547–561. doi:10.1057/jors.2012.23.

French, S., Maule, J., & Papamichail, N. (2009). *Decision behaviour, analysis and support*. New York: Cambridge University Press.

Grant, K. A. (2007). Tacit knowledge revisited—we can still learn from Polanyi. *The Electronic Journal of Knowledge Management, 5*(2), 173–180.

Gunnells, K., Hale, J. E., & Hale, D. P. (2006). Revisiting knowledge, skills, and abilities needed for development and delivery project staff. In M. Murray (Ed.), *Proceedings of the 2006 Southern Association for Information Systems Conference (March 11–12)* (pp. 217–223). Jacksonville, FL: Southern Association for Information Systems.

Jaques, E. (1989). *Requisite organization: A total system for effective managerial organization and managerial leadership for the 21st century amended* (2nd ed.). Arlington, VA: Cason Hall & Company.

Klein, G. A., Moon, B., & Hoffman, R. R. (2006). Making sense of sensemaking 1: Alternative perspectives. *IEEE Intelligent Systems, 21*(4), 70–73.

Klein, G. A., Phillips, J. K., Rall, E. L., & Peluso, D. A. (2007). A data/frame theory of sensemaking. In R. R. Hoffman (Ed.), *Expertise out of control: Proceedings of the sixth international conference on naturalistic decision making* (pp. 113–155). New York: Lawrence Erlbaum Associates.

Kurtz, C. F., & Snowden, D. J. (2003). The new dynamics of strategy: Sense-making in a complex-complicated world. *IBM Systems Journal, 42*(3), 462–483.

Leedom, D. K. (2001). *Report on 2001 Sensemaking symposium*. Retrieved from McLean, VA.

Mish, F. C. (Ed.). (2009). *Merriam-Webster's collegiate dictionary* (11th ed.). Springfield, MA: Merriam-Webster, Incorporated.

Nonaka, I. (1991). The knowledge-creating company. *Harvard Business Review, 69*(6), 96–104.

Nonaka, I. (1994). A dynamic theory of organizational knowledge. *Organization Science, 5*(1), 14–37.

Nonaka, I. (2007). The knowledge-creating company. *Harvard Business Review, 85*(7/8), 162–171.

Nonaka, I., & Toyama, R. (2003). The knowledge-creating theory revisited: knowledge creation as a synthesizing process. *Knowledge Management Research & Practice, 1*(1), 2–10.

Nonaka, I., von Krogh, G., & Voelpel, S. (2006). Organizational knowledge creation theory: Evolutionary paths and future advances. *Organization Studies, 27*(8), 1179–1208. doi:10.1177/0170840606066312.

Polanyi, M. (1962). *Personal knowledge: Towards a post-critical philosophy*. Chicago: University of Chicago Press.

Polanyi, M. (2009). *The tacit dimension*. Chicago: University of Chicago Press.

Renck, R., Kahn, E. L., & Gardner, B. B. (1969). *Continuing education in R & D careers*. Retrieved from Washington, DC.

Rumsfeld, D. (2002). DoD news briefing by the secretary of defense on February 12, 2002 [Press release].

Russell, D. M., Stefik, M. J., Pirolli, P., & Card, S. K. (1993). The cost structure of sensemaking. In S. Ashlund (Ed.), *Proceedings of the INTERACT '93 and CHI '93 conference on human factors in computing systems* (pp. 269–276). Amsterdam, The Netherlands: IOS Press.

Snowden, D. J. (2000). The social ecology of knowledge management. In C. Despres & D. Chauvel (Eds.), *Knowledge horizons: The present and the promise of knowledge management* (pp. 237–266). Woburn, MA: Butterworth Heinemann.

Snowden, D. J. (2002). Complex acts of knowing: paradox and descriptive self-awareness. *Journal of Knowledge Management, 26*(2), 100–111. doi:10.1108/13673270210424639.

Snowden, D. J., & Boone, M. E. (2007). A leader's framework for decision making. *Harvard Business Review, 85*(11), 69–76.

Vitalari, N. P. (1985). Knowledge as a basis for expertise in systems: An empirical study. *MIS Quarterly, 9*(3), 221–241.

Weick, K. E. (1993). The collapse of sensemaking in organizations: The Mann Gulch disaster. *Administrative Science Quarterly, 38*(4), 628–652.

Weick, K. E. (1995). *Sensemaking in organizations*. Thousand Oaks, CA: Sage Publications.

Weick, K. E., Sutcliffe, K. M., & Obstfeld, D. (2005). Organizing and the process of sensemaking. *Organization Science, 16*(4), 409–421. doi:10.2307/25145979.

Welford, A. T. (1968). *Fundamentals of skill*. London: Methuen.

Winterton, J., Delamare-Le Deist, F., & Stringfellow, E. (2006). *Typology of knowledge, skills and competences: Clarification of the concept and prototype*. Luxembourg: Office for Official Publications of the European Communities.

Zadeh, L. A. (1965). Fuzzy sets. *Information and Control, 8*(3), 338–353. doi:10.1016/S0019-9958(65)90241-X.

Chapter 11
The *When* of Systemic Thinking

Abstract The *when* question of systemic thinking attempts to determine the appropriate time for interacting with our mess in an effort to increase our understanding about it. Recalling the TAO of systemic thinking, we must think before we act on (and observe) our mess. The understanding gained from our thinking informs when (and if) we decide to intervene in our mess. In order to discern the appropriate time for action, we explore two criteria of our messes, its *maturity* and its *stability*. These two criteria will first be explored by investigating life cycles and their relevance to the maturity of our mess. We will then explore the phenomena of evolution, as it pertains to both biological systems and to purposeful systems. Then, we will discuss entropy as it relates to evolution. Finally, we develop a framework to address the *when* as it applies to any efforts at intervention in our mess.

11.1 Life Cycles and Maturity

There are many similarities between biological systems and purposeful systems, but perhaps none is more fundamental than the basic life cycle each follows. Although there are more complex models for both in the biological and systems literature, we can summarize biological systems as comprising a "birth-growth-aging and death life cycle" (Sage & Armstrong, 2000, p. 7). Blanchard (2004) discusses a purposeful system's life cycle, saying it

> …includes the entire spectrum of activity for a given system, commencing with the identification of need and extending through system design and development, production and/or construction, operational use and sustaining maintenance and support, and system retirement and material disposal. (p. 13)

Succinctly, and in terms analogous to the phases associated with a biological life cycle, we may describe purposeful man-made systems as having a life cycle consisting of a definition (birth), development (growth), use (aging), and retirement (death). A depiction juxtaposing both life cycles is shown in Fig. 11.1.

P.T. Hester and K.M. Adams, *Systemic Decision Making*, Topics in Safety, Risk, Reliability and Quality 33, DOI 10.1007/978-3-319-54672-8_11

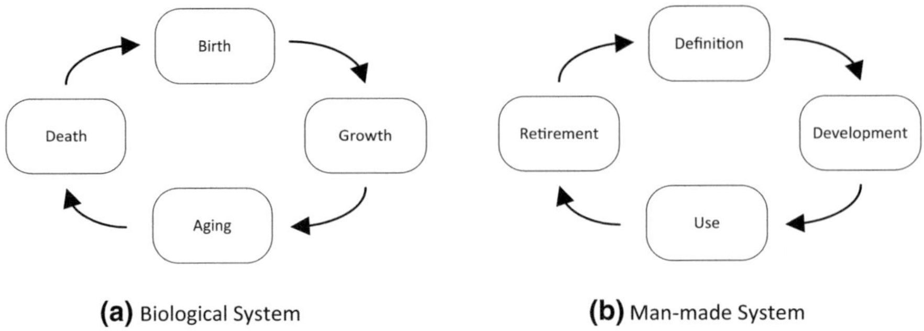

Fig. 11.1 Depiction of biological (**a**) and human-made system (**b**) life cycles

A short description of each stage as it pertains to purposeful man-made systems is as follows:

- *Definition*: Our system is *born* here. We begin to conceptualize it here by identifying a need that is to be satisfied by our system and determining the constraints on our system. As it concerns a mess, definition is an artificial construct. We define the context and environment of our system (see Chap. 8 for further guidance). We define the elements that comprise the mess as a construct of convenience; they likely have no real abstraction at the level we choose to analyze them. A perfect example is the education system in the USA. Our level of abstraction is subjective and purposeful; whether we wish to explore the national education system or the education afforded to the children in our home influences the lens through which we view the problem.
- *Development*: Our system begins to take shape. It matures and *grows* through iterative development and evolution. It may require resources to take a form that is either useful or recognizable to us.
- *Use*: Our system is in use. It requires maintenance and effort to sustain its performance at a level that is acceptable to its users. At this point, consideration and maintenance of our system's entropy (discussed at length in Sect. 11.4) become paramount to its continued viability.
- *Retirement*: Our system has fulfilled its intended purpose (and thus, it may be retired from service) or surpassed its expected life (and thus, it *dies* organically). In the context of a mess, this element is problematic as not all components will have the same timescale or life expectancy. Thus, we may need to invest resources into our mess in an effort to artificially extend its useful life.

The two cycles in Fig. 11.1 show significant similarity between the basic life cycles of biological and purposeful man-made systems. However, when we think about *messes*, which occur as a result of system operation and human involvement and are not purposefully *designed*, the conceptualization of a life cycle becomes a little less clear and orderly. Most notably, the *birth* and *death* of a mess are nebulous constructs. When does a traffic problem in a locality become a mess? When a second mode of transportation (i.e., public transportation) becomes available?

When it has to cross traditional jurisdictional boundaries (i.e., city, county, state, or country)? There are certainly several explanations for the birth of said mess that may be reasonable, and yet, none may be of any value. A more fundamental question may be whether or not the specific birth or death of our mess is a construct that is of any value to its observers. How it came into being (be it by our own purposive behavior or otherwise) and how it will cease to exist (be it by forced retirement, simply run out its expected life, or evolve into an entirely different mess of an unrecognizable nature) is likely of little value. More importantly, it is of interest to us to understand the *life* of our mess, and thus, we should primarily focus on the development and use of it, or to use biological terms, its growth, and aging. In concerning ourselves with its birth and death, we are likely to get mired in trivialities that are of no value. We must undertake a holistic consideration of the life of our mess. Blanchard (2004) agrees, noting

> The past is replete with examples in which major decisions have been made in the early stages of system acquisition based on the "short term" only. In other words, in the design and development of a new system, the consideration for production/construction and/or maintenance and support of that system was inadequate. These activities were considered later, and, in many instances, the consequences of this "after-the-fact" approach were costly. (pp. 14–15)

Noted systems engineer Derek Hitchins offers a unique, but complementary perspective which may help us. His principle of cyclic progression offers a lens to view our system's development through

> Interconnected systems driven by an external energy source will tend to a cyclic progression in which system variety is generated, dominance emerges, suppresses the variety, the dominant mode decays or collapses, and survivors emerge to regenerate variety. (Hitchins, 1993, p. 633)

This principle can be depicted graphically and annotated with the phases of the biological cycle discussed earlier as shown in Fig. 11.2. We can see the cyclic

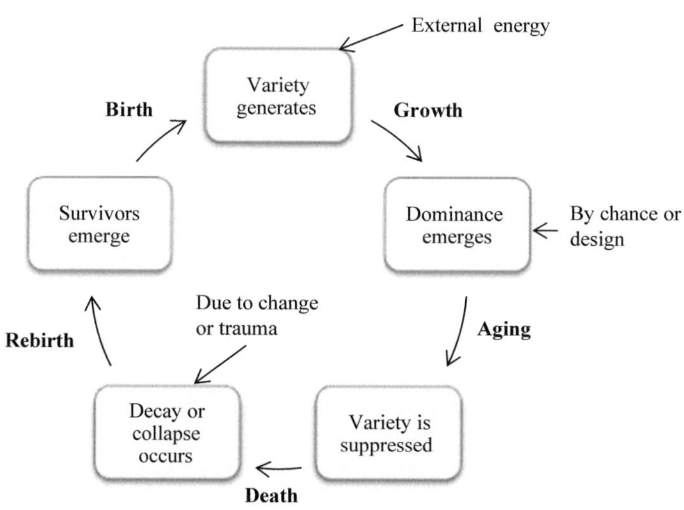

Fig. 11.2 Illustration of cyclic progression (adapted from Fig. 2.9 in Hitchins, 2007, p. 58)

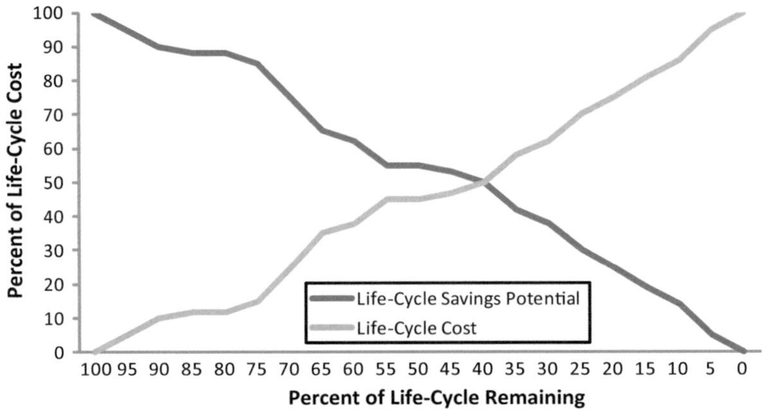

Fig. 11.3 Closed system life-cycle cost and benefit

nature of the life cycle as it is juxtaposed with Hitchins' illustration of cyclic progression.

The question then becomes, at what point in the life of our mess should we intervene? We can look to Fig. 11.3 to give us a clue. All closed systems have a finite life. Without intervention from external sources, our system will cease to exist (more on this and its related element of entropy are found later in this chapter). Thus, as the life of our system progresses, the cumulative costs associated with it increase and the potential for savings decrease. While the exact shapes of the curves shown in Fig. 11.3 vary depending on the circumstances, we know that the total cost is monotonically increasing (i.e., it never goes down), and the savings potential is monotonically decreasing (i.e., it never increases).

Thus, for any given system, every day that passes has the potential to incur more cost for us and present less opportunity for savings. So, should we just invest as early as possible? The answer is not so clear.

To answer this question, we can adapt the notion of a basic cost–benefit analysis (CBA). Traditionally in CBA, alternatives are designed for a system and we trade off their respective benefits (typically in terms of dollars) with their costs (also typically in dollars) as a ratio expressed in Eq. 11.1.

$$C/B = \frac{\text{Cost}}{\text{Benefit}} \tag{11.1}$$

The alternative with the lowest *C/B* is chosen as the preferred option to pursue. However, with a mess being so inherently unpredictable, it may not be advantageous for us to use cost and benefit in this sense. More importantly, we may consider the trade-off between cost and benefit as a litmus test of feasibility for considering whether or not to intervene in our mess (and thus, to commit resources). For such an analysis, we can invert Eq. 11.1 and consider the following relationship in Eq. 11.2.

$$\max(B/C) \geq 1 \tag{11.2}$$

Utilizing this inequality, we try to conceptualize if *any* option exists for intervention in our system that provides a larger benefit than its associated cost. This is of course a simplifying assumption in that it typically equates cost in dollars to benefit in dollars, but we can abstract the discussion to any relevant measure of merit.

Let us take a biological example. It would be difficult for a doctor to endorse an urgent heart transplant for a 95-year-old patient regardless of the circumstances (i.e., even if death is certain without the operation). The benefit of the operation may be conceptualized in a number of ways. For instance,

- Five years of additional life or alleviated pain for the patient can be compared to the cost associated with it, or
- The actual cost of the operation, the expected survival rate of the patient, or the risk of not providing the donor heart to a more viable (and arguably more deserving) patient.

It seems fairly straightforward that the inequality represented by Eq. 11.2 is not met. Complicating this scenario is its likely status as a mess. Maybe the patient would pay cash for the operation alleviating insurance concerns. Alternatively, perhaps there is a clearly more deserving patient (although it may seem abhorrent to some, merit-based rankings of individuals seeking a donor organ can be generated). These and other concerns make this quite a difficult scenario to understand. If we determine that the *B/C* ratio is not sufficient for this alternative, we can conceive of other options. One such alternative is to utilize hospice care for the patient in an effort to allow him to die with dignity. In this case, the cost is minimal (at least from the medical expenditure perspective, the *cost* to the world of losing the individual is another debate entirely, and one we would not dare explore) and the benefit is arguably justified by the cost. Thus, we have found a proposed solution that satisfies Eq. 11.2. In this way, we have satisfied the *maturity* concern associated with the when of systemic thinking. It is in this way that we should think of maturity.

If we take the ratio of the benefit and cost curves in Fig. 11.3 and plot them against the inequality of Eq. 11.2, we can generate the curves shown in Fig. 11.4. This graphic demonstrates that early on in our system development, there is a high potential for a high benefit to be realized from intervening in our system, given the significant expected life left in our system. Additionally, early in the development, it is cheap to change our system. At some point, when the curves cross, it is no longer advantageous to intervene in our system.

Figure 11.4 must be taken with two caveats as they pertain to a mess:

1. Messes exist in open systems. Open systems interact with their environment. As a result, they are unstable such that Fig. 11.4 can be recalibrated by interjecting resources into the mess. Thus, the *B/C* curve (and its underlying components of cost and benefit) can be improved or worsened by expending resources in the form of additional mechanisms (the focus of Chap. 10) on the mess. In doing so,

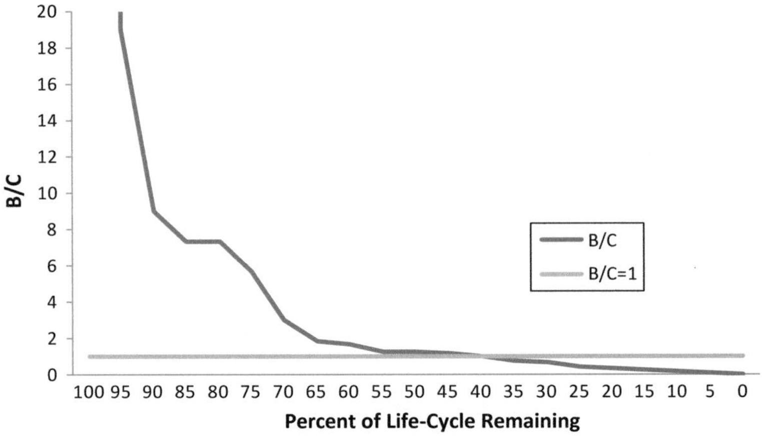

Fig. 11.4 *B/C* as a function of time

we have transitioned our mess, perhaps to a form that is unrecognizable to us (and hopefully to an improved state). Such potential transitions are illustrated in Fig. 11.5.

2. Figure 11.4 illustrates a system, not a mess. Messes are unpredictable. They are likely not to possess a clear crossover point. Thus, our understanding of the mess is likely to coincide with a range of options, such as those denoted by the improved and worsened curves in Fig. 11.5. This is largely due to the unpredictability of the system and due to the first caveat, i.e., our ability to make adjustments based on our limited understanding.

Thus, the guidance provided in this section is to be taken as a heuristic. The key takeaway of the maturity discussion is for us to consider the relative cost (monetary

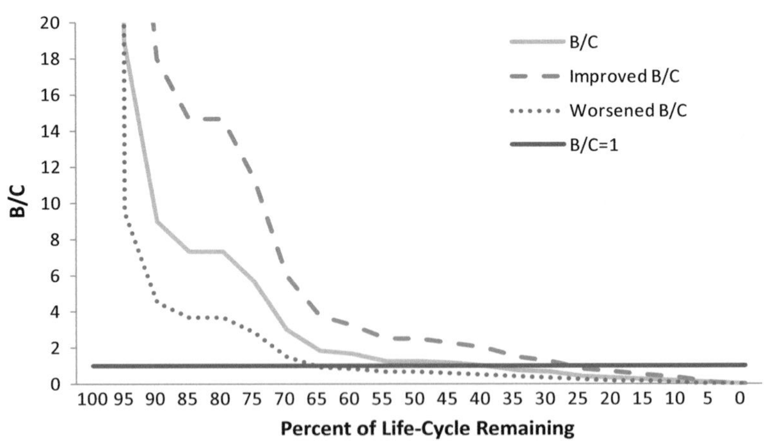

Fig. 11.5 Moving the *B/C* curve

or otherwise) and expected benefits resulting from increasing our understanding, especially if this increased understanding requires the expenditure of resources, before investing our time, money, and efforts. We must aim for the region above the *break-even* point for our mess (when $B = C$), but this is not the only concern. Given the unpredictable nature of our mess, we must also consider its evolution. Sage and Armstrong (2000) illustrate the linkage between life-cycle and evolution concerns: "This life-cycle perspective should also be associated with a long-term view toward planning for system evolution, research to bring about any new and emerging technologies needed for this evolution, and a number of activities associated with actual system evolution..." (p. 7). Indeed, the *stability* of our system, a measure equally as important as its *maturity*, must be considered by exploring its evolution and development.

11.2 Evolution

The development and evolution of our mess are continual. Understanding the mechanism of evolution and determining an appropriate time to intervene in our mess are a significant endeavor and yet one we are tasked with. First, we can explore the notion of evolution within biological, or living, systems. Many definitions for evolution exist. Several definitions taken from the biological complexity domain include

- Biological evolution is the process of gradual (and sometimes rapid) change in biological forms over the history of life (Mitchell, 2009, p. 72).
- Evolution here is simply robustness to (possibly large) changes on long timescales (Csete & Doyle, 2002, p. 1666).
- Evolution is the historical process that leads to the formation and change of biological systems (Johnson & Lam, 2010, p. 880).
- The word evolution comes from the Latin evolvere, "to unfold or unroll"–to reveal or manifest hidden potentialities. Today "evolution" has come to mean, simply, "change." (Futuyma, 2005, p. 3)

Each of the above definitions connotes change; however, only one, Csete and Doyle's, addresses the purposeful notion of change (they support that evolution exists to maintain system functionality despite uncertainty). As systemic thinkers, we support the notion of purposeful change in systems and we believe the following discussion will bear out a historical belief in this notion as well. Thus, for our purposes, we define evolution succinctly as *purposeful change in system structure or behavior*.

Jean-Baptiste Lamarck [1744–1829] developed arguably the most famous pre-Darwin theory of evolution, the idea that living organisms can pass characteristics they acquired throughout their lifetime on to their offspring. These acquired characteristics, or adaptations, were "changes for the better, or at least, for the more complex" (Mitchell, 2009, p. 73).

As Charles Darwin [1809–1882] and others rose to prominence, it was clear that the notion of acquired characteristics in biological systems was false. Darwin's voyages to the Galapagos Islands aboard the H.M.S. *Beagle* survey ship led to his empirical observations about the gradual development and adaptation of finches. His observations led to his belief in the idea of *gradualism*, the notion that small factors, extended over significant time horizons, could have long-reaching effects, and his publication of *The Origin of Species* (Darwin, 1859). Two major premises arose from this work, as summarized by (Futuyma, 2005):

- The first is Darwin's theory of **descent with modification**. It holds that all species, living and extinct, have descended, without interruption, from one or a few original forms of life….Darwin's conception of the course of evolution is profoundly different from Lamarck's, in which the concept of common ancestry plays almost no role.
- The second theme of The Origin of Species is Darwin's theory of the causal agents of evolutionary change…This theory is a VARIATIONAL THEORY of change, differing profoundly from Lamarck's TRANSFORMATIONAL THEORY, in which individual organisms change. (p. 7)

Mitchell (2009) adds one additional point of note regarding Darwin's beliefs:

- Evolutionary change is constant and gradual via the accumulation of small, favorable variations. (p. 79)

This theory was in sharp contrast, at least in the eyes of the early adherents of both, to Gregor Mendel's [1822–1884] *mutation theory*. Mutation theory stated that variation in organisms was due to mutations in offspring which drive evolution, with natural selection unnecessary to account for origin of species. Mendel's perspective evolved into the Evolutionary Synthesis or Modern Synthesis movement, which provided its own set of principles of evolution. Describing the development of its underlying theory, Futuyma (2005) notes

> Ronald A. Fisher and John B.S. Haldane in England and Sewall Wright in the United States developed a mathematical theory of population genetics, which showed that mutation and natural selection together cause adaptive evolution: mutation is not an alternative to natural selection, but is rather its raw material. (p. 9)

The idea of gradualism was questioned in the 1960s and 1970s, when paleontologists Stephen Jay Gould and Niles Eldridge began to challenge it as "very rare and too slow, in any case, to produce the major events of evolution" (Gould & Eldredge, 1977, p. 115). Instead, they proposed a theory of *punctuated equilibrium* (Eldredge & Gould, 1972) which instead hypothesized that "Most evolutionary change, we argued, is concentrated in rapid (often geologically instantaneous) events of speciation in small, peripherally isolated populations (the theory of allopatric speciation)" (Gould & Eldredge, 1977, pp. 116–117).

Despite this challenge, evolutionary synthesis remains crucial to our understanding of evolution today. "The principal claims of the evolutionary synthesis are the foundations of modern evolutionary biology…most evolutionary biologists today accept them as fundamentally valid" (Futuyma, 2005, pp. 9–10). While this consensus persists, many questions remain concerning the complexities of modern evolution. The presence of holistic connections in living systems complicates our understanding of biological organisms: "The complexity of living systems is largely due to networks of genes rather than the sum of independent effects of individual genes" (Mitchell, 2009, p. 275).

At this point, then, most of science believed that evolution alone, in one form or another, was responsible for the complexity inherent in biological systems. This perspective was in sharp contrast to that of theoretical biologist Stuart Kauffman; in studying complex biological systems, Kauffman has developed remarkable theories about evolution and complexity. Arguably, his most fundamental point is that biological complexity does not necessarily arise from a process of natural selection.

> Most biologists, heritors of the Darwinian tradition, suppose that the order of ontogeny is due to the grinding away of a molecular Rube Goldberg machine, slapped together piece by piece by evolution. I present a countering thesis: most of the beautiful order seen in ontogeny is spontaneous, a natural expression of the stunning self-organization that abounds in very complex regulatory networks. We appear to have been profoundly wrong. Order, vast and generative, arises naturally…much of the order in organisms may not be the result of selection at all, but of the spontaneous order of self-organized systems.…If this idea is true, then we must rethink evolutionary theory, for the sources of order in the biosphere will now include both selection and self-organization. (Kauffman, 1993, p. 25)

Further, Kauffman's *fourth law* introduced the notion that "life has an innate tendency to become more complex, which is independent of any tendency of natural selection" (Mitchell, 2009, p. 286). Kauffman's book *The Origins of Order* (1993) talks at length about this concept.

Astrophysicist Erich Jantsch [1929–1980] contrasted internal and external self-organizing systems as those that change their internal organization and those that adapt their way of interacting with their environment, respectively. Jantsch (1972) discussed three types of internal self-organizing behavior useful to our study:

- mechanistic systems do not change their internal organization;
- adaptive systems adapt to changes in the environment through changes in their internal structure in accordance with preprogrammed information (engineering or genetic templates); and
- inventive (or human action) systems change their structure through internal generation of information (invention) in accordance with their intentions to change the environment (p. 476)

The systems we are concerned with reside in the adaptive or inventive classification. For our purposes, we are concerned with order and stability and what we may learn of purposeful systems by studying biological systems. If we can summarize, we may conceive of two major streams of evolutionary thought (1) those who believe natural selection is primary, be it via gradual means (e.g., Darwin) or punctuated means (e.g., Gould and Eldredge); and (2) those that believe that self-adaptation and self-organization have arisen via emergent behavior of biological systems (e.g., Kauffman). We may describe evolution by natural selection as being "conceived using data at the macroscopic level" (Johnson & Lam, 2010, p. 879) and thus as a meta-theory of the development of systems, whereas we may think of self-organization as "essentially present, but..not well controlled" (Johnson & Lam,

2010, p. 882) and thus an emergent, inherent property of both the system and its circumstances. It is our belief that these two perspectives may be complementary given their presence on differing levels of logical abstraction, and, in fact, both perspectives have implications for how we may seek to understand problems and messes. If we accept the parallelism of biological and purposeful system life cycles, then perhaps, it is not much of a stretch to understand the increasing complexity of both biological and purposeful systems. What drives this increasing complexity? Is it evolution *or* self-organization? We contend that a system that is to maintain its viability (Beer, 1979) must be allowed to evolve *and* self-organize. How to ascertain if our mess has evolved or is evolving; what about self-organizing? More fundamentally perhaps is, does it even matter? The answer, if we are to effect change, is yes. The answer in how to identify the opportunity for this change lies in the concept of entropy, to which we now turn.

11.3 Entropy

How do patterns emerge in systems and in nature? As if appearing to occur by some magical *slight of hand*, structure and patterns emerge in systems without external interference (i.e., they self-organize). This behavior is seemingly illogical, but some investigation will clarify how independent elements arrange themselves in an ordered and purposeful pattern. Understanding this phenomena and its role in systemic thinking requires that we first understand the second law of thermodynamics, which says that entropy (the property of matter that measures the degree of randomization or disorder at the microscopic level) can be produced but never destroyed (Reynolds & Perkins, 1977). The potential energy of our system, which is inversely proportional to its entropy, will decrease without the application of energy to our system. Stated another way, it states that "in a closed system, entropy always increases" (Bertalanffy, 1968, p. 144). But, as Mitchell points out, "nature gives us a singular counterexample: Life…According to our intuitions, over the long history of life, living systems have become vastly more complex and intricate rather than more disordered and entropic" (Mitchell, 2009, p. 71). The key is that living systems are *open systems*.

The second law of thermodynamics is true of all *closed systems*, those systems that exchange no materials with their environment. A car's fuel stores its potential energy; without refueling, the car will have a finite driving range. Similarly, our bodies store our potential energy; without consuming calories will cease to be able to function and eventually we will die. The flow of this energy maintains order and continued existence. There is no such thing as a perpetual motion machine; all systems are less than 100% efficient, and thus, they consume resources, requiring intervention from external entities, to remain viable. Open systems solve this entropy conundrum by exchanging matter with their environment. As a result, they can exhibit the equifinal behavior where "If a steady state is reached in an open system, it is independent of the initial conditions, and determined only by the system parameters, i.e., rates of reaction and transport" (Bertalanffy, 1968, p. 142).

If no energy enters or leaves a closed system, the potential energy of the system dissipates with time (i.e., its entropy increases). We can express this notion mathematically. If we designate entropy as S, then the change in entropy of a closed system can be expressed as follows:

$$\Delta S_C = S_{\text{final}} - S_{\text{initial}} \geq 0 \quad (11.3)$$

where ΔS_C = change in closed system entropy

S_{final} = final system entropy

S_{initial} = initial system entropy.

Open systems behave much differently, owing to their ability to transport matter in and out of the system. Their change in entropy, then, can be denoted as follows:

$$\Delta S_O = \Delta S_{\text{transport}} + \Delta S_{\text{reactions}} \quad (11.4)$$

where ΔS_O = change in open system entropy

$\Delta S_{\text{transport}}$ = change in entropy transport (either positive or negative) in and out of the system

$\Delta S_{\text{reactions}}$ = the production of entropy due to internal processes such as chemical reactions, diffusion, and heat transport.

The relevance of these two conceptualizations is that open systems can reach the same final state from different initial conditions due to exchanges with the system's environment (i.e., the principle of equifinality). This is directly relevant to us as we assess messes, which are open and involve significant matter (and information) exchange across their system boundaries.

The concept of entropy may be generalized to other contexts. Arguably, the most famous beside the thermodynamics perspective is physicist Ludwig Boltzmann's [1844–1906] statistical entropy (Boltzmann, 1905), which shows the relationship between entropy and the number of ways the atoms or molecules of a thermodynamic system can be arranged. Boltzmann's formula is as follows:

$$S = k_b \ln W \quad (11.5)$$

where S is entropy, as before, k_b is the Boltzmann's constant equal to 1.38×10^{-23} J/K, and W is conceptualized as the *thermodynamic probability* of a particular macro-state for some distribution of possible micro-level states of a thermodynamic system.

In a thermodynamic system where each state may have an unequal probability, it is useful to utilize a reformulation of this concept developed by J. Willard Gibbs [1839–1903] in his seminal work (Gibbs, 1902):

$$S = -k_b \sum_i p_i \ln p_i \tag{11.6}$$

where p_i refers to the probability that a given micro-state can occur. Claude Shannon [1916–2001], the father of information theory, adapted these concepts to the analysis of entropy in information, stating:

> That information be measured by entropy is, after all, natural when we remember that information, in communication theory, is associated with the amount of freedom of choice we have in constructing a message. (Shannon & Weaver, 1949, p. 13)

Shannon's conceptualization of information entropy, then, can be defined as follows:

$$H = -\sum_i p_i \log_b p_i \tag{11.7}$$

where H is the information entropy, b is the base of the logarithm used (typically taken to be 2 due to the predominant use of binary logic in information theory), and p is the probability associated with each of the symbols in each discrete message i. It is worth noting that this formula is maximized when all state probabilities are equal (i.e., for a two-state system, $p_1 = p_2 = 1/2$). In this case, the most uncertainty possible is present in the system.

The question is, how is this energy handled by our system, be it information, thermodynamic, or statistical entropy? The short answer lies in the exploration of the concept of self-organization. Self-organization is a well-established phenomena in chemistry, physics, ecology, and sociobiology (Nicolis & Prigogine, 1977) defined as "the spontaneous reduction of entropy in a dynamic system" (Heylighen & Joslyn, 2003, p. 155). Recall our discussion of the second law of thermodynamics stating that entropy can be produced but not destroyed. How, then, is entropy in a system reduced?

Ilya Prigogine [1917–2003] received the 1977 Nobel Prize in Chemistry for his investigation, starting in the 1950s, of the case where self-organizing systems do not reach an equilibrium state. Nicolis and Prigogine (1977) were studying structures that they referred to as dissipative; these were structures that exhibited dynamic self-organization. As such, these open systems generated energy, which was dissipated to their environment. Thus, they were able to self-organize (i.e., decrease their entropy) by increasing the disorder (and thus, the entropy) of their environment. This is the key to survival for living systems; they reduce their internal entropy to avoid disorder and chaos prescribed by the second law of thermodynamics (and only true for closed systems). As such, these dissipative systems are able to maintain a dynamic equilibrium (D'Alembert, 1743) by dissipating their energy to the environment in an effort to create a reproducible steady state. This steady state can arise through multiple means, be it by system evolution, manufactured means, or a combination of the two. Examples of these systems range from purposeful systems such as climate control systems (i.e., heaters and air conditioners) to natural systems such as convection, hurricanes, and cyclones, to all living systems.

While these numerous examples illustrate the prevalence of self-organization, they do little to explain how or why self-organization occurs. The varying entropic perspectives of Nicolis, Prigogine, Boltzmann, Gibbs, and Shannon and Weaver are complemented by work in control theory and cybernetics. The term cybernetics was coined by Norbert Wiener in his seminal book whose title defined it as *the study of control and communication in the animal and the machine* (Wiener, 1948). Heylighen and Joslyn (2003), in a discussion of cybernetic control, speak of basins (Varghese & Thorp, 1988) and their relationship to self-organization:

> An attractor y is in general surrounded by a basin B(y): a set of states outside y whose evolution necessarily ends up inside: $\forall$ s $\in$ B(y), s $\notin$ y, n such that $f^n(s) \in$ y. In a deterministic system, every state either belongs to an attractor or to a basin. In a stochastic system there is a third category of states that can end up in either of several attractors. Once a system has entered an attractor, it can no longer reach states outside the attractor. This means that our uncertainty (or statistical entropy) H about the system's state has decreased: we now know for sure that it is not in any state that is not part of the attractor. This spontaneous reduction of entropy or, equivalently, increase in order or constraint, can be viewed as a most general model of self-organization. (Heylighen & Joslyn, 2003, p. 165)

The attractors described by Heylighen and Joslyn (2003) will end up in a state of dynamic equilibrium. This arrangement of elements and emergence of order are what W. Ross Ashby [1903–1972] called the *principle of self-organization* (Ashby, 1947). This self-organization results in a lowered entropy for our system as uncertainty has decreased within our system. Heinz von Foerster [1911–2002] devised the principle of *order from noise* (1960). Self-organization can be expedited by the presence of noise; the larger the random perturbations (*noise*) of a system, the more entropy exists in the system, and thus, the more quickly it will become ordered.

So, what does all of this mean? Our system changes, and maintains stability, as a result of mechanisms involving both evolution and self-organization. The order that emerges (both through evolution on longer time horizons and self-organization on shorter time horizons) is essential for our system to maintain its continued viability. We can enhance this viability through mechanisms such as those described by Beer (1979, 1981) in his Viable Systems Model. A self-organizing system achieves this viable equilibrium state by random exploration, with purposeful systems being aided by control mechanisms (recall Checkland's (1993) control principle), which reduce the feasible solution space (i.e., the variety) for these systems to explore. Ashby (1947), von Foerster (1960), and von Foerster and Zopf (1962) further postulate that this process can be expedited by increasing variation or noise into the system, thereby increasing system entropy and accelerating the systems search's for an equilibrium state. This process is confirmed by Prigogine's theory of dissipative structures, which increase their variation (and thus, entropy) until it is unsustainable and then dissipate this energy back into the environment.

What does this all mean for the systemic thinker? In theory, it provides a mechanism for determining when to interfere in our system; we should interact with it before its natural tendency to dissipate (or in Hitchins' terms, to decay or collapse) in an effort to expedite its search for equilibrium. In practice, this undertaking is not so straightforward as self-organizing systems, by definition exhibit behavior, as described by the *principle of homeostasis* (Cannon, 1929) in an effort to regulate their internal environment. Thus, the most practical approach for us is to identify application points or individual properties where a small change may result in a large, predictable effect. Accordingly, we turn to analysis of an approach which will enable us to determine an appropriate time for intervention in our system.

11.4 Another View of Sensemaking

Because complexity is such an important characteristic of systems, a number of frameworks have been developed for understanding the relationship between complexity and systems. One such framework is the Cynefin framework presented in Chap. 10.

Another way to look at the Cynefin framework is by the types of systems' connections expected to exist in each of the domains depicted in Fig. 11.6. Kurtz and Snowden (2003) discuss these connections:

> On the side of order, connections between a central director and its constituents are strong, often in the form of structures that restrict behavior in some way—for example, procedures, forms, blueprints, expectations, or pheromones. On the side of un-order, central connections are weak, and attempts at control through structure often fail from lack of grasp or

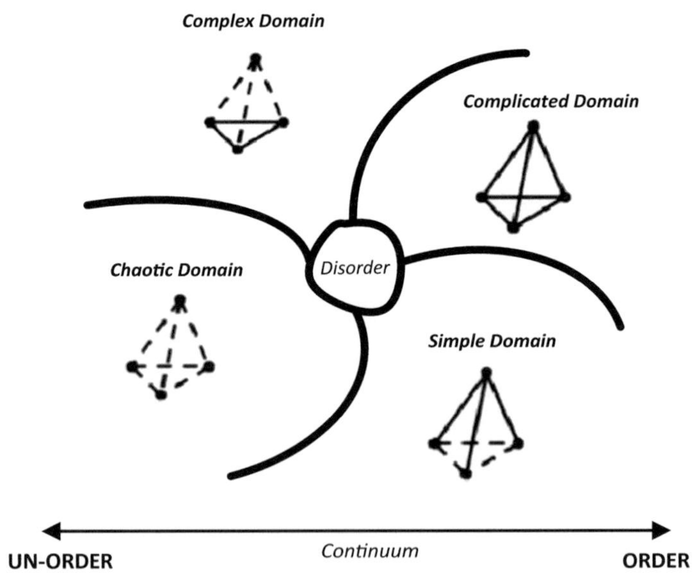

Fig. 11.6 Connection strength of Cynefin domains (adapted from Kurtz & Snowden, 2003, p. 470)

> visibility. In the complex and knowable domains, connections among constituent components are strong, and stable group patterns can emerge and resist change through repeated interaction, as with chemical messages, acquaintanceship, mutual goals and experiences. The known and chaotic domains share the characteristic that connections among constituent components are weak, and emergent patterns do not form on their own. (p. 470)

It is problematic for us to try to interfere in messes that reside primarily in the unorder domain (complex and chaos), both due to their weak central connections (in our terms, at the mess level) and their unpredictable and unperceivable relationships. It is our goal in these regimes, at best, to shift to an ordered domain. Here, we are invoking the *principle of relaxation time* (see Chap. 4), which sets the requirement for stability as a precursor to analysis and the need to avoid messes during periods of instability. Most importantly, we should concentrate on utilizing our resources to effect changes in the order domain, if possible. Kauffman (1993) echoes the difficulty in intervening in chaotic systems:

> Deep in the chaotic regime, alteration in the activity of any element in the system unleashes an avalanche of changes, or damage, which propagates throughout most of the system (Stauffer, 1987). Such spreading damage is equivalent to the butterfly effect or sensitivity to initial conditions typical of chaotic systems. The butterfly in Rio changes the weather in Chicago. Crosscurrents of such avalanches unleashed from different elements means that behavior is not controllable. Conversely, deep in the ordered regime, alteration at one point in the system only alters the behavior of a few neighboring elements. Signals cannot propagate widely throughout the system. Thus, control of complex behavior cannot be achieved. Just at the boundary between order and chaos, the most complex behavior can be achieved. (p. 302)

An alternative way of conceptualizing conditions for interaction is presented in Fig. 11.7. This figure shows the relationship of entropy and self-organization when

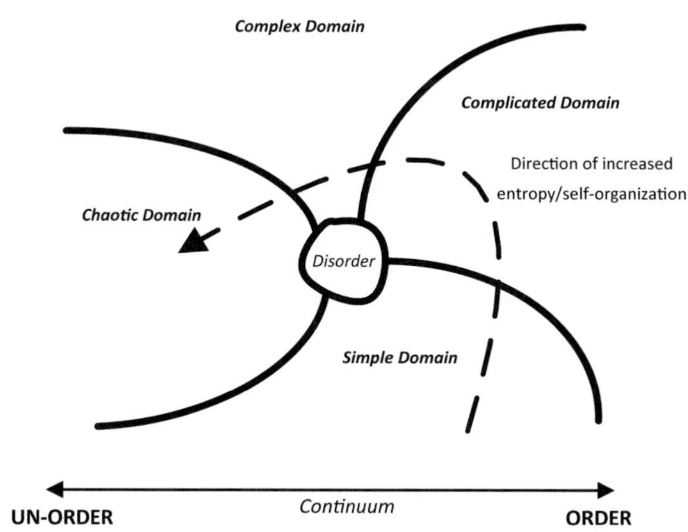

Fig. 11.7 Entropy and self-organization as applied to the Cynefin framework

compared to each Cynefin domain. As the underlying complexity of a situation increases, its entropy increases. This entropy feeds self-organizing behavior, which makes intervention problematic. Thus, it is advantageous for us to intervene in our system in the less entropic states (and set up conditions for self-organization, such as feedback mechanisms and regulators, in more entropic states).

How, then, should we intervene? This is the focus, in large part, of Chap. 10. *When* should we intervene in our system? We need to balance our desire for intervention (i.e., our bias for action) with consideration of the efficacy of our actions. For an answer to this question, we develop a decision flowchart for assessing intervention timing in the next section.

11.5 Decision Flowchart for Addressing *When* in Messes and Problems

Figure 11.8 shows our proposed decision flowchart for assessing if and when we should intervene in our mess in an effort to increase understanding about it. A discussion of the flowchart's elements is as follows:

Element 1 urges us to ask, *Is max(B/C) $\geq$ 1 for our problem*? Put another way, is our system too mature? This question arises from the material presented in

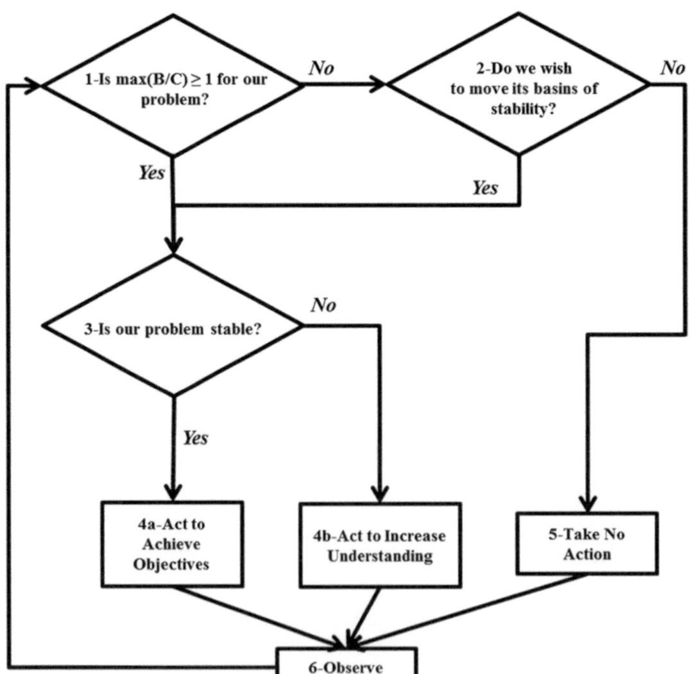

Fig. 11.8 Decision flowchart for assessing intervention timing

Sect. 11.1. The key here is asking whether or not our system has sufficient life remaining to warrant us expending resources to intervene in it. If it is too mature ($B/C < 1$), then we move on to Element 2. If not, we move to Element 3.

Element 2 serves as a follow-up to Element 1. If we have deemed the system too mature under its current configuration, the question we must ask ourselves is, recalling Varghese and Thorp (1988), *do we wish to move its basins of stability?* That is, do we wish to shift the system in a manner that perhaps renders it unrecognizable to observers previously familiar with it (see Fig. 11.5 and its shifted *B/C* curves to conceptualize the potential result of a shift in the system's basins, keeping in mind that intervention in a mess may result in either a positive or negative result). If the answer is no, we move to Element 5. If we do wish to alter it, we move to Element 3.

Element 3 encourages us to ask, *Is our problem stable*? While it is possible that no mess will ever exist here, we may decompose it further and explore its constituent problems. Stability can be thought of in the terms presented in Chap. 5; namely, if it exhibits simple or complicated behavior, then it is stable (or ordered, in Cynefin terms). If it exhibits complex or chaotic behavior, it is not (unordered in Cynefin terms). This can be checked by estimating current (i.e., unchanged) parameter values in our current scenario (using our FCM representation) and assessing scenario stability using a trivalent transfer function. If the scenario is stable, we should move to Element 4a. If it is not stable, we should move to Element 4b.

Element 4 (both 4a and 4b) represents our decision to act. Arriving here compels us to do *something*. Our resultant action is dependent on what effect we are trying to achieve, which is in turn influenced by the problem's stability. If we have a stable problem, then we can reasonably act to achieve our problem's objectives (*Element 4a*). If we have an unstable problem, we should act to increase our understanding about our problem (*Element 4b*). This action and its mechanisms are described in Part III of the text, starting with Chap. 12. While we offer no prescriptive advice regarding what action is to be taken at this point, we assert that an individual arriving at this element in the framework is compelled to do *something*. Failing to act, given the factors that led to this point, is likely to result in a Type V error (inaction when action is warranted). After acting, we move to Element 6.

Element 5 represents our decision not to act. If we have arrived here, our system, in its current form, is beyond help or we simply do not wish to try to salvage it. Thus, we choose to not act in order to avoid committing a Type IV error (taking inappropriate action to resolve a problem). This does not mean we are done with our mess; it merely means we will move on to observing without interfering with it. This stage continues to Element 6.

All elements eventually lead to *Element 6*. Element 6 asks us to observe. After acting (or not) based on the factors associated with our problem, we must observe the effects of our decisions. This may include waiting to see whether our mess becomes more orderly or attempting to realize the benefits of a programmed intervention in our system. Regardless of why we have arrived here, it is important to observe our system before the framework compels us to return to Element 1 and begin anew.

11.6 Framework for Addressing *When* in Messes and Problems

Addressing the *when* perspective in our messes and problems requires that we complete the following two steps for an identified problem:

1. Assess the problem FCM to ensure all concepts operate on the same timescale. If necessary, adjust causal weights to synchronize time steps.
2. Use the decision flowchart (Fig. 11.8) for addressing intervention timing to determine the appropriateness of intervening in a problem and to document the accompanying rationale.

A note on Step 1 is necessary before illustrating this framework on an example problem. Step 1 asks us to ensure that all causal weights are being assigned based on the same temporal scale (i.e., one day, one week, etc.) and adjust if necessary. We can investigate this notion by listing all concepts and their associated time period for change (i.e., a week, month, year, etc.). If all concepts do not change in the same time period, we can adjust *incoming* weights for those that do not synchronize them. We focus on adjusting incoming weights as they influence the speed at which a concept changes in our FCM. We can make adjustments as a rough order of magnitude by adjusting weights according to a reference point (e.g., the minimum time horizon in which a change in any concept in the FCM is observable).

More sophisticated techniques for dealing with temporal inconsistencies can be found in Park and Kim (1995), who add intermediate concepts to synchronize time steps between concepts and Hagiwara (1992), who provides techniques for incorporating nonlinear weights, conditional weights, and time delays between concepts. Each approach complicates the FCM development significantly and can be pursued if more sophisticated analysis is desired.

11.7 Example Problem

We return to our real estate example pictured in Fig. 11.9.

11.7.1 Timescale Assessment

First, we must analyze our FCM to ensure all concept transitions occur on the same timescale. We can list all of our concepts and their accompanying time horizon for change to ensure that they change at the same rate. This information is found in

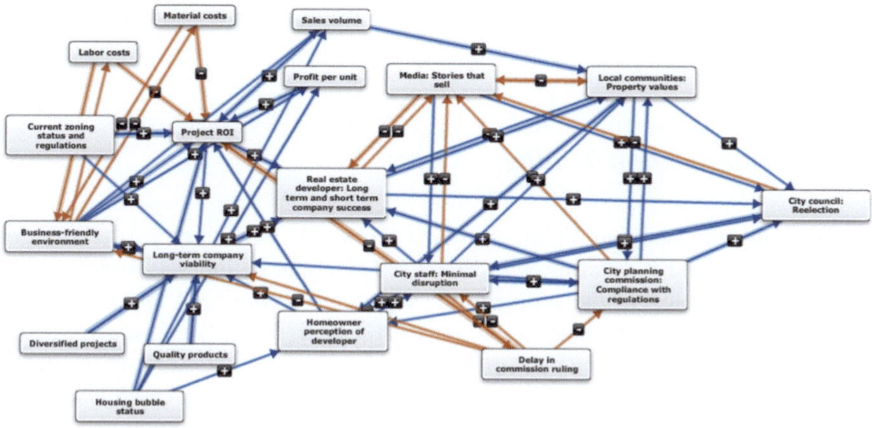

Fig. 11.9 Real estate example FCM

Table 11.1 Assessment of concept time horizons

Concept	Time period for change	Proposed change
City staff: minimal disruption	Weekly	None
Homeowner perception of developer	Weekly	None
Media: stories that sell	Weekly	None
Business-friendly environment	Monthly	–
City planning commission: compliance with regulations	Monthly	–
Current zoning status and regulations	Monthly	–
Delay in commission ruling	Monthly	–
Diversified projects	Monthly	–
Labor costs	Monthly	–
Material costs	Monthly	–
Profit per unit	Monthly	–
Project ROI	Monthly	–
Quality products	Monthly	–
Sales volume	Monthly	–
City council: reelection	Yearly	–
Housing bubble status	Yearly	–
Local communities: property values	Yearly	–
Long-term company viability	Yearly	–
Real estate developer: long-term company success and short-term company success	Yearly	–

Table 11.1. Note that proposed changes indicate whether the total magnitude of a weight should be increased (+) or decreased (−). An indication of two or more plus or minus values indicates a stronger temporal adjustment is necessary.

We can now adjust our causal weights using the information found in Table 11.1.

11.7.2 Intervention Timing

Armed with our modified FCM, we must work our way through the decision flowchart in Fig. 11.8. Starting with Element 1, we can definitively conclude that the benefit remaining in the problem (as it pertains to financial return) certainly outweighs the cost of intervention. So, max(B/C) $\geq$ 1. Next, we must ask whether or not our problem is stable (Element 3). This requires us to consider initial values for our concepts as the status quo. In this case, we believe *compliance with regulations* is at 1.0, while all remaining concepts are at 0 (absent any further information). The results of this analysis are shown in Fig. 11.10.

Clearly, the scenario is stable, but complicated. In this case, we move to Step 4a, *Act to achieve objectives*. Thus, overall, we can conclude that the problem has sufficient time to act and is stable enough to warrant action to resolve its objectives.

Fig. 11.10 Stability analysis of real estate example

11.8 Summary and Implications for Systemic Thinking

This chapter discussed the *when* question of systemic thinking. Thinking about this compels us to determine the appropriate time for us to intervene in our system, if ever. In order to develop an approach for determining the appropriate time for intervention in our mess, we developed an approach to assess the *maturity* and *stability* of our mess. The maturity discussion focused on life-cycle concerns and on evaluating the cost-to-benefit ratio of mess intervention, while our stability perspective focused on a discussion of system evolution and self-organization, leading to a method for classifying and understanding our system's state. We then combined these concepts into a six-element framework to serve as a guide for individuals interested in increasing understanding about their mess. After reading this chapter, the reader should

1. Be able to assess the maturity and stability of a problem; and
2. Understand the appropriateness of intervening in a given mess.

References

Ashby, W. R. (1947). Principles of the self-organizing dynamic system. *Journal of General Psychology, 37*(1), 125–128. doi:10.1080/00221309.1947.9918144.
Beer, S. (1979). *The heart of the enterprise*. New York: Wiley.
Beer, S. (1981). *Brain of the firm*. Chichester, UK: Wiley.
Bertalanffy, Lv. (1968). *General system theory: Foundations, development, applications* (Revised ed.). New York: George Braziller.
Blanchard, B. S. (2004). *Systems engineering management* (3rd ed.). Hoboken, NJ: Wiley.
Boltzmann, L. (1905). *The 2nd law of thermodynamics*. Dordrecht: Reidel.
Cannon, W. (1929). Organization for physiological homeostasis. *Physiological Reviews, 9*(3), 399–431.
Checkland, P. B. (1993). *Systems thinking, systems practice*. New York: Wiley.
Csete, M. E., & Doyle, J. C. (2002). Reverse engineering of biological complexity. *Science, 295* (5560), 1664–1669.
D'Alembert, J. (1743). *Traité de Dynamique*. Paris: David l'Ainé.
Darwin, C. (1859). *On the origin of species by means of natural selection, or the preservation of favored races in the struggle for life*. New York: Modern Library.
Eldredge, N., & Gould, S. J. (1972). Punctuated equilibria: An alternative to phyletic gradualism. In T. J. M. Schopf (Ed.), *Models in paleobiology* (pp. 82–115). San Francisco: Freeman, Cooper & Company.
Foerster, H. V. (1960). On self-organizing systems and their environments. In M. Yovits & S. Cameron (Eds.), *Self-organizing systems* (pp. 30–50). London: Pergamon.
Foerster, H. V., & Zopf, G. (Eds.). (1962). *Principles of self-organization*. New York: Pergamon.
Futuyma, D. J. (2005). *Evolution*. Sunderland, MA: Sinauer Associates.
Gibbs, J. W. (1902). *Elementary principles in statistical mechanics: Developed with especial reference to the rational foundation of thermodynamics*. New Haven, CT: Yale University Press.
Gould, S. J., & Eldredge, N. (1977). Punctuated equilibria: The tempo and mode of evolution reconsidered. *Paleobiology, 3*(2), 115–151. doi:10.2307/2400177.

Hagiwara, M. (1992). *Extended fuzzy cognitive maps*. Paper presented at the IEEE International Conference on Fuzzy Systems, San Diego, CA.

Heylighen, F., & Joslyn, C. (2003). Cybernetics and second-order cybernetics. In R. A. Meyers (Ed.), *Encyclopedia of physical science and technology* (3rd ed., pp. 155–169). New York: Academic Press.

Hitchins, D. K. (1993). A unified systems hypothesis. *Systems Practice, 6*(6), 613–645. doi:10.1007/BF01059481.

Hitchins, D. K. (2007). *Systems engineering: A 21st century systems methodology*. Chichester, UK: Wiley.

Jantsch, E. (1972). Forecasting and the systems approach: A critical survey. *Policy Sciences, 3*(4), 475–498.

Johnson, B. R., & Lam, S. K. (2010). Self-organization, natural selection, and evolution: Cellular hardware and genetic software. *BioScience, 60*(11), 879–885.

Kauffman, S. (1993). *The origins of order: Self-organization and selection in evolution*. New York: Oxford University Press.

Kurtz, C. F., & Snowden, D. J. (2003). The new dynamics of strategy: Sense-making in a complex-complicated world. *IBM Systems Journal, 42*(3), 462–483. doi:10.1147/sj.423.0462.

Mitchell, M. (2009). *Complexity: A guided tour*. New York: Oxford University Press.

Nicolis, G., & Prigogine, I. (1977). *Self-organization in nonequilibrium systems: From dissipative structures to order through fluctuations*. New York: Wiley.

Park, K. S., & Kim, S. H. (1995). Fuzzy cognitive maps considering time relationships. *International Journal of Human-Computer Studies, 42,* 157–168.

Reynolds, W., & Perkins, H. (1977). *Engineering thermodynamics* (2nd ed.). New York: McGraw-Hill.

Sage, A. P., & Armstrong, J. E. (2000). *Introduction to systems engineering*. Hoboken, NJ: Wiley.

Shannon, C. E., & Weaver, W. (1949). *The mathematical theory of communication*. Champaign, IL: University of Illinois Press.

Stauffer, D. (1987). Random Boolean networks: Analogy with percolation. *Philosophical Magazine, 56*(6), 901–916.

Varghese, M., & Thorp, J. S. (1988). An analysis of truncated fractal growths in the stability boundaries of three-node swing equations. *IEEE Transactions on Circuits and Systems, 35*(7), 825–834. doi:10.1109/31.1829.

Wiener, N. (1948). *Cybernetics: Or control and communication in the animal and the machine*. Cambridge: MIT Press.

Part III
Acting Systemically

With Part III, we have now transitioned from the think to act phase in the systemic decision making process (as shown in Fig. 1, originally from Chap. 3).

We have now thought extensively about each of our constituent problems using each of our six perspectives; that is, we have *thought systemically* about our problems. Now, we must combine these perspectives into a comprehensive understanding of our situation in order to prepare for our next phase, action. At this point, we are eager to improve our situation and to increase our understanding about it. This section provides the reader the necessary tools to guide analysis of appropriate decisions for intervention in our mess. This part of the text is comprised of Chaps. 12–14. Chapter 12 provides guidance on combining our constituent problem representations for holistic understanding of our mess. Chapter 13 discusses the anatomy of a decision, including how to make a decision regarding appropriate intervention in your mess. Chapter 14 addresses decision implementation and how to avoid errors when implementing our chosen course of action. After you have read Part III of the text, you will be equipped with both a plan for action and a set of strategies for executing the plan.

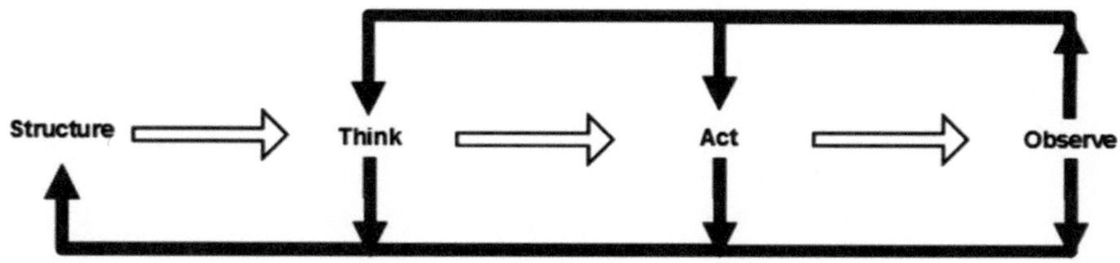

Fig. 1 Systemic decision making process with feedback

Chapter 12
Systemic Action

Abstract We have come a long way together. The assumption at this point is that the reader has read through the first eleven chapters of this book and understands how to analyze a singular problem from each of the six perspectives presented in Chaps. 6–11. Now we are ready to take action. To this end, this chapter addresses putting the pieces back together (i.e., mess reconstruction) in order to understand our mess systemically. Two meta-perspectives, the *what is* and the *what ought-to-be*, are proposed to represent our current and idealized states. Generation of these meta-perspectives is demonstrated on the real estate example we have carried throughout the text.

12.1 Mess Reconstruction

Recall Fig. 3.7, which presented a basic illustration of the systemic decision making process, presented here as Fig. 12.1 with additional annotation regarding topics covered since our discussion in Chap. 3.

The primary focus of this chapter is on mess reconstruction (the synthesis element shown in Fig. 12.1) as a vehicle for increasing understanding and as preparation for taking action. Messes, of course, are a construct of convenience. They are envisioned and constructed in a somewhat arbitrary manner by each of us (as an observer) and yet, in identifying a mess and deconstructing it as we did in Chap. 2, and then analyzing its elements as we did in Chaps. 6–11 (we were acting as an analyst), we have placed a responsibility on ourselves to reconstitute these pieces into a coherent whole to allow for systemic understanding of our mess. Of course, in order to have a mess to reconstruct, we must have analyzed at least two constituent problems. Problem 1 (the initial real estate problem) analysis has been addressed in the previous six chapters. Appendix A provides analysis for Problem 2 (the 2nd real estate problem) of this mess. The reader would be fine to continue without reading Appendix A; however, a more thorough understanding of the material will result from doing so.

P.T. Hester and K.M. Adams, *Systemic Decision Making*, Topics in Safety, Risk, Reliability and Quality 33, DOI 10.1007/978-3-319-54672-8_12

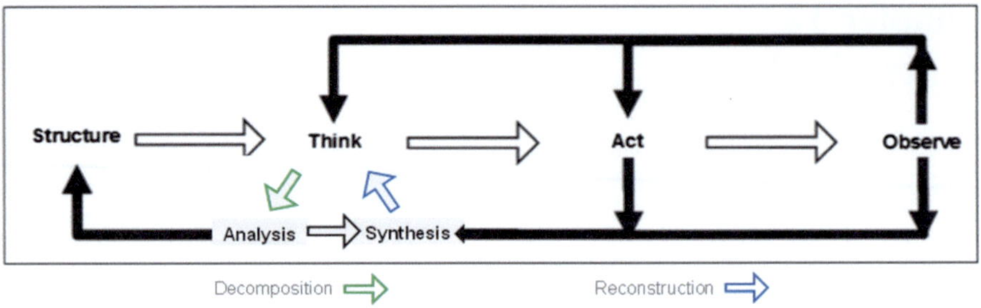

Fig. 12.1 Illustration of mess decomposition and reconstruction (adapted from Fig. 3.7)

12.2 The *What Is* Meta-Perspective

To start on this journey, we must begin to reconstruct our mess. We wish to generate a singular mess-level graphic depicting elements from each of our six perspectives, a depiction known as the *what is* meta-perspective. This combined representation provides us a mess-level understanding of our stakeholders, objectives, feedback, context, boundaries, timing issues, mechanisms, and their interactions as we believe they currently exist. Of course, this representation is limited by the number of perspectives we have available and our limited capacity for understanding. The question then is, how do we create this what is meta-perspective?

At this point, we have a single fuzzy cognitive map (FCM) generated for each of our problems of interest. They must be combined in order to holistically view our mess (and to support action). In order to do so, we first check for model inconsistencies between FCMs (i.e., concept names that are similar and should be reconciled or those that are misspelled and require fixing). We can then use the *merge* function within Mental Modeler software package to merge our FCM files. Alternatively, we can manually merge the files, although this will require significantly more effort. A mess-level depiction of the real estate *what is* meta-perspective is shown below in Fig. 12.2.

The natural question arises as to what has changed by fusing the two (or more) problem representations. Now that we are interested in understanding our situation at the mess level, we are operating at a different level of abstraction than before. This leads to a perception of *increased complexity* (as a mess observer) as compared to our constituent problems. For our real estate example, Problem 1 has 19 concepts and 61 connections, Problem 2 has 16 concepts and 34 connections, and the mess representation has 30 concepts and 94 connections. Of course, the underlying complexity of the situation has not changed in any form. Rather, the perceived change in complexity arises from the increased number of *interactions* among the components and the sheer volume of information we are now faced with. We are now overwhelmed with information and have trouble processing it in its entirety. This invokes notions of the *principle of requisite parsimony* (Miller, 1956) and information overload.

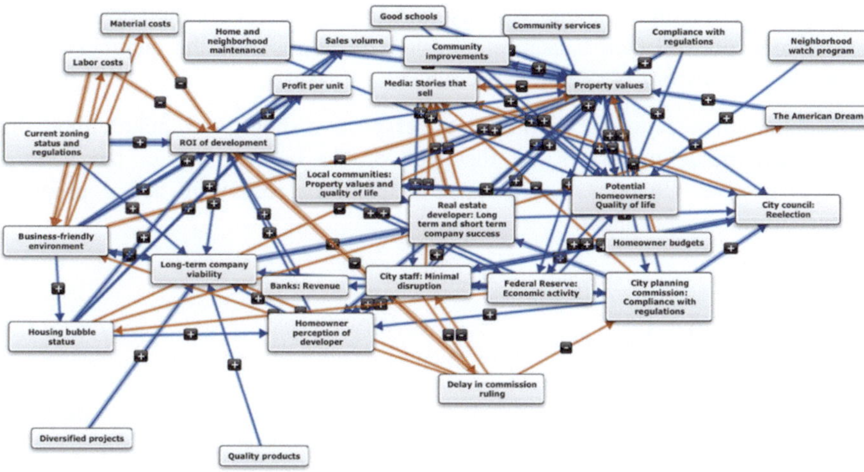

Fig. 12.2 Illustration of real estate what is meta-perspective

This perspective on complexity is known as subjective complexity (Aulin, 1987). Aulin (1987) gives the example of chess as exhibiting subjective complexity, noting, "In Chess, and in other complex games, the rules are always simple, and the complexity follows merely from the too great a number of different possibilities for a player to keep simultaneously in his mind" (p. 114).

Once we have generated our *what is* meta-perspective, we can review it for the following interactions as they pertain to each perspective:

- Stakeholder interactions (the who).
- Fundamental objective interactions (the what).
- Feedback interactions (the why).
- Context and boundary interactions (the where).
- Mechanism interactions (the how).
- Timing interactions (the when).

Each of these elements helps provide improved qualitative understanding of the mess we are facing.

12.3 The *What Ought-to-Be* Meta-Perspective

Once we have created the *what is* meta-perspective, we may wish to investigate proposed changes to the situation it represents. In order to capture these changes, we should create a second representation (i.e., FCM) known as the *what ought-to-be* meta-perspective. This second meta-perspective should include:

- Our proposed feedback modifications captured during our *why* analysis.
- Any context or boundary changes captured during our *where* analysis.

- Any other structural changes we may wish to propose. These include changes to any concepts and their causal relationships resulting from the qualitative evaluation of our *what is* meta-perspective.

Note that changes to our *what is* meta-perspective should not include modeling errors. Any modeling errors we discover should be rectified in the *what is* meta-perspective. Changes incorporated at this stage represent actual proposed changes to the system itself.

12.4 Example Analysis

Investigating our real estate mess, using Fig. 12.2 as a baseline, we can make the following changes as a result of boundary/context analysis:

- *Developer's balance sheet* influence reduced from 1 to 0.5.
- *Current zoning status and regulations*' influence reduced from −1 to −0.5.
- *Homeowner perception of developer* influence reduced from −0.5 to −0.25.
- *Housing bubble status* influence reduced from −0.5 to 0.25.
- "*The American Dream*" influence raised from 0.5 to 1.
- *Business-friendly environment* influence raised from −1 to −0.5.
- *Homeowner budgets'* influence raised from −0.5 to −0.25.

No other changes were made, including as a result of feedback analysis. This new depiction, known as the *what ought-to-be meta-perspective*, is shown in Fig. 12.3.

Changes in the meta-perspectives are present at the weight level, and thus, difficult to discern from a casual glance at the two figures.

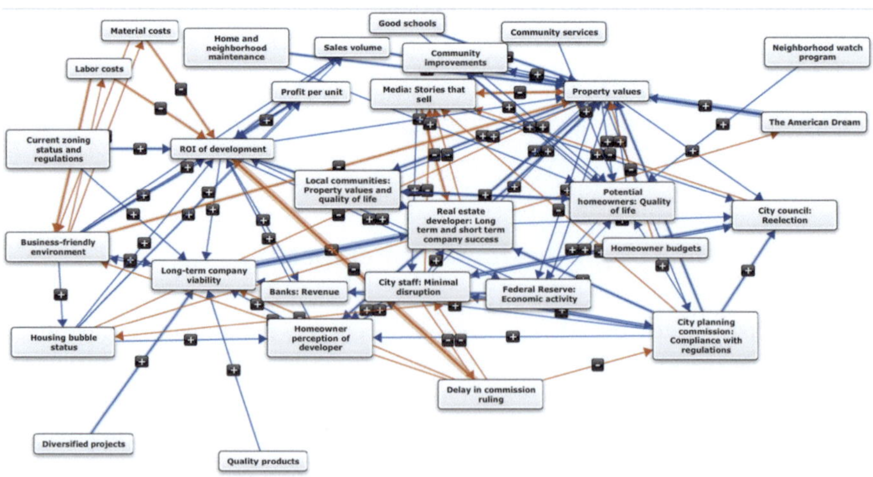

Fig. 12.3 Illustration of real estate what ought-to-be meta-perspective

12.5 Iteration

Even after we have reassembled our mess and formulated both our *what is* and *what ought-to-be* meta-perspectives, we are not really *finished*. To be fair, there is really no *finished* what it comes to a mess. We will use our two meta-perspectives to inform the *Act* stage, the focus of the next two chapters. Completion of the *Act* stage will lead to the *Observation* stage, which in turn will lead back to *Thinking,* and so on. As we gain more information and increase our understanding, we can make better choices, invest our resources more wisely, ask better questions, use better mechanisms, and truly, think systemically about our mess. This will force us to reassess our choices and assumptions about what we know; it will ask us to refine the perspectives we have of our mess.

12.6 Summary

This chapter introduced two meta-perspectives, the *what is* and the *what ought-to-be* perspectives, in an effort to make sense at the mess level of our problem-level analyses. This guidance provided a framework for systemic understanding of our mess. As we said early on in this book, everyone's got problems (and messes). How we think about them determines whether or not we'll be successful in understanding and addressing them. This chapter provided a discussion of how to combine problem representations to create a mess-level representation of our situation. We are setting ourselves up for the upcoming chapters (and for action and observation).

After reading this chapter, the reader should:

1. Understand how our six perspectives work together to provide systemic understanding of a mess;
2. Be able to articulate, using our two meta-perspectives, the current and desired states for a mess.

References

Aulin, A. (1987). Cybernetic causality III: The qualitative theory of self-steering and social development. *Mathematical Social Sciences, 14,* 101–140.

Miller, G. A. (1956). The magical number seven, plus or minus two: Some limits on our capability for processing information. *Psychological Review, 63*(2), 81–97. doi:10.1037/h0043158.

Chapter 13
Anatomy of a Decision

Abstract By now we all know that our mess is composed of multiple problems which all interact in some capacity. This mess can be thought of in terms of its actual state and its desired state. We also know that if the desired and actual states are the same, we have no decision to make as we do not have a problem. If, however, there is a delta between the two, we need to invoke a decision process in an attempt to bridge the gap. This involves creating a model of our mess utilizing the actual state of our system as best we know (as we have done in Chaps. 6–12). This modeled state, along with a desired state, feeds into a decision process. The output of this decision process is a decision or set of decisions which provide feedback through action (and the result of our action) to our mess and its constituent problems. This set of relationships is shown in Fig. 13.1. This chapter aims to elaborate on the decision process element in Fig. 13.1. We first begin with discussing some basics of decision analysis and decision science. We then proceed to discuss the decision process and how we can execute it. This leads to a framework for action in our mess. This framework is demonstrated on our real estate problem, and some additional concerns for decision making are discussed.

13.1 Introduction

At this point, we have moved on from thinking about our problem and we are ready to act. Reviewing our errors' taxonomy shown in Table 13.1, we have now moved on from avoiding the Type III error (the focus of the thinking systemically section of the text). The action stage requires two major steps: (1) We must decide what to do (and aim to avoid the Type IV error of wrong action and the Type V error of inaction) and (2) we must implement this action (and aim to avoid the Type VIII error of incorrect implementation).

So, we have to *choose the correct action* and *implement it correctly*. The former is the focus of this chapter; the latter is the focus of the next chapter.

P.T. Hester and K.M. Adams, *Systemic Decision Making*, Topics in Safety, Risk, Reliability and Quality 33, DOI 10.1007/978-3-319-54672-8_13

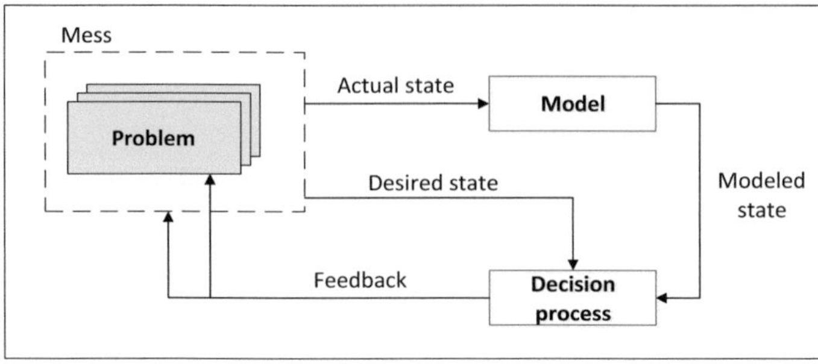

Fig. 13.1 Modeling and decision making

Table 13.1 Taxonomy of systems errors (adapted from Adams & Hester, 2012)

Error	Definition	Issue
Type III (γ)	Solving the wrong problem precisely	Wrong problem
Type IV (δ)	Inappropriate action is taken to resolve a problem as the result of a correct analysis	Wrong action
Type V (ε)	Failure to act when the results of analysis indicate action is required	Inaction
Type VIII (η)	Incorrectly implementing the correctly decided action	Incorrect implementation
Type I (α)	Rejecting the null hypothesis when the null hypothesis is true	False positive
Type II (β)	Failing to reject the null hypothesis when the null hypothesis is false	False negative
Type VI (θ)	Inferring causation when only correlation exists	Unsubstantiated inference
Type VII (ζ)	An error that results from a combination of the other six error types, often resulting in a more complex problem than initially encountered	System of errors

13.2 Roles

Before diving into the specifics of decision analysis, we must distinguish between the roles of individuals associated with its execution. It is important to distinguish between an analyst, an owner, and a decision maker. An *analyst* performs an assessment of a problem (i.e., a decision analyst), but this individual does not necessarily control any resources (or care about the outcome, that is, he/she may simply be a hired hand). We defined an *owner* in Chap. 2 as someone who wishes to see a problem resolved and allocate resources to do so (although he/she may not control these resources). Finally, a *decision maker* has authority over a decision (and has the associated resources to implement a decision).

Thus, a decision maker can be the problem's owner, but he/she does not have to be. When the owner and decision maker are not the same individual, we have an interesting situation as we have someone who does not necessarily wish to see a problem resolved and yet has the authority to allocate resources to resolve it. This harkens back to our discussion of Chap. 6 with regard to stakeholder analysis and management. If we *own* a problem but we are not the decision maker, we must utilize appropriate stakeholder management techniques to convince the decision maker to help with our plight. Sometimes, this is easier said than done. The remainder of this chapter will ignore this potential conflict and assume the reader is serving in the role of both decision maker and owner.

13.3 Decision Analysis

Some background on the field of decision analysis is necessary to set the context for this chapter's discussion. Keeney and Raiffa (1976) describe decision analysis as a "*prescriptive* approach…to think hard and systematically about some important real problems" (p. vii). Thus, it is a description of how individuals *should* solve problems and not necessarily how they actually solve problems. Further complicating this prescriptive discussion is the introduction of subjective evaluations. Keeney and Raiffa (1976) further elaborate on the nature of decision analysis and subjectivity:

> It is almost a categorical truism that decision problems in the public domain are very complex…It would be nice if we could feed this whole mess into a giant computer and program the super intellect to generate an 'objectively correct' response. It just can't be done!…We believe that complex social problems–and, for that matter, complex business problems–demand the consideration of subjective values and tradeoffs. (p. 12)

Characterization of a problem is a necessary first step in undertaking the decision process depicted in Fig. 13.1. Keeney and Raiffa (1976) discuss a taxonomy of problems composed of the following four types, which we shall call cases:

1. Single attribute, no uncertainty.
2. Single attribute, uncertainty present.
3. Multiple attributes, no uncertainty.
4. Multiple attributes, uncertainty present.

Case 1 is trivial, at least conceptually. It simply requires that we contrast all feasible alternatives and choose the one with the best possible objective function value. However, few, if any, realistic problems reside within this decision space. Generalizing to cases of uncertainty (Case 2), multiple attributes (Case 3), or both (Case 4) requires additional thought and structure. However, multiple attributes and uncertainty are a fact due to the principles underlying a mess (recall our discussion from the Preface and Chap. 1), and they characterize all but the most trivial of problems. Thus, it is only of interest for us to consider Case 4. Table 13.2 lists those

Table 13.2 Systems principles demanding consideration of uncertainty

Principle	Rationale
Emergence	Accounting for emergence means accepting that there will be uncertain, unpredictable phenomena occurring within our mess
Darkness	We can never truly know a system completely. As such, there will always be uncertainty surrounding those elements of our mess that are unknown
Equifinality	Since there are multiple paths that may lead to a singular end point, it may be difficult for us to predict what trajectory a problem will take in its resolution. Further, it is uncertain as to whether or not it matters. The phrase "the end justifies the means" is appropriate here
Multifinality	Once our system has departed from its initial conditions, and given the uncertainty present in our system, it may follow many trajectories to radically disparate paths. These end states are uncertain
Self-organization	Order arises out of seemingly independent and unrelated components of a system. Especially when we are assessing a sociotechnical system, and humans are part of the equation, this self-organization is rather uncertain and difficult to predict

Table 13.3 Systems principles requiring multiple attributes

Principle	Rationale
Complementarity	Complementarity ensures that no singular, unified, wholly correct perspective of a system exists. Thus, consideration of multiple attributes is necessary to capture these divergent perspectives
Requisite saliency	Multiple attributes are necessary to fully capture the complexity of a mess. While each contributes to our understanding, the principle of requisite saliency informs us that each of these attributes is likely to have its own relative importance (weight) which contributes to the overall system and its goals
Suboptimization	Suboptimization requires us to understand that competing objectives exist within our system. If we optimize any singular objective, we in turn suboptimize the entire system. Thus, consideration of all relevant attributes ensures we don't suboptimize our system
Hierarchy	On one level, i.e., at the system level, a particular set of attributes may be rolled up or decomposed as appropriate. This structure requires multiple attributes to capture the nuances of these hierarchical relationships

principles demanding consideration of uncertainty, which drive those concerned with systemic thinking toward problems categorized by Case 4.

Similarly, Table 13.3 lists those principles requiring multiple attributes which drive those concerned with systemic thinking toward problems categorized by Case 4.

Systemic consideration of a mess requires us to think using a paradigm that supports both multiple objectives and uncertainty. While none of this information is novel to the reader at this point in the text, it sets the tone for the discussion of the scientific underpinnings of decision science to follow.

13.4 Decision Science

The ontological and epistemological constructs for decision science may be shown as a continuum between *idealism*, or the subjective school, and *realism*, or the objective school. In the decision analysis discipline, there are four subdisciplines which span this continuum and are easily positioned along it. The formal, empiricist school of decision analysis, labeled Classical Decision Making (CDM), is positioned at the realism end of the continuum with the objectivist school. Recognizing that decision making is inherently flawed due to human error, the Judgment and Decision Making (JDM) school of decision analysis shifts to the left on the continuum and falls within the rationalist school of thought. The Naturalistic (NDM) and Organizational Decision Making (ODM) schools of decision making belong to the naturalistic paradigm (Lipshitz, Klein, & Carroll, 2006) and are placed farther left on the continuum. The placement, with respect to their ontological assumptions and epistemological stances, of the four subdisciplines of decision analysis is depicted in Fig. 13.2.

Each of the four recognized subdisciplines within decision analysis, their focus, and associated scholarly journals is:

1. *Classical Decision Making (CDM)*: Focuses on the development and study of operational decision making methods. The primary scholarly publication is the journal *Decision Analysis* [ISSN 1545-8490].
2. *Naturalistic Decision Making (NDM)*: Emphasizes psychological approaches and methods in decision processes. While somewhat ethnographic in nature, it is based on careful descriptions of how experts actually make choices in complex, real-world situations. The primary scholarly publication is the *Journal of Behavioral Decision Making* [ISSN 0894-3257].
3. *Organizational Decision Making (ODM):* Focuses on decision making as an element of organizational behavior, specifically decision making behaviors in individuals when acting as a member of an organization. The primary scholarly publications are the journals *Organization Science* [ISSN 1047-7039] and *Management Science* [ISSN 0025-1909].

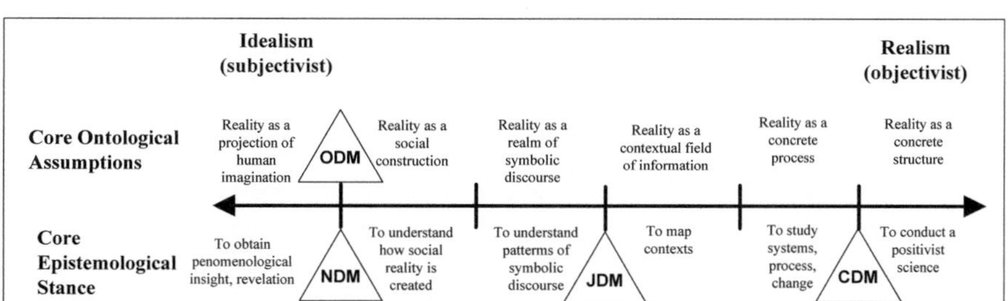

Fig. 13.2 Ontological assumptions and epistemological stances of decision analysis subdisciplines

4. *Judgment and Decision Making (JDM)*: Emphasizes normative, descriptive, and prescriptive theories of human judgments and decisions. The primary scholarly publication is the *Judgment and Decision Making* [ISSN 1930-2975].

Given the presence of multiple objectives and uncertainty, and the pluralistic nature of messes, it behooves us to elicit multiple perspectives in order to invoke the *principle of complementarity* (Bohr, 1928). Keeney and Raiffa (1976) agree, elaborating:

> In many situations, it is not an individual but, instead, a group of individuals who collectively have the responsibility for making a choice among alternatives. Such a characterization is referred to as a group decision problem. With each group decision, there is the crucial metadecision of selecting a process-oriented strategy by which the group decision is to be made. (p. 26)

As we incorporate multiple perspectives as a necessary requirement for understanding messes, objectivity becomes increasingly complex, if not impossible. Thus, mess-level decision making tends to reside on the left side of the continuum depicted in Fig. 13.2.

13.5 The Decision Process

Now that we've realized that we're going to make a decision and that our decision will be subjective, at least to some degree, it is time to evaluate our options using a structured decision process. But how do we compare the numerous potential paths at our disposal? Gilboa (2011) offers that all potential courses of action (i.e., decisions) must be *feasible* and *desirable*.

Feasibility is determined by whether or not we could execute the action. "A choice is *feasible* if it is possible for the decision maker, that is, one of the things that she *can* do" (Gilboa, 2011, p. 191). Desirability is determined, ultimately, by our fundamental objectives. The most desirable decision is the one that has the highest value in terms of our fundamental objective achievement. "An outcome is *desirable* if the decision maker wishes to bring it about" (Gilboa, 2011, p. 191). Gilboa (2011) elaborates on the two concepts:

> Typically, feasibility is considered a dichotomous [i.e., binary] concept, while desirability is continuous: a choice is either feasible or not, with no shades in between; by contrast, an outcome is desirable to a certain degree, and different outcomes can be ranked according to their desirability. (p. 191)

It is important to note that actions do not guarantee our desired outcomes, as all potential future circumstances are speculative and subject to uncertainty. After all, if this were not the case, decision making would be trivial. We would simply choose the course of action with the highest guaranteed outcome. Ultimately, our decision's quality is not determined by the decision outcome. A simple example comes from the world of gambling. It would be silly to accept an even money bet (winning

Table 13.4 Intersection of popularity and activity (From Chap. 6)

		Popularity	
		Low	High
Activity	High	Important and easy to influence	Important but hard to influence
	Low	Not important but easy to influence	Not important and hard to influence

$1 for every $1 you bet) that two unloaded die would roll *snake eyes*, or two 1's, even if it did. Why? The odds of such a roll are 1/6 * 1/6 = 1/36. So, a fair bet would be odds of 36:1. Even the shadiest of bookies should offer you odds better than even money (1:1). We must remember that hindsight is always 20/20. When we make decisions, we must focus on the future. We are looking for a *defensible mechanism* by which to make a decision. In doing so, potential outcomes, uncertainties, and the inherent value to the decision maker (and other relevant stakeholders) at the time of the decision must be taken into account. These concerns must be balanced with the mechanisms necessary to undertake our actions in an effort to conserve scarce resources. This notion invokes the idea of *rationality* as we aim to make a rational decision, or a decision that we would repeat if given the same information at a later date.

We can apply the concepts of popularity and activity from Chap. 6 in order to help us choose between potential courses of action, as shown in Table 13.4. We wish to find those concepts that are high activity/low popularity in order to achieve the highest return on our actions (i.e., maximize our scarce resources).

In order to determine a priority for intervention in our mess, we first calculate activity and popularity for all concepts using Eqs. 6.6 and 6.7, respectively. Then, we sort all concepts first by activity (in descending order), and then, if necessary, by popularity (in ascending order). This sorts each of our concepts by largest influence and easiest to influence. This order reflects the order of consideration for action.

Now we can ask, is the concept feasible to influence? Recall that we are taking on the role of the decision maker. We have resources at our disposal, and those resources can be used to accomplish a set of objectives. So, we should ask ourselves, in a binary fashion, whether or not influence on a particular concept is achievable. A good place to start in identifying feasible concepts is our means objectives, context and barrier elements, and mechanisms. For those concepts that we answer *yes* to, we can then evaluate these for their desirability. Using our real estate mess, pictured in Fig. 13.3, we can calculate the highest priority concepts for engagement. These concepts can be evaluated based on the feasibility of influencing them.

The next element, then, is to evaluate the desirability of our actions. In order to do this, we need to discuss the notion of performance, a concept to which we now turn.

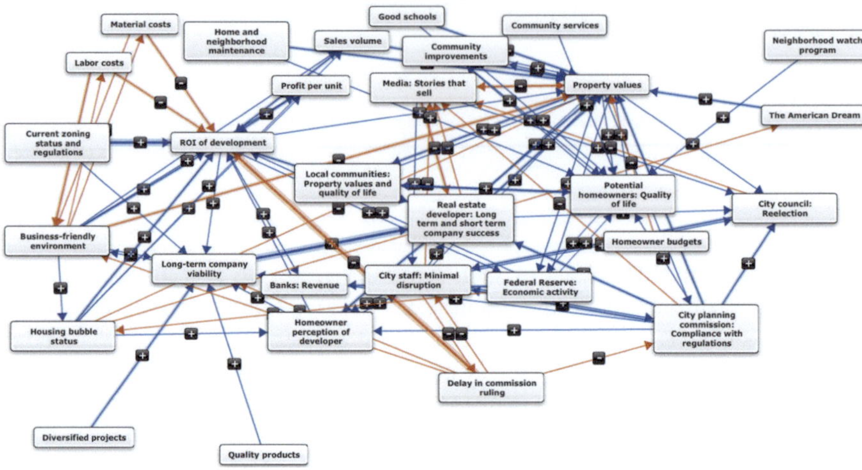

Fig. 13.3 Real estate mess

13.5.1 Measuring Performance

In order to evaluate the desirability of our actions, we need a quantitative measure for the effect of our actions. We provide a set of definitions that will permit understanding of the relationship between learning, performance, and productivity that build upon the foundation level terms of actuality, capability, and potentiality. These definitions come from cybernetician and management scientist, Stafford Beer [1926–2002]. Beer (1981) succinctly describes his views on performance as being "about *both* short- and long-term viability" (p. 163). He continues, "The notion that cost should be minimized or profit maximized within a fixed epoch leaves right out of the count other factors which are vital to the future viability of the business contained within the division" (Beer, 1981, p. 163). Beer defines performance mathematically as:

> The ratio of Actuality to Potentiality is called Performance (Beer, 1979, p. 293).

Beer was a holistic thinker who viewed the world as an interrelated collection of systems which are arranged hierarchically, with each operating in a recursive fashion, evoking both the *principles of hierarchy* (Checkland, 1999) and *recursion*. Beer felt that it was important to provide systemic definitions that provided linkage to the wider systems structure within which all organizations and processes belong. Consequently, his definitions for both *productivity* and *performance* emphasized two interrelated operational aspects: (1) learning and (2) improvement. Beer (1981) defined three foundational terms:

- **Actuality**: What we are managing to do now, with existing resources, under existing constraints.

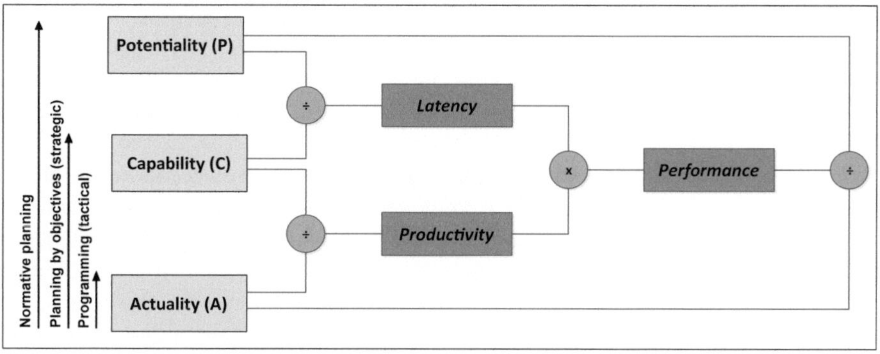

Fig. 13.4 Mathematical constructs for latency, performance, and productivity (adapted from Beer, 1981, p. 164)

- **Capability**: This is what we could be doing (still right now) with existing resources, under existing constraints, if we really worked at it.
- **Potentiality**: This is what we ought to be doing by developing our resources and removing constraints, although still operating within the bounds of what is already known to be feasible. (p. 163)

From these terms, Beer was able to express, mathematically, constructs for (1) latency, (2) performance, and (3) productivity. All three constructs are functions of the foundational terms and have mathematical operations as depicted in Fig. 13.4.

The three mathematically derived constructs in Fig. 13.4 are defined as follows:

- **Latency**: The ratio of capability and potentiality.
- **Productivity**: The ratio of actuality and capability.
- **Performance**: The ratio of actuality and potentiality, and also the product of latency and productivity. Defined as: *The ratio of (1) what we are managing to do now, with existing resources, under existing constraints [actuality] to (2) what we ought to be doing by developing our resources and removing constraints, although still operating within the bounds of what is already known to be feasible [potentiality].* (Adams & Bradley, 2012, p. 18)

Beer (1981) expands on the types of plans which are associated with each of three measures of performance:

> Planning on the basis of actuality I call programming. Planning on the basis of capability I call planning by objectives. Planning on the basis of potentiality I call normative planning. The first of these three is simply a programme because it accepts the inevitable short-comings of the situation, and does not admit that anything can be imminently done about them. Programming is a tactical ruse. We move to genuine planning only when we set new objectives and try to achieve them. This is the strategic planning level. Normative planning sets potentiality as its target – and incurs major risks and penalties, although it also offers major and perhaps decisive benefits. (p. 163)

We can relate these differing types of plans to our meta-perspectives from the previous chapter. Actuality (and tactical planning), as well as capability (and strategic planning), is achieved through our what is meta-perspective. Potentiality (and normative planning) requires our what ought-to-be meta-perspective to achieve.

To clarify these concepts, an example is in order. Consider a scenario in which it takes 50 full-time equivalent (FTE) employees to achieve a level of success in accomplishing a task. Thus, our actuality is 50. With some training, this number can be reduced to 20 FTEs. So, our capability is 20. Thus, productivity = capability/actuality = 20/50 = 0.4. If we invest in new equipment (and thereby change our underlying system), the number of FTEs can be reduced even further to 5. Thus, latency = potentiality/capability = 5/20 = 0.25. Further, we can calculate performance = latency * productivity = potentiality/actuality = 0.25 * 0.4 = 5/50 = 0.10.

The natural next question concerns the improvement of our performance in this situation. If we consider our current business practices (actuality) and long-term company vision (potentiality) unchanged, and a scenario in which division management wishes to improve capability by optimizing a product line, we can evaluate the resultant effect on our performance. So, actuality = 50 and potentiality = 5. Let us say that suggested improvements can reduce our capability FTE to 15. We can recalculate our values as:

Latency = potentiality/capability = 5/15 = 0.33
Productivity = capability/actuality = 15/50 = 0.30
Performance = latency * productivity = potentiality/actuality = 0.33 * 0.30 = 5/50 = 0.10

A curious phenomenon has occurred. Our latency has improved, while our productivity has worsened. Overall, our performance has remained the same. A preferred approach would be to improve all three indices if possible. So how, then, do we determine the desirability of influence on our concepts using Beer's three measures? For each of our fundamental objectives, we should seek to maximize their performance using feasible interventions in our system using the following guidelines:

1. We can first explore the *actuality* of a scenario in an attempt to assess where we currently operate.
2. If it is maximized, then we can try to minimize resources in an effort to improve *capability*. This involves exploration of the what is meta-perspective.
3. If it is not maximized, then we can reason, using our mess-level understanding (e.g., boundaries and mechanisms), about its *potentiality*. This involves exploration of the what ought-to-be meta-perspective.

Thus, using our prioritized list of concepts and our performance indices (actuality, capability, and potentiality), we can decide how to intervene in our mess.

13.6 Framework for *Action* in Messes and Problems

Summarizing the discussion thus far, taking action requires that we complete the following steps for our mess:

1. Sort concepts by priority of engagement, using popularity and activity.
2. Assess the feasibility of concepts.
3. Explore scenarios using feasible concepts to assess desirability (i.e., actuality, capability, and (if necessary) potentiality) of fundamental objectives.

This framework is demonstrated in the next section on our real estate example problem.

13.7 Example Action Analysis

Our action analysis investigates the real estate mess that we have discussed throughout the text and that is pictured in Fig. 13.3. First, activity and popularity are calculated for this mess for all concepts using Eqs. 6.6 and 6.7, respectively. The top ten concepts (in terms of order of engagement), as well as an assessment of their feasibility, are listed in Table 13.5.

Assessing the top ten concepts (in terms of their order of engagement), we see there are only three feasible concepts for influence. *Compliance with regulations* is directly able to be influenced; the real estate developer can continue to follow the law and not build on its parcel of land until it receives a zoning variance. Next, *minimal disruption* is feasible; both the local communities and real estate developer can attempt to cooperate without disruptive protests or other commotion. Finally, *stories that sell* can be influenced by both parties. If the real estate developer follows zoning laws and the local communities do not cause a disruption, then the likelihood of generating stories that sell is minimal.

Table 13.5 Assessment of feasibility for real estate mess

Concept	Activity	Popularity	Order of engagement	Feasible?
Property values	4.97	9.59	1	N
Compliance with regulations	4.69	1.50	2	Y
Business-friendly environment	4.47	2.24	3	N
Minimal disruption	3.50	2.96	4	Y
Housing bubble status	3.24	1.00	5	N
Delay in commission ruling	3.16	0.50	6	N
Economic activity	2.74	1.50	7	N
ROI of development	2.74	7.42	8	N
Stories that sell	2.45	2.50	9	Y
Quality of life	2.24	4.50	10	N

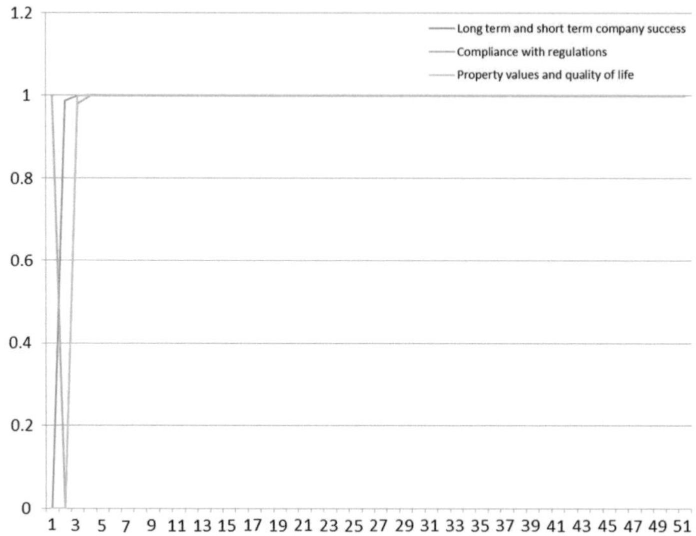

Fig. 13.5 Comply with regulations set to +1 initially

We can now explore scenarios to achieve our objectives. *Compliance with regulations* is the highest ranked feasible element for intervention. Under current circumstances, the real estate company is complying with regulations; thus, it can be said to have a value of +1. Results of this scenario are shown in Fig. 13.5.

This scenario results in both *long-term and short-term company success* and *property values and quality of life* being +1 in a steady state. From the real estate company's viewpoint, they can choose to take two actions—continue to comply or fail to comply with regulations. Continued compliance maximizes our performance at unity (potentiality/actuality = 1/1 = 1) and is low effort.

Alternatively, we can explore what happens in the real estate company fails to comply. This result is shown in Fig. 13.6.

Failure to comply with regulations reduces both *long-term and short-term company success* and *property values and quality of life* to −1, yielding a performance value for each of 0 (potentiality/actuality = 0/1 = 0) or the worst possible performance. If we suspect the real estate company is insistent on failing to comply, perhaps due to frustration with persistent delays on behalf of the commission's ruling, we can explore alternative scenarios for maximizing performance. For example, a scenario of *minimal disruption* set to 1 simultaneously with *compliance with regulations* and *stories that sell* set to −1 also yields maximum performance (see Fig. 13.7). However, this scenario does not seem *feasible* as failure to comply with regulations will no doubt generate stories that sell.

A summary of the three scenarios that were investigated is shown in Table 13.6.

The best course of action for both fundamental objectives is for the real estate company to comply with regulations. This represents the status quo, or *no action*. It is important to keep *no action* as a viable option, especially if it is the choice that

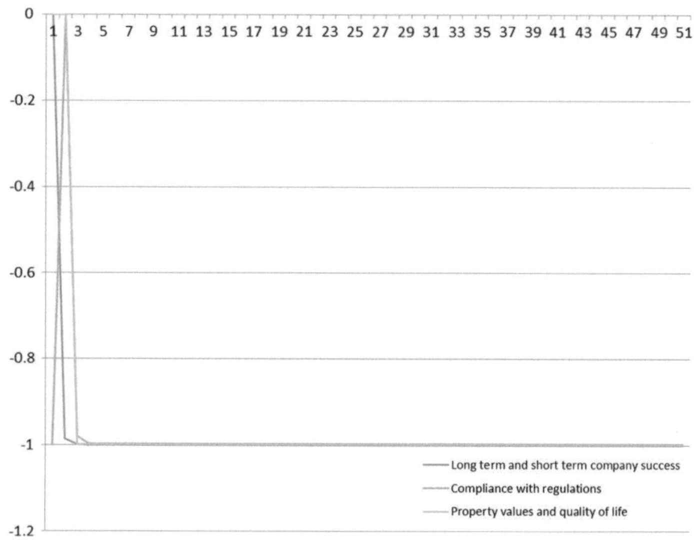

Fig. 13.6 Comply with regulations set to −1 initially

Fig. 13.7 Minimal disruption set to 1 and compliance with regulations and stories that sell set to −1

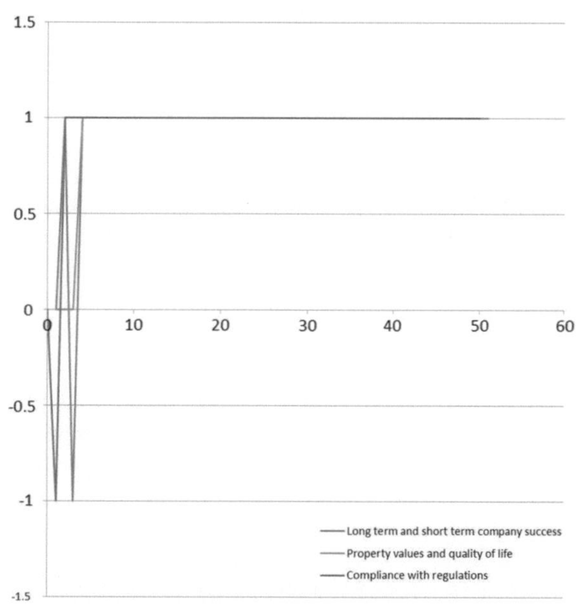

maximizes our performance. Acting when unwarranted also poses the opportunity for us to commit a Type IV error. In this case, it is important to note that we have deliberately chosen *no action* rather than simply refusing to act (and thereby committing a Type V error).

Table 13.6 Summary of mess scenario exploration

Potential course of action	Feasible?	Performance (desirability)
Comply with regulations = −1	Yes	Worst performance on both fundamental concepts
Comply with regulations = −1, Minimal disruption = 1, Stories that sell = −1	No	Best performance on both fundamental concepts
Comply with regulations = 1	Yes	Best performance on both fundamental concepts

13.8 Additional Concerns

This chapter thus far has developed an approach for evaluating and choosing an action based on the concepts of feasibility and desirability (measured by actuality, capability, and potentiality). This section addresses a number of concerns that may warrant additional thought during the decision process, namely decision robustness and decision optimality.

13.8.1 Decision Robustness

Once we have made a decision, a natural follow-up question may be how sure we are of our choice. An underlying assumption of FCM analysis is determinism, which is not true in a mess. In order to evaluate our decision's robustness, we can explore variability in our weights using Monte Carlo simulation (MCS). Monte Carlo simulation is a valuable tool that can allow individuals with a basic understanding of simulation and probability to investigate sophisticated problems and the uncertainty associated with them. Typically, it is employed as a computerized mathematical technique which allows individuals to account for uncertainty in decision making. It is useful when data are not available, difficult to obtain, expensive, or impractical.

In Monte Carlo simulation, we simulate a series of unknown random variables in our system many times, generating random outcomes for a scenario. We are simulating virtual trials of the phenomenon under study. For example, if we are choosing between two generators for purchase, each with a specified mean horsepower and standard deviation, we can use MCS to simulate their characteristics in order to compare their relative performance. Our resultant analysis can tell us, for example, which one is more likely to meet a minimum horsepower threshold or which has the least variability under load.

But how do we do this? The first step in Monte Carlo analysis is to generate a series of random values for each variable. Then, we convert these raw values to the appropriate distribution under study. Examples of variable transformations include the following:

- *Uniform*: $y = a + x(b - a)$, where x is the uniform random variable, b is the upper limit of the uniform distribution, a is the lower limit of the uniform distribution, and y is the uniformly distributed variable.
- *Exponential*: $y = -\ln(1 - x)/\lambda$, where x is the uniform random variable, and λ is the rate parameter.

Alternatively, we can use a software package such as Microsoft Excel, which has built-in functions for assisting with Monte Carlo simulation. Excel distributions include the following:

- BETAINV(), which returns the inverse of the beta distribution.
- LOGINV(), which returns the inverse of the log distribution.
- NORMINV(), which returns the inverse of the normal distribution.

Applying these ideas to our real estate mess, we can use a random distribution and add it to our original weight matrix, W, to experiment with our decision's robustness. A snapshot of our baseline adjacency weight matrix (W) for this problem is shown in Fig. 13.8.

Using this baseline adjacency matrix (with no variability), we can simulate our scenario 25 times and look at the performance of our two fundamental objectives. The result of this analysis is shown in Fig. 13.9. Given the deterministic nature of FCMs, this analysis produces a predictable result, namely average values of unity (as before) and zero standard deviation for the two fundamental objectives.

We can also explore random behavior in our weight matrix to investigate the effects of uncertainty on our decision. An example Monte Carlo scenario is shown in Fig. 13.10. This scenario represents uncertainty incorporated into the weight matrix.

While uncertainty quantification is not the focus of this text, we can explore a number of illustrative scenarios to investigate the effects of variability on our recommended courses of action. First, we can explore the effect of 10% variance in weights. That is, we are applying a uniform random perturbation to our initial weight matrix defined on $[-0.1,0.1]$. The results of this analysis are shown in Fig. 13.11. In this case, the average value of our 25 runs of *property values and quality of life* is 0.92, while the average value of *long-term and short-term company*

	Property v	Long term	Complian	Reelectior	Stories th	Homeowr	Quality of	Property v	Economic	Communi	Minimal d	Near-term	Delay in c	ROI of dev	Revenue	Long-term	Business-f	Sales volu	Profit per	Material c L
Property values	0	0.25	0.25	0.25	-0.25	0.5	0.25	0.5	0.25	0.25	0	0	0	0	0	0	0	0	0	0
Long term and sh	0.5	0	0	0.25	-0.25	0	0	0	0	0	0	0	0	0	0	0	0	0	0	0
Compliance with r	0.25	0.5	0	0.5	-0.25	0.25	0.25	0	0	0	0.5	0.25	0	0	0	0	0	0	0	0
Reelection	0	0	0	0	-0.25	0	0	0	0	0	0.5	0	0	0	0	0	0	0	0	0
Stories that sell	-0.5	-0.5	0	0.25	0	0	0	0	0	0	0.25	0	0	0	0	0	0	0	0	0
Homeowner perc	0	0	0	0	0	0	0	0	0	0	0.25	0.25	0	0.25	0	0	0	0	0	0
Quality of life	0.25	0	0	0	0	0	0	0.5	0.25	0.25	0	0	0	0	0	0	0	0	0	0
Property values ar	0.25	0	0	0	0	0	0.25	0	0	0	0	0	0	0	0	0	0	0	0	0
Economic activity	0.25	0	0	0	0	0	0.25	0	0	0	0	0	0	0.25	0.5	0	0	0	0	0
Community imprc	0.25	0	0	0	0	0	0.25	0	0	0	0	0	0	0	0	0	0	0	0	0
Minimal disruptio	0.25	0.25	0.25	0.25	-0.25	0.25	0	0	0	0	0	0	-0.25	0	0	0	0	0	0	0
Near-term Financi	0	0	0	0	0	0	0	0	0	0	0	0	0	0	0	1	0	0	0	0
Delay in commissi	0	0	-0.25	0	0	0	0	0	0	0	-0.25	-0.25	0	-1	0	0	-0.25	0	0	0
ROI of developme	0.25	0.5	0	0	0	0	0	0	0.25	0	0	0.25	0	0	0.25	0	0	0	0	0
Revenue	0	0	0	0	0	0	0	0	0	0	0	0	0	0	0	0	0	0	0	0
Long-term compa	0	1	0	0	0	0	0	0	0	0	0	0	0	0	0	0	0.25	0	0	0
Business-friendly	-0.5	0	0	0	0	0	0	0	0	0	0	0.25	0	0.5	0	0	0	0.25	0.25	-0.25
Sales volume	0.5	0	0	0	0	0	0	0	0	0	0	0	0	0.5	0	0	0	0	0	0
Profit per unit	0	0	0	0	0	0	0	0	0	0	0	0	0	0.5	0	0	0	0	0	0
Material costs	0	0	0	0	0	0	0	0	0	0	0	0	0	-0.5	0	0	-0.25	0	0	0

Fig. 13.8 Baseline adjacency matrix

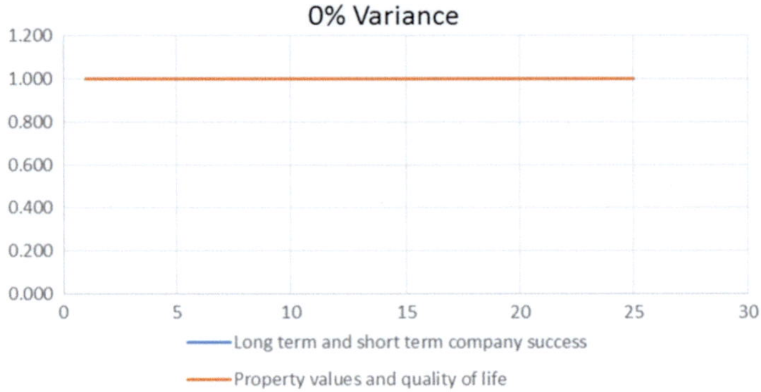

Fig. 13.9 Baseline variance assessment

	Property v	Long term	Complian	Reelection	Stories tha	Homeowr	Quality of	Property v	Economic	Communi	Minimal d	Near-term	Delay in c	ROI of dev	Revenue	Long-term	Business-f	Sales volu	Profit per	Material c
Property values	0.011616	0.263645	0.337252	0.25	-0.20918	0.469943	0.299553	0.468299	0.25	0.28965	-0.03435	0.018083	0.057679	-0.07817	-0.08028	0	-0.08453	0.071538	0.005952	0
Long term and sh	0.5	0.004171	-0.05305	0.191131	-0.25125	0.070291	-0.00401	0.034184	0.001992	0.056263	0.031748	0	0	0.034553	-0.03878	0.018123	-0.0992	0.041297	-0.02723	0
Compliance with r	0.25	0.5	0	0.5	-0.25	0.348707	0.25	-0.05904	-0.02756	0	0.498837	0.270015	-0.08368	-0.05702	-0.06168	0	0	0	0	0.035392
Reelection	-0.05162	-0.00269	0.040405	-0.01417	-0.20265	0	0.045476	0.023478	0	-0.04688	0.468356	0	0.026489	-0.02115	0	0.068496	-0.016	-0.07481	0	0.027027
Stories that sell	-0.49714	-0.43218	0	0.309996	0	0.033632	0.097637	-0.03492	-0.06223	0.085498	0.244131	0.063445	0.08108	0	-0.08338	0	0	-0.04649	-0.01318	0
Homeowner perc	-0.00041	0	0	-0.03072	0	0	0	0.009318	0	-0.06344	0.190124	0.25	0	0.25	0	0.066235	0	-0.05468	0.038741	0
Quality of life	0.248887	0	-0.07916	-0.07749	0.069736	0.092411	-0.0121	0.540818	0.214659	0.246401	-0.08368	0.047836	0	-0.00609	-0.09241	-0.07326	-0.09114	0	-0.02042	0.084442
Property values a	0.25	0	0.021309	-0.06649	0	-0.05354	0.154099	0	0.044628	0	0	0.073508	0.018318	0.010819	-0.08085	0	0	-0.07894	0	0.081802
Economic activity	0.346523	0	0.023647	0	0	0.079611	0.25	0	0.095466	-0.05866	-0.04685	0.014101	-0.05661	0.217506	0.5	-0.026	0.020381	0	0	0.047877
Community impr	0.340153	0	0.018087	0.098629	-0.05518	0	0.25	0	-0.07164	-0.02435	0.016424	0	-0.07615	0.073137	-0.00514	0	-0.02117	0.027943	0.070584	-0.07988
Minimal disruptio	0.154276	0.299469	0.25	0.157061	-0.17919	0.25	0	0	0.009737	-0.0546	0	0.037409	-0.25	0.021804	0	0.008068	-0.08001	0	0	0.061889
Near-term Financi	-0.04791	0.051794	0.027914	0	0.052545	0.049112	-0.01188	0	0	-0.08621	-0.01457	-0.02623	0.076885	0.019723	-0.00594	1	0	-0.05105	0.07771	-0.02476
Delay in commissi	0.049608	0	-0.25	0.007974	-0.00121	0	-0.08232	0.088628	0.092229	0	-0.25	-0.29798	0.041766	-1	0	0.006332	-0.31625	0.018088	-0.00644	0
ROI of developme	0.183139	0.5	0.086719	0.018312	-0.00545	-0.08018	0.082876	0	0.282932	-0.04604	0.088877	0.153147	0	0	0.241823	0	-0.03151	0.068979	-0.02261	0
Revenue	-0.08994	-0.04698	-0.08235	0	0.083783	0	0.093247	-0.0118	0	-0.07012	0.015207	0	-0.05504	-0.03879	-0.05626	0.086613	0	0.006875	0.088519	0
Long-term compa	0	1	0	0	0.095943	0.019561	0	-0.07818	0	0	0	0.0075	0.062269	-0.09312	0.020234	-0.09928	0.21492	0.059044	0	0.042229
Business-friendly	-0.5	0.043172	-0.03032	0.036241	-0.09433	0.069276	-0.03174	-0.00207	-0.0325	-0.04783	0	0.311341	-0.08923	0.485324	-0.08875	0.030093	0.038775	0.215916	0.333095	-0.19379
Sales volume	0.43111	-0.03176	-0.01449	0.078084	0.093859	0.015297	0	0.016846	-0.09529	-0.0094	-0.08478	0.032034	0	0.434397	-0.03091	-0.02981	0.026231	-0.03179	-0.03727	0
Profit per unit	-0.08502	0	-0.08581	0.011053	-0.05858	0	0.03612	0.092188	0.067489	0.05832	-0.09027	0.017902	0.044773	0.520422	0	-0.09277	-0.06104	0.023129	0.000414	0.071453
Material costs	0	0	0.003558	0.092275	0.027347	0	0.056695	0.048972	0	0	0	-0.01417	0	-0.50411	0	0.077746	-0.32525	-0.08892	-0.00104	-0.00833

Fig. 13.10 Example Monte Carlo simulation matrix

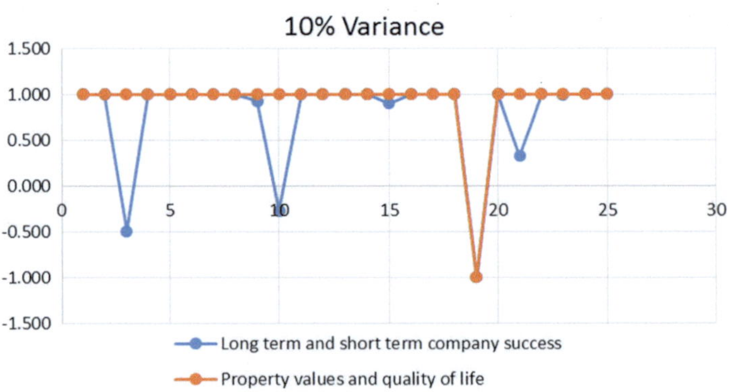

Fig. 13.11 10% weight variance assessment

success is 0.775. The standard deviation of *property values and quality of life* is 0.4, while the standard deviation of *long-term and short-term company success* is 0.543. Thus, we may say that there is a fair amount of variation in the *long-term and short-term company success*, and a moderate amount of variation in *property values*

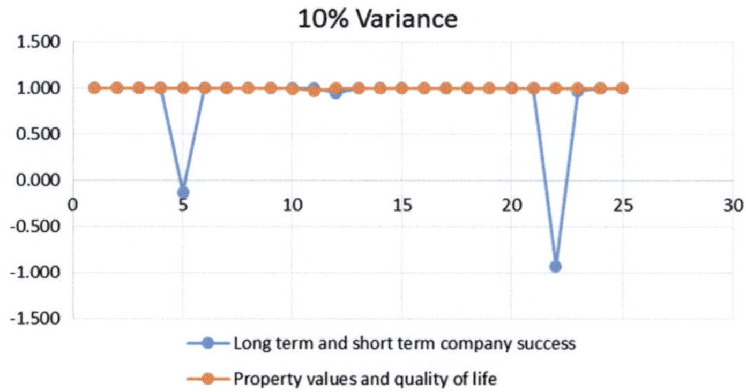

Fig. 13.12 10% weight variance assessment, while activating minimal disruption

and quality of life. It should be noted that this analysis is illustrative, as 25 Monte Carlo trials are insufficient for substantive conclusions to be drawn from.

Next, we can explore the effect of 10% variance in weights, while also activating minimal disruption. That is, we are applying a uniform random perturbation to our initial weight matrix defined on [−0.1,0.1]. The results of this analysis are shown in Fig. 13.12. In this case, the average of *property values and quality of life* is 0.998, while the average value of *long-term and short-term company success* is 0.874. The standard deviation of *property values and quality of life* is 0.006, while the standard deviation of *long-term and short-term company success* is 0.438. Under this scenario, we may conclude that there remains a fair amount of variation in the *long-term and short-term company success*, but now there is a minimal amount of variation in *property values and quality of life*. It should be noted again that this analysis is illustrative, as we again ran 25 Monte Carlo trials.

Further assessment of these two scenarios allows us to draw additional conclusions. It appears as though the *property values and quality of life* objective are quite robust, while the development company may wish to explore *minimal disruption* to more assuredly achieve its objective. However, even in the second case, there is still significant variability, leading to the conclusion that the real estate development company may wish to seek additional actions in order to more assuredly achieve its objective. We can also explore variability in our initial starting vector, although minor variations in starting concept activation levels are typically more robust to variability than the weights in FCMs.

13.8.2 Decision Optimality

An additional consideration that we should think about when deciding on interventions is the notion of optimality and its implications on our actions and expected results. Specifically, two principles must be considered when evaluating, and trying to improve, the performance of a system with respect to its objective(s): (1) the *Pareto principle* and (2) the *satisficing principle*.

In any mess, we will have parts that are not of equal importance (Boulding, 1966) and this importance is a dynamic characteristic of our mess. The *Pareto principle* (Pareto, 1897) states, simply, that 80% of the outcomes of a system will be achieved through 20% of the means. The corollary of this principle is that 20% of the results of a system absorb 80% of the resources. This becomes an issue of diminishing returns. We must seek to concentrate on the most productive 20% of our effort and avoid the limited marginal gains that accompany trying to wholly maximize our production (i.e., to reach 100%). A disproportionately large amount of resources is necessary for this minimal gain. While the 80/20 distribution is a heuristic, it represents a general rule of thumb that highlights the highly nonlinear generalized power law relationship of cause and effect or resources and outcomes to be expected when dealing with complex systems. It also informs us that while we may be able to achieve greater performance using the notion of potentiality, it may not be worth the additional effort to do so. This principle is exacerbated when considered in concert with the notion of optimality.

One could calculate a set of potential decisions whose performance lies along what is known as an isopreference curve, along which all configurations have equal subjective value (i.e., we are indifferent between any solutions on this curve). This set consists of a set of possible answers for the multiobjective mess that we are evaluating. Theoretically, this set can yield an infinite number of solutions that decision makers can choose from. All points on the curve shown in Fig. 13.13, for example, lie on an isopreference curve. This notion is problematic if we are searching for *a* singular optimal solution to our problem, or if we are trying to optimize at all, as multiple, equally valued solutions may exist.

Given that the behavior of a mess involves emergent behavior and structure, the search for an optimal solution to a mess is problematic. "The structure as well as the parameters of problem situations continually change. Because optimal solutions are very seldom made adaptive to such changes, their optimality is generally of short duration" (Ackoff, 1977, p. 1). While a singular evaluation may represent an approximately optimal solution, its utility will be fleeting. Ackoff (1977) agrees,

Fig. 13.13 Illustration of an isopreference curve

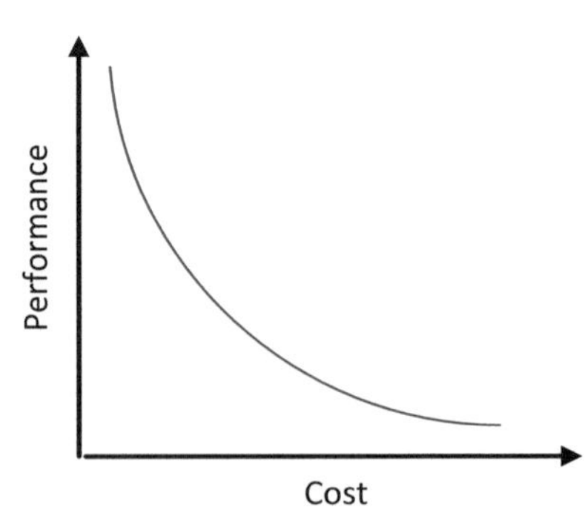

noting, "The effectiveness of a solution that is claimed to be optimal at the time of its implementation tends to deteriorate over time" (p. 2).

The inability for us to determine an optimal solution may be troublesome to some, while others may find it a stop-work of sorts, yet others may find it liberating. We side with the final camp. The *principle of satisficing* allows for a practical solution to our stakeholders' problem in the face of a desire for optimality. Satisficing is a term coined by Nobel Laureate Simon (1955, 1956) to describe how individuals make rational decisions between available alternatives in a constrained environment. Simon argued that individuals rarely, if ever, obtain all necessary information to analyze a decision scenario. Thus, they work with a limited scope of information in an effort to reach an acceptable compromise (they satisfice, i.e., satisfy and suffice) rather than attempt to obtain a globally optimal solution to a problem. Ackoff (1977) agrees with the absurdity of attempting to optimize a mess, noting:

> It is silly to look for an optimal solution to a mess. It is just as silly to look for an optimal plan. Rather we should be trying to design and create a process that will enable the system involved to make as rapid progress as possible towards its ideals, and to do so in a way which brings immediate satisfaction and which inspires the system to continuous pursuit of its ideals. (p. 5)

Satisficing uses bounded rationality to select an alternative. Brown and Sim (2009) elaborate on Simon's bounded rationality:

> One of the key principles from Simon's (1955) bounded rationality model is that, rather than formulating and solving complicated optimization problems, real-world agents often can choose the first available actions, which ensure that certain aspiration levels will be achieved. In other words, given the computational difficulties in the rational model paradigm, a more sensible (and descriptively accurate) approach may in fact be to view profit not as an objective to be maximized, but rather as a constraint relative to some given aspiration level. (p. 71)

Hester (2012) elaborates on the appropriateness of satisficing as a mechanism for analyzing complex problems:

> Satisficing, as a mechanism for evaluating a [mess] is not to be feared, to be regarded as a "less than optimal" solution to a problem, but rather it should be viewed as an approach to be embraced. Bounded rationality can be useful when a decision must be made and the decision maker does not have an eternity to exhaustively compare all alternatives and their resultant consequences …In this case, individuals gather information for a finite period of time and make a decision based on this subset of information (rather than exhaustively collecting all information regarding the decision). Additionally, bounded rationality allows us to incorporate decision costs into our approach to decision making. Sometimes, gathering information is detrimental. It is often said that "time is money). (pp. 273–274)

When we combine this perspective with our use of performance parameters (actuality, capability, and potentiality) in evaluating system behavior, we can see that this is a satisficing approach to our mess. Thus, we can adjust the parameters of our decisions in an effort to achieve a satisfactory system rather than naively search for an optimal one.

13.9 Summary

At this point, we have thought our mess to death. Now is the time for action! This chapter introduced the basics of decision analysis and provided a structured approach to determine an appropriate course of action for our mess. Additional concerns such as our decisions' robustness and optimality were also addressed. Armed with a plan, we can now take the appropriate action for our problem.

After reading this chapter, the reader should:

1. Be capable of evaluating feasibility and desirability for a potential course of action;
2. Select a preferred course of action for intervention in our mess; and
3. Understand additional concerns that may affect our decision process.

References

Ackoff, R. L. (1977). Optimization + objectivity = opt out. *European Journal of Operational Research, 1*(1), 1–7.

Adams, K. M., & Bradley, J. M. (2012). *Productivity, performance and availability and intermediate level maintenance*. Norfolk, VA.

Adams, K. M., & Hester, P. T. (2012). Errors in systems approaches. *International Journal of System of Systems Engineering, 3*(3/4), 233–242.

Beer, S. (1979). *The heart of enterprise*. New York: Wiley.

Beer, S. (1981). *Brain of the firm*. Chichester, UK: Wiley.

Bohr, N. (1928). The quantum postulate and the recent development of atomic theory. *Nature, 121* (3050), 580–590.

Boulding, K. E. (1966). *The impact of social sciences*. New Brunswick, NJ: Rutgers University Press.

Brown, D. B., & Sim, M. (2009). Satisficing measures for analysis of risky positions. *Management Science, 55*(1), 71–84. doi:10.2307/40539128.

Checkland, P. B. (1999). *Systems thinking, systems practice*. New York: Wiley.

Gilboa, I. (2011). *Making better decisions: Decision theory in practice*. West Sussex, UK: Wiley.

Hester, P. T. (2012). Why optimization of a system of systems is both unattainable and unnecessary. *International Journal of System of Systems Engineering, 3*(3/4), 268–276.

Keeney, R. L., & Raiffa, H. G. (1976). *Decisions with multiple objectives: Preferences and value tradeoffs*. New York: Wiley.

Lipshitz, R., Klein, G., & Carroll, J. S. (2006). Naturalistic decision making and organizational decision making: Exploring the intersections. *Organization Studies, 27*(7), 917–923.

Pareto, V. (1897). *Cours D'économie Politique Professé à l'Université de Lausanne*. Luzerne University of Luzerne.

Simon, H. A. (1955). A behavioral model of rational choice. *Quarterly Journal of Economics, 69* (1), 99–118. doi:10.2307/1884852.

Simon, H. A. (1956). Rational choice and the structure of the environment. *Psychological Review, 63*(2), 129–138. doi:10.1037/h0042769.

Chapter 14
Decision Implementation

Abstract Once we've decided on a course of action, we must implement it. This requires human intervention in some capacity, even if it's initializing a computer algorithm or turning on a machine. In inserting a human into our process, we create an opportunity for human error. This chapter discusses the classification, management, and prevention of these errors as it focuses on decision implementation.

14.1 Introduction

The theme of this chapter is captured by the old saying *the best-laid plans of mice and men often go awry*. That is, despite our best planning efforts, our actions often don't go as we had anticipated. After we decide, we must act; otherwise, all our planning efforts are for naught. Remember, though, that action does not connote direct intervention in our mess (and a propensity for a bias for action); rather, it can encompass a cognizant decision on the part of the problem owner for no action. In this case, our chosen action is to avoid trying to fix our situation and observe to increase our understanding of a complex situation.

If we choose to intervene, however, we also invite the potential for errors in implementation. These errors, introduced by human intervention in our mess, are the focus of this chapter. We will discuss the recognition of human error, its classification, management, and prevention, all with an aim toward improved decision implementation.

14.2 Human Error Classification

The formal study of human error was pioneered by psychologist and human factors researcher James Reason. When considering errors in the context of implementation, we must first consider the role of intention. Why? "The notions of intention and error are inseparable. Any attempt at defining human error or classifying its

P.T. Hester and K.M. Adams, *Systemic Decision Making*, Topics in Safety, Risk, Reliability and Quality 33, DOI 10.1007/978-3-319-54672-8_14

forms must begin with a consideration of the varieties of intentional behavior" (Reason, 1990, p. 5).

The study of intention is key to understanding of human errors, as Reason (1990) explains:

> …the term error can only be applied to intentional actions. It has no meaning in relation to nonintentional behavior because error types depend critically upon two kinds of failure: the failure of actions to go as intended (slips and lapses) and the failure of intended actions to achieve their desired consequences (mistakes). (p. 7)

"The notion of intention comprises two elements: (a) an expression of the end-state to be attained, and (b) an indication of the means by which it is to be achieved" (Reason, 1990, p. 5). Element "a" is a fundamental element of the problem itself, recalling a problem as an *undesirable situation without a clear resolution that an individual or group wishes to see resolved* (Chap. 2). Element "b" is the result of rigorous decision analysis techniques introduced in the previous chapter. Thus, if we are to assume that we have: (1) defined the problem correctly and (2) chosen the correct course of the action, the next potential for error arises when we elect to execute our desired actions.

Reason (1990) provides a useful set of questions for distinguishing between different types of intentional behavior by answering the following three questions in yes–no fashion:

- Were the actions directed by some prior intention?
- Did the actions proceed as planned?
- Did they achieve their desired end? (p. 5)

Reason's framework for determining intention is shown graphically in Fig. 14.1. Each element is elaborated on in the paragraphs that follow.

In order to understand intention, we must first ask, was there a prior intention to act? That is, was there planning involved? This is important as it is necessary to distinguish between "prior intentions" and "intentions in action." Searle (1980) elaborates:

> All intentional actions have intentions in action but not all intentional actions have prior intentions. I can do something intentionally without having formed a prior intention to do it, and I can have a prior intention to do something and yet not act on that intention. (pp. 52, 53)

If there was not a prior intention to act, then we can ask, *was there intention in action*? That is, did you mean to do it? If no, we can classify the action as an involuntary or a nonintentional solution. These include, for example, having a seizure while driving a car, losing control, and killing someone in an accident. This is a tragic circumstance, no doubt, but it was unintentional. Such concerns are often the focus of criminal liability cases. If we meant to do it, i.e., there was intention in action, yet no prior intention, we can classify this as a spontaneous or subsidiary action. Once again returning to our legal context, this may represent the distinction

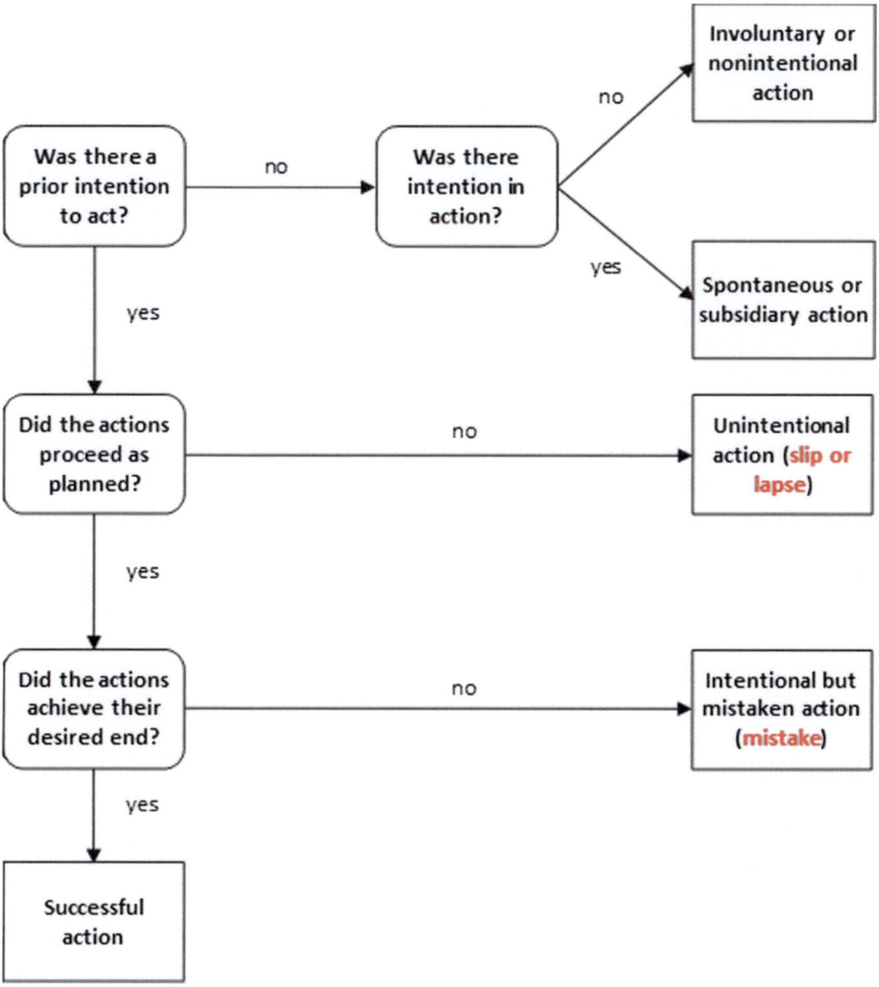

Fig. 14.1 Framework for distinguishing intentional behavior (adapted from Reason, 1990, p. 6)

between involuntary manslaughter (no intention in action), voluntary manslaughter (no prior intention, but intention in action) where, for example, an individual is acting intentionally in self-defense, or murder (where there is prior intention to act and thus a deliberate planning effort that is undertaken). Given the purposeful nature of our mess intervention, that is, our prior intention to act when using the guidelines laid out in the previous chapter, we will not explore actions that lack an intention to act any further. Thus, we will assume going forward that interventions in our mess involve some level of planning (and thus intentional behavior).

If we have prior intention to act (i.e., planning has occurred) and the results of our actions do not conform to our expectations, then we have committed an implementation error. Implementation errors carry unique connotations when compared with their counterparts in the thinking stage. As such, they require unique definitions in the context of human error, where an *error* is defined as "a generic term to encompass all those occasions in which a planned sequence of mental or physical activities fails to achieve its intended outcome, and when these failures cannot be attributed to the intervention of some chance agency" (Reason, 1990, p. 9). More specifically, Rasmussen (1982) discusses the notion of human error:

> If a system performs less satisfactorily than it normally does - due to a human act or to a disturbance which could have been counteracted by a reasonable human act - the cause will very likely be identified as a human error. (p. 313)

The next question in Reason's framework involves assessment of the results of our actions. *Did the actions proceed as planned*? If the answer is no, then we have committed an unintentional action (a so-called slip or lapse). *Slips and lapses* are "errors which result from some failure in the execution and/or storage stage of an action sequence, regardless of whether or not the plan which guided them was adequate to achieve its objective" (Reason, 1990, p. 9). In our terms, this would be seen as an execution failure (an implementation, or Type VIII, error). In the case, we did not execute our plan appropriately, irrespective of the reason.

If we answer that the actions proceeded as planned, then we are faced with a third and final question regarding intentional behavior. That is, *did the actions achieve their desired end*? If we answer no, we have committed an intentional but mistaken action, or simply, a mistake. Mistakes can be defined as:

> …deficiencies or failures in the judgmental and/or inferential processes involved in the selection of an objective or in the specification of the means to achieve it, irrespective of whether or not the actions directed by this decision-scheme run according to plan. (Reason, 1990, p. 9)

A mistake is the result of a planning failure. That is, we have chosen the wrong intervention to address our mess. Mistakes cannot be known a priori. After all, if we anticipate a mistake, why would we execute the planned action? Thus, the incidence of a mistake is evaluated during the observation stage when we determine whether or not an action had its desired efficacy. If we deem that it has, indeed, achieved its desired end, we can declare our action successful.

Thus, ultimately, we can distinguish between the two basic error forms (mistakes, slips, and lapses) as "*planning failures* (mistakes) and *execution failures* (slips and lapses)" (Reason, 1990, p. 8). Slips and lapses represent Type VIII errors, whereas mistakes represent errors in decision making, potentially resulting from a number of disparate sources, to include faulty *thinking* about our problem. Mistakes are evaluated during the *observation* phase, although they are committed during the *thinking* and *action* phases when we assess and ultimately choose an intervention strategy. It is this temporal disconnect that allows mistakes to linger and ultimately cause larger issues within our mess, a topic we will return to later in this chapter.

14.3 Classification and Performance Levels

Each of the aforementioned categories of human error (slips, lapses, or mistakes) can be further categorized according to the type of behaviors that produce them under the skill–rule–knowledge framework (Rasmussen, 1976, 1980). This framework attributes different task types at each of the three levels (skill-, rule-, and knowledge-based) and different types of errors as a result. Skill-based performance involves well-known situations and retrieval of automatic decision making processes. Rule-based performance involves learned solutions that take the form of if (*state*) *X*, then (*action*) *Y*, such as if *fire alarm goes off*, then *call Fire Department*. Errors here are due to misclassification of the situation at hand leading to incorrect rule implementation. Finally, knowledge-based performance is associated with the ad hoc generation of solutions to novel situations. Errors here are due to our inability to adapt our knowledge to these new or unique situations.

Typically, skill-based performance is associated with slips and lapses, while rule- and knowledge-based performance is associated with mistakes. As the focus of this chapter is on implementation errors (execution failures), we will focus on slips and lapses due to poor skill-based performance. A summary of the characteristics of skill-based errors is provided below (Reason, 1990):

- Involve routine actions
- Focus is on something other than the task at hand
- Are largely predictable
- Constitute a small proportion of the total number of opportunities for error
- Are influenced by intrinsic factors more than situational factors
- Are easily and rapidly detected (p. 62)

Skill-based failure typically occurs based on inattention or overattention. These failures are often influenced by our biases and heuristics, as we attempt to use automated decision making in an effort to reduce cognitive effort. Zeroing in on skill-based failures, we can investigate error management (as it pertains to implementation).

14.4 Human Error Management

It should come as no surprise that the increased complexity of messes in modern society has led to an increased potential for human error. But why is that potential more likely to be realized? We are often faced with the need to multitask, and this multitasking is influenced by the principles of requisite parsimony and requisite saliency. The *principle of requisite parsimony* limits the amount of information we can process and, thus, our response capabilities (Miller, 1956). The *principle of requisite saliency* says we can deal with this information limitation by affording differing levels of importance to this overabundance of information (Boulding,

1966). Given these factors, application of a systemic perspective reveals that human errors are likely inevitable. Reason (2000) offers:

> The basic premise in the system approach is that humans are fallible and errors are to be expected, even in the best organisations. Errors are seen as consequences rather than causes, having their origins not so much in the perversity of human nature as in "upstream" systemic factors. (p. 768)

Kontogiannis and Malakis (2009) agree, discussing the prevalence of errors in the context of increasingly complex circumstances:

> With the increasing complexity of technical systems, however, there has been a realization that total elimination of human error may be difficult to achieve. There are always bound to be complex situations in which errors may creep up due to high workload, decision making under stress, and poor team coordination. What seems to be more important in these situations is preventing or containing adverse consequences through the detection and correction of errors rather than prevention of errors in the first place. (p. 693)

Thus, the idea of error management becomes fundamental. "Error management has two components: limiting the incidence of dangerous errors and—since this will never be wholly effective—creating systems that are better able to tolerate the occurrence of errors and contain their damaging effects" (Reason, 2000, p. 769). The notion of managing is perhaps best captured by Reason's model of human error causation commonly known as the *Swiss cheese* model. A depiction of the Swiss cheese model is shown in Fig. 14.2.

A mishap or accident requires a perfect storm of *latent failures*, represented by organizational influences, unsafe supervision, and preconditions for unsafe acts, and

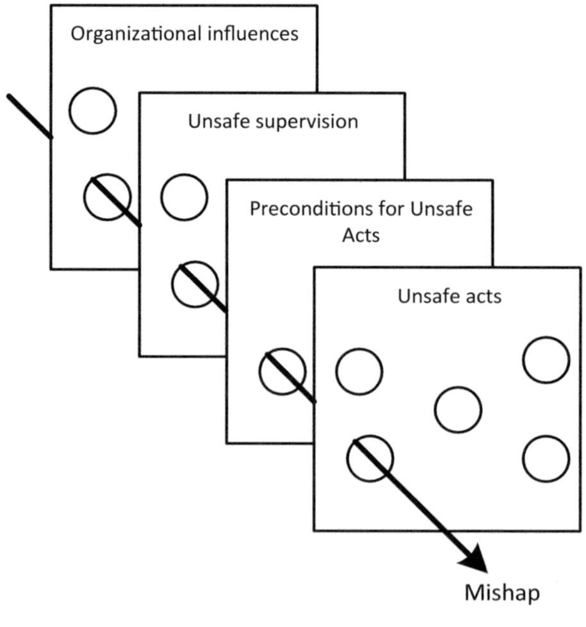

Fig. 14.2 *Swiss cheese* model of human error causation (adapted from Reason, 1990)

active failures, represented by unsafe acts. Unsafe acts include, slips, lapses, mistakes, and a new class of actions and *violations*. Violations are "willful disregard for the rules and regulations…" (Shappell and Wiegmann, 2000, p. 3). For the purposes of this discussion, violations are not considered. It is assumed that the individual executing an intervention in a mess is doing so with appropriate intentions and thus will execute the task at hand to the best of his or her ability.

14.5 Latent and Active Failures

Ramanujam and Goodman (2003) define latent errors (failures) as "uncorrected deviations from procedures and policies that potentially can contribute to adverse organizational consequences" (p. 815). Characteristics of latent failures are defined by Ramanujam and Goodman (2003) as: (1) a set of organizational expectations; (2) a deviation from these expectations; and (3) an absence of any direct consequence. "We emphasize the absence of immediate consequences because nothing inherent in a deviation can automatically generate adverse consequences" (Ramanujam & Goodman, 2003, p. 817). They emphasize that other factors or triggers (active failures in Reason's terms) must interact with a deviation in order to produce any adverse outcome for the system. They give the following example:

> …the failure to follow procedures for shutting off safety valves in a nuclear power plant or the failure to monitor for gas in a coal mine represents deviations that have the potential for causing adverse outcomes such as an explosion. These deviations by themselves might not lead to an explosion either immediately or even over time. Other factors or triggers need to interact with the deviations to produce adverse outcomes. (Ramanujam & Goodman, 2003, p. 818)

Ramanujam and Goodman (2003) provide a framework regarding latent and active failures, context, and mishaps shown in Fig. 14.3.

Several elements of this framework require elaboration. Organizational and situational context creates an environment, due to inherent deviations in the factors we've identified, for latent failures to take place. As latent failures increase, so does the potential for a mishap. Two feedback systems contribute to the effects of latent failures, the positive feedback system and the negative feedback system.

"Positive feedback systems play a pivotal role in accelerating the increase in latent [failures]" (Ramanujam & Goodman, 2003, p. 825). They lead to mislearning, escalation of commitment, and reduced task attention, conditions which both have the potential to increase deviations and exacerbate latent failures. Ramanujam and Goodman (2003) discuss the three terms:

> Mislearning refers to the adoption of deviation-induced behaviors…escalation of commitment represents another process that leads to a build-up of latent errors where decision-makers, faced with the negative results of their decisions, continue to invest resources in these decisions in an effort to recoup their investment…lowered availability of attention reduces vigilance and contributes to latent errors. (pp. 826, 827)

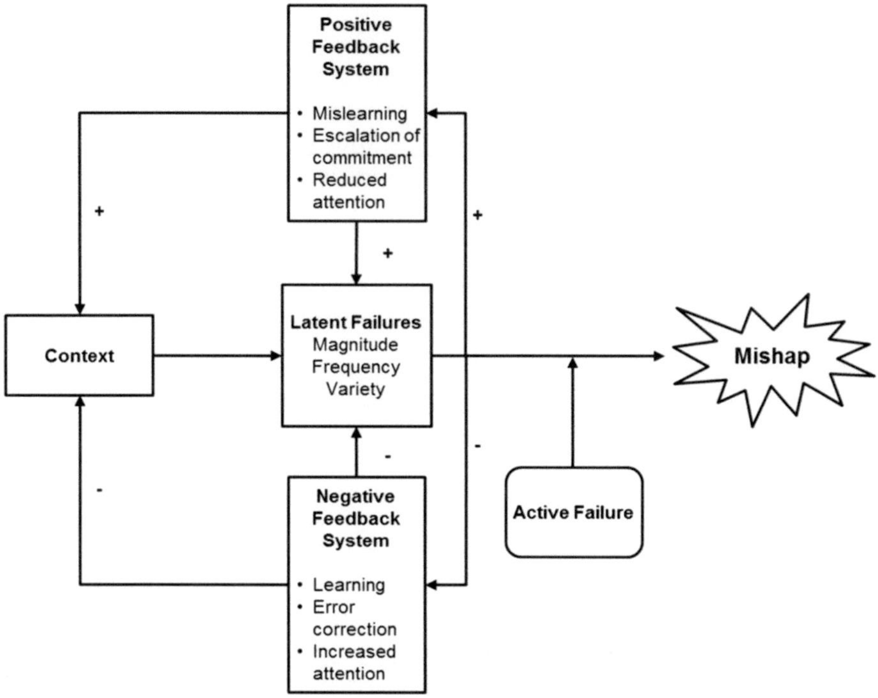

Fig. 14.3 Failures, context, and mishaps (adapted from Ramanujam & Goodman, 2003, p. 824)

Negative feedback systems, on the other hand, can help to mitigate the effects of latent failures. Ramanujam and Goodman (2003) elaborate:

> Negative feedback systems exert a different effect on latent errors. These systems have pre-specified standards of performance. Deviations from these standards initiate a feedback mechanism that leads to corrective action. The deviation is corrected and the system returns to a state of equilibrium. The necessary elements in a negative feedback system include standards that are operational, a monitoring system that detects deviations from these standards, reporting systems that provide information about deviations to appropriate organizational units and managers, and organizational capabilities to initiate and implement corrective actions. (p. 827)

Ultimately, though, latent failures require an active failure (an implementation error) to *connect the dots* or to create a mishap, or a negative consequence. This situation is depicted by the arrow in Fig. 14.2 which has somehow found a way to bypass all of our fail-safes and protections against human error. But how do we prevent these errors? We now turn our attention to this question.

14.6 Human Error Prevention

It should be clear by now that a perfect storm is necessary to commit an implementation error which, combined with several latent failures, leads to a serious consequence. How, though, does this happen? Figure 14.4 summarizes Reason's factors that contribute to fallible, high-level decision making. Decision makers apply resources to meet goals. Attainment of these goals is assessed through feedback, which is fed through defensive filters (which may also mitigate their effect) back to the decision makers. The decision makers then adjust their behaviors accordingly and the cycle begins again.

But this lens is applied at too high a level to be useful for implementation assessment purposes. This process can be localized to individual actions by simply replacing the *goals* element with *actions*, thereby providing a more granular perspective on action accomplishment and error avoidance. In terms of committing a Type VIII error, application of resources to undertake actions provides opportunities for decision makers to commit errors.

Ultimately, elimination of latent failures inherent in a system is a design function that precedes most systemic decision making efforts (i.e., typically, our mess is already operational). However, two elements are key from our discussion this chapter, and that is (1) the central role of feedback and (2) the fact that implementation errors don't always lead to mishaps. Feedback and a lack of negative consequences are key to the frameworks shown in Figs. 14.3 and 14.5 and yet lacking in Reason's initial framework regarding intentional behavior (Fig. 14.1). Feedback is the key to preventing future human error occurrence (or at least mitigating it in the present). This should not come as a shock to us, given Chap. 8's

Fig. 14.4 Factors that contribute to fallible, high-level decision making (adapted from Reason, 1990, p. 204)

Fig. 14.5 Factors that contribute to fallible, localized decision making (adapted from Reason, 1990, p. 204)

discussion of motivation. Thus, we have altered the Reason (1990, p. 5) framework as shown in Fig. 14.6.

The updated framework makes two important changes to the original. First, we have introduced a new connection between unintentional actions and an assessment of whether or not the actions achieved their desired end. We can commit an error in execution and yet still achieve our objectives. This is an important element to include in our implementation assessment. Second, all paths lead to a (potentially) revised plan. This reflects the importance of feedback in our decision making and a recognition that it is an iterative undertaking (i.e., we make a decision, implement it, receive feedback, and then reassess). This new framework mirrors the generic control theory of motivation that we introduced in Chap. 8, and its basic elements should come as no surprise to the reader by now.

Incorporation of feedback is essential to appropriate decision implementation. As it pertains to the limitation of skill-based slips and lapses (our main focus in this chapter), there is good news:

> In the skill-based mode, recovery is usually rapid and efficient, because the individual will be aware of the expected outcome of his or her actions and will therefore get early feedback with regard to any slips that have occurred that may have prevented this outcome being achieved. This emphasizes the role of feedback as a critical aspect of error recovery. In the case of mistakes, the mistaken intention tends to be very resistant to disconfirming evidence. People tend to ignore feedback information that does not support their expectations of the situation… (Center for Chemical Process Safety, 2004, p. 76).

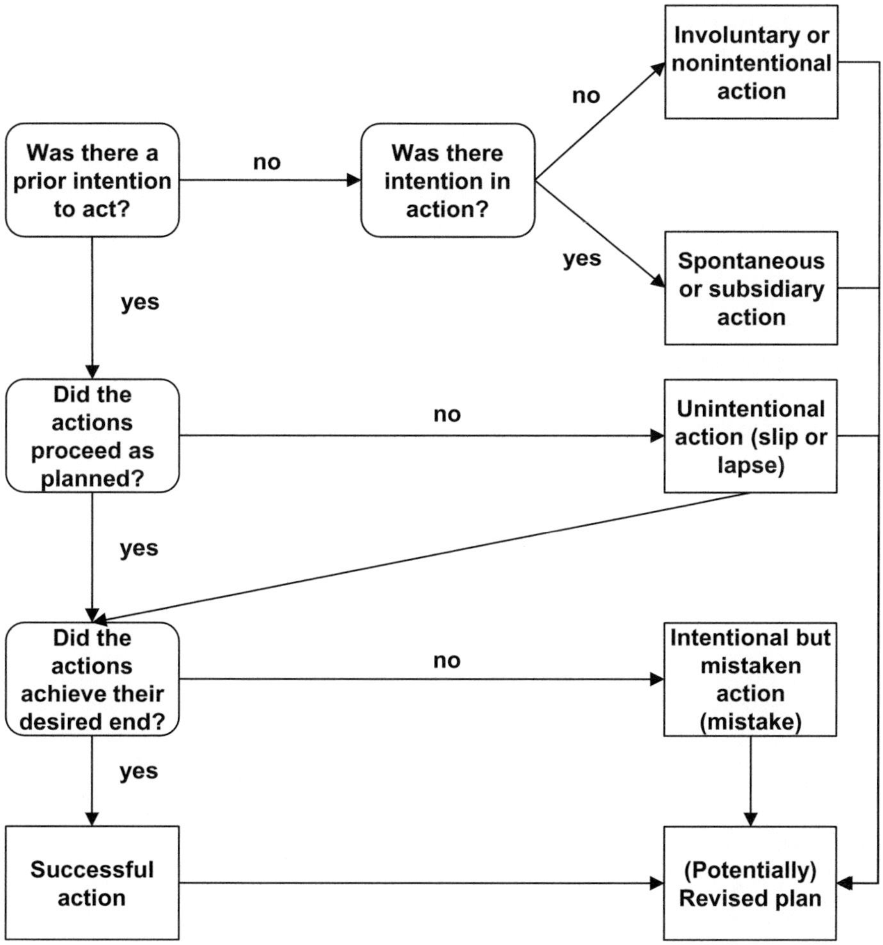

Fig. 14.6 Updated framework for distinguishing intentional behavior (adapted from Reason, 1990, p. 6)

Feedback is an effective mechanism for preventing future active failures. After all, it is much simpler to determine what happened after an event (hindsight is 20/20) than to predict human behavior due to the vast number of potential outcomes. Human error prevention, then, is composed of two elements:

(1) Maintain feedback to both the system (using cybernetic principles) and the decision maker to ensure actions are executed properly; and
(2) If possible, work to incorporate feedforward procedures, that is, to anticipate future effects and act accordingly, in an effort to prevent active failures in situ.

14.7 Summary

The potential for human error is an inevitable part of decision implementation. Being able to classify, manage, and prevent human error helps us to appropriately implement our decisions. Classification helps us to understand which errors are driven by planning (mistakes) and which are driven by implementation (slip and lapses). Management helps us to understand that both latent and active failures lead to slips and lapses. Finally, prevention is accomplished through proper feedback (and feedforward) mechanisms.

After reading this chapter, the reader should:

1. Be capable of classifying human errors;
2. Distinguish between latent and active failures; and
3. Utilize feedback to prevent human error.

References

Boulding, K. E. (1966). *The impact of social sciences*. New Brunswick, NJ: Rutgers University Press.

Center for Chemical Process Safety. (2004). Human performance understanding and error. *Guidelines for preventing human error in process safety*. Hoboken, NJ: John Wiley & Sons.

Kontogiannis, T., & Malakis, S. (2009). A proactive approach to human error detection and identification in aviation and air traffic control. *Safety Science, 47*(5), 693–706.

Miller, G. A. (1956). The magical number seven, plus or minus two: Some limits on our capacity for processing information. *Psychological Review, 63,* 81–97.

Ramanujam, R., & Goodman, P. S. (2003). Latent errors and adverse organizational consequences: A conceptualization. *Journal of Organizational Behavior, 24*(7), 815–836.

Rasmussen, J. (1976). Outlines of a hybrid model of the process operator. In S. A. Johannsen (Ed.), *Monitoring behaviour and supervisory control*. New York: Plenum Press.

Rasmussen, J. (1980). What can be learned from human error reports? In K. D. Duncan, M. M. Gruneberg, & D. Wallis (Eds.), *Changes in working life*. Hoboken, NJ: John Wiley & Sons.

Rasmussen, J. (1982). Human errors: A taxonomy for describing human malfunction in industrial installations. *Journal of Occupational Accidents, 4,* 311–333.

Reason, J. (1990). *Human error*. Cambridge, UK: Cambridge University Press.

Reason, J. (2000). Human error: Models and management. *BMJ, 320,* 768–770.

Searle, J. R. (1980). The intentionality of intention and action. *Cognitive Science, 4,* 47–70.

Shappell, S., & Wiegmann, D. (2000). *The human factors analysis and classification system—HFACS*. Washington, DC: Office of Aviation Medicine.

Part IV
Observing Systemically

With Part IV, we have now transitioned from the act to the observe phase in the systemic decision making process (as shown in Fig. 1 below, originally from Chap. 3).

We have now thought extensively about each of our constituent problems using each of our six perspectives; that is, we have *thought systemically* about our problems. We have also *acted systemically*. We carefully considered our options, chose a course of action, and implemented it. Now, all that is left is to *observe systemically*. This phase helps us to understand whether or not our actions have had the efficacy we desired. In short, it allows us to learn from all of our efforts thus far. This part of the text is comprised of Chaps. 15–18. Chapter 15 discusses the general process of observation and what can go wrong. Chapter 16 discusses systemic learning to focus on what insights we can gain from this process as it pertains to our mess. Chapter 17 provides a holistic case study outlining all the materials presented thus far. Finally, Chap. 18 provides some concluding thoughts to tie the text together.

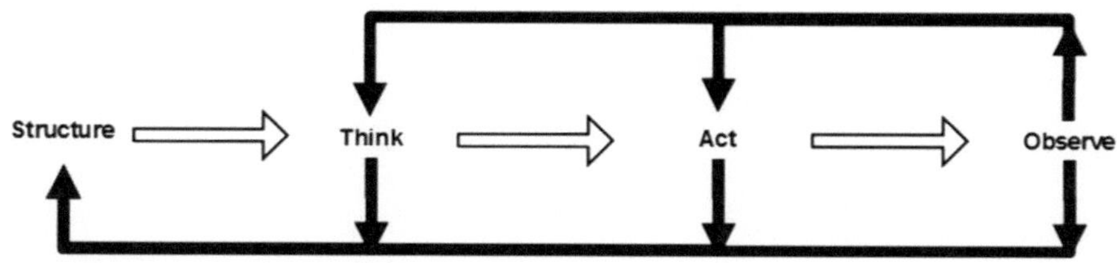

Fig. 1 Systemic decision making process with feedback

Chapter 15
Observation

Abstract Observation is the central method by and through which human beings engage with the real-world. Observation is the source of virtually all empirical evidence for science. In this chapter, we will be underscoring the inevitable impact of the observer's role in the process of observation. The central point we wish to emphasize is that the human observer impacts each and every observation during the systemic decision making process. This is crucial as we wish to avoid committing a Type I and Type II error. We will discuss the process of observation and suggest a model for understanding observation and will also be describing how empirical observations are subject to the theories and ideas possessed by the observer. Additional sections will describe a model for situations where technological systems are employed as part of the observation process, the role of measurement during observations, and how bias and heuristics affect observations.

15.1 Introduction

At this point, we have thought about and acted on our problem. Reviewing our errors' taxonomy shown in Table 15.1, we have now moved on from avoiding the Type IV, V, and VIII errors (the focus of the acting systemically section of the text). The observation stage requires two major processes: (1) we must observe and interpret our observations carefully (and avoid committing a Type I or Type II error), and (2) we must carefully learn from our observations (and avoid jumping to conclusions and committing a Type VII error).

So, we have to *observe* and *learn from our observations*. The former is the focus of this chapter; the latter is the focus of the next chapter.

P.T. Hester and K.M. Adams, *Systemic Decision Making*, Topics in Safety, Risk, Reliability and Quality 33, DOI 10.1007/978-3-319-54672-8_15

Table 15.1 Taxonomy of systems errors (adapted from Adams & Hester, 2012)

Error	Definition	Issue
Type III (γ)	Solving the wrong problem precisely	Wrong problem
Type IV (δ)	Inappropriate action is taken to resolve a problem as the result of a correct analysis	Wrong action
Type V (ε)	Failure to act when the results of analysis indicate action is required	Inaction
Type VIII (η)	Incorrectly implementing the correctly decided action	Incorrect implementation
Type I (α)	Rejecting the null-hypothesis when the null-hypothesis is true	False positive
Type II (β)	Failing to reject the null-hypothesis when the null-hypothesis is false	False negative
Type VI (θ)	Inferring causation when only correlation exists	Unsubstantiated inference
Type VII (ζ)	An error that results from a combination of the other six error types, often resulting in a more complex problem than initially encountered	System of errors

15.2 Avoiding the Type I and Type II Errors

Recall that the observation process is prone to Type I and Type II errors. How do we use proper observation to avoid or lessen the incidence of these errors? Type I and Type II error minimization is a matter of proper experimental design and execution. Table 15.2 contains a representation of and definitions for the Type I and Type II errors framed in terms of a hypothesis test.

When we wish to conduct a hypothesis test, we choose the *level of significance*, also called α. This value tells us the probability of committing a Type I error. Our *confidence* in a given hypothesis may be expressed as $1 - \alpha$. We can also choose the *power* of a test which is represented by $1 - \beta$, or the probability that we will not commit a Type II error. β is affected by the sample size, α, and true parameter value. In the end we cannot avoid a Type I and Type II error entirely as our sample size will be limited. We can be aware of their possibility and act accordingly. However, proper observation goes a long way toward minimizing their incidence. For this reason, the remainder of this chapter focuses on observation and how to conduct it properly.

Table 15.2 Type I and Type II errors

	Actual condition	
Test result	Positive	Negative
Positive	True positive $p = 1 - \alpha$	False positive $p = \alpha$
Negative	False negative $p = \beta$	True negative $p = 1 - \beta$

15.3 Observation

Observation is the central source of knowledge gained from exposure to the real world. This is true whether the knowledge is being generated in a controlled laboratory or in a natural setting.

> Observation is being understood in a very broad way here, to include all kinds of sensory contact with the world, all kinds of perception. (Godfrey-Smith, 2003, p. 156)

In this section, we elaborate on the notion of observation as it pertains to systemic thinking in general. The two sub-sections that follow will review observation as a process with a descriptive model and how all observation is laden with theory that is imbedded within the notions and ideas of the observer.

15.3.1 A Model for the Process of Observation

Observation is the operation where raw sensory inputs are filtered by the human thought process. The physiological capacity for sensory perception in humans is limited by the five senses: (1) hearing, (2) sight, (3) smell, (4) taste, and (5) touch. Over time, raw perceptions are converted by the human thought process and begin to form impressions, which are stored for future use. Stored impressions and their relationships with one another are formed into constructs that permit the individual to develop more complex implications and associations from the sensory inputs.

In a literature too vast to summarize here, theorists have argued that observation is already cognition and that we cannot describe a fact without implying something more than the fact. As a result, Clyde H. Coombs [1912–1988] proposed that the term *data* be used for observations already interpreted in some way. The diagram in Fig. 15.1 depicts the scope of Coombs' theory of data (1964).

Figure 15.1 depicts how an observer's interpretation of the universe of all possible observations can lead to logical inferences as a result of four distinct phases conducted during the process of observation. The graphic has additional importance when considered with the following statement from Coombs (1964) pertaining to those phases after Phase 0:

> The scientist enters each of these three phases in a creative way in the sense that alternatives are open to him and his decisions will determine in a significant way the results that will be obtained from the analysis. Each successive phase puts more limiting boundaries on what the results might be. At the beginning, before phase 1, there are perhaps, no limits on the potential conclusions; but each phase then constrains the universe of possible inferences that can be ultimately drawn from the analysis. (p. 5)

It is important to note that the observer depicted in Fig. 15.1 directly influences the data in many ways. Table 15.3 provides a glimpse of the how the observer influences the observations during the four phases and associated stages.

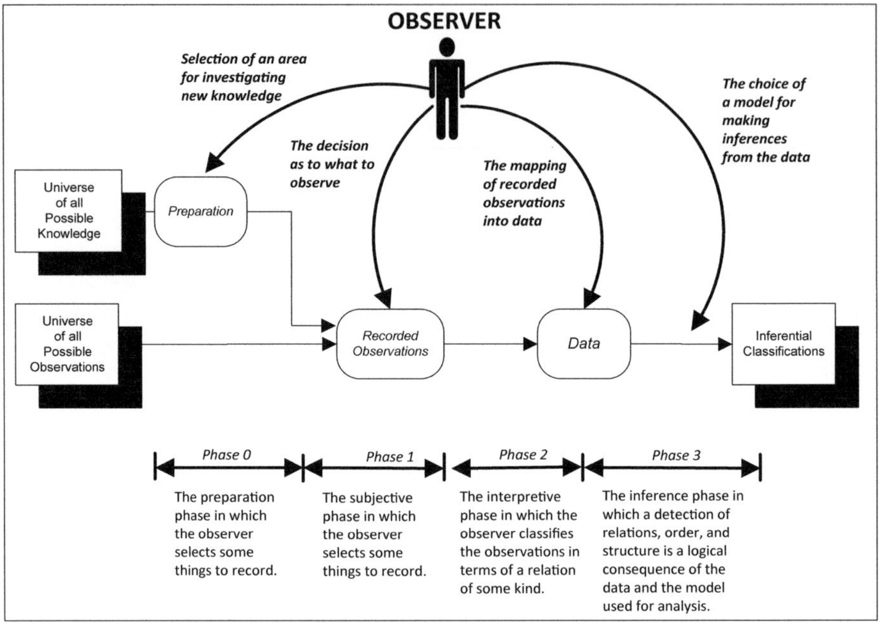

Fig. 15.1 Flow diagram of observable to inference

Table 15.3 How and where an observer exhibits influence during observation

Phase	Stage	Description
0—Preparatory	Knowledge area	Selection of an area for investigating new knowledge
	Preparation	Preparatory reading in the area's existing body of knowledge
1—Subjective	Selection	Selection of things to observe
	Method	The sensors and methods used to record and measure the observation
2—Interpretive	Analysis	The observer interprets the data
	Classification	The observer classifies the observations
3—Inferential	Inference	The observer makes an inference based on the order structure and model used in analysis and classification
	Publication	The observer reports the interpretation of the new knowledge

Table 15.3 demonstrates that the potential to influence observations is problematic and must be mitigated during the conduct of all research and problem solving efforts. Thus, in terms of the stages of observation and their relation to the systems errors discussed in Chap. 1, we must be careful to avoid the Type I and II errors (described in Sect. 1.3.5) in Phase II and the Type VI error (described in Sect. 1.3.6) in Phase III.

This leads the discussion to the notion that all observation is impacted by the observer's personal beliefs in what is termed *theory-laden observation.*

15.3.2 Theory-Laden Observation

Based upon Coombs' notion that observation has already been subjected to analysis, a number of major scholars in the field of Philosophy of Science have argued that all observation is theory-laden (Feyerabend, 1962; Kuhn, 1996). Specifically,

> Observation cannot function as an unbiased way of testing theories (or larger units like paradigms) because observational judgments are affected by the theoretical beliefs of the observer. (Godfrey-Smith, 2003, p. 156)

Feyerabend (1962) cautions all observers of empirical data to separate the observation from the consequent description:

> We must carefully distinguish between the 'causes' of the production of a certain observational sentence, or the features of the process of production, on the one side, and the 'meaning' of the sentence produced in this manner on the other. More especially, a sentient being must distinguish between the fact that he possesses certain sensation, or disposition to verbal behavior, and the interpretation of the sentence being uttered in the presence of this sensation, or terminating this verbal behavior. (p. 94)

Observations that may not be true are associated with the term *judgment,* which takes on a very specific meaning:

> An observation of doubtful truth is sometimes called a "judgment." Judgments are identified by a non-negligible incidence of disagreement between independent observers that persists under the best of conditions for observation. If for analytic purposes one asserts that all observations are subject to error, then all observations are properly called "judgments" - this fact is not unrecognized in the use of the term but ordinarily it refers to observations that are most obviously fallible. (Ellson, 1963, p. 41)

So, how are the truths associated with human observations rendered reliable? Reliability is an important dimension associated with all empirical research. Ellson (1963) reports that there are a number of methods for improving or establishing the reliability of observation which include: (1) personal observation; (2) instrumentation; (3) repetition; (4) inter-observer agreement through statistical correlation and frequency of agreement; and (5) the use of competent observers.

We would be remiss if we did not address the acquisition of knowledge without the benefit of direct observation. Fodor (1984) discusses belief fixation via *inference* from beliefs previously held as a method used by human observers. His example is the notion of Martian fauna (i.e., Martian life forms) that is held by many which cannot have been acquired through empirical observation, but may only be acquired through some belief system to which the observer belongs.

Many theories and models exist for a further exploration of the concepts of awareness, observation, and cognition. While this subject area is beyond the scope of this text, the reader is referred to literature on situation awareness (Endsley,

1995), the recognition-primed decision (RPD) model (Klein, Calderwood, & Clinton-Cirocco, 1986), and gestalt psychology (Ellis, 1997) for further guidance on the topic. The next section will address how data from observations is processed into information and how information is formed into knowledge, which is utilized in decision making.

15.3.3 Data, Information, Knowledge and Observation

The data, information, knowledge, decisions, and metrics (DIKDM) model described in Sect. 9.2.6 will be utilized to understand how observation is the essential process responsible for the processing of empirical data into information, and information into knowledge. A brief review is in order.

Most pieces of *data* are of limited value until it they are processed into a useable form. Processing data into useable forms requires human intervention, most often accomplished with the use of an information system. Data are transformed into information in the following ways (Davenport & Prusak, 2000):

- Contextualized: we know for what purpose the data was gathered.
- Categorized: we know the units of analysis or key components of the data.
- Calculated: the data may have been analyzed mathematically or statistically.
- Corrected: errors may have been removed from the data.
- Condensed: the data may have been summarized in a more concise form. (p. 4)

Like data, *information* has little utility without additional processing. Processing information into useful elements is a higher-order process that requires a purposeful human involvement and intervention. Information is transformed into knowledge by conducting the following actions (Davenport & Prusak, 2000):

- Comparison: how does information about this situation compare to other situations we have known?
- Consequences: what implications does the information have for decisions and actions?
- Connections: how does this bit of knowledge relate to others?
- Conversation: what do other people think about this information? (p. 6)

It is important to note that data and information can always be made explicit and codified while some knowledge is tacit (i.e., understood or implied without being stated). The resulting structure may be characterized as data-information-knowledge (DIKDM) and is depicted in Fig. 15.2.

Figure 15.2 shows context as an essential element in the information to knowledge transformation process. Context serves as the *wrapper* that must be supplied, most often as explanations, to enable knowledge. Human involvement, through intervention, is an essential requirement in this process because “an explanation always takes place relative to a space of alternatives that require

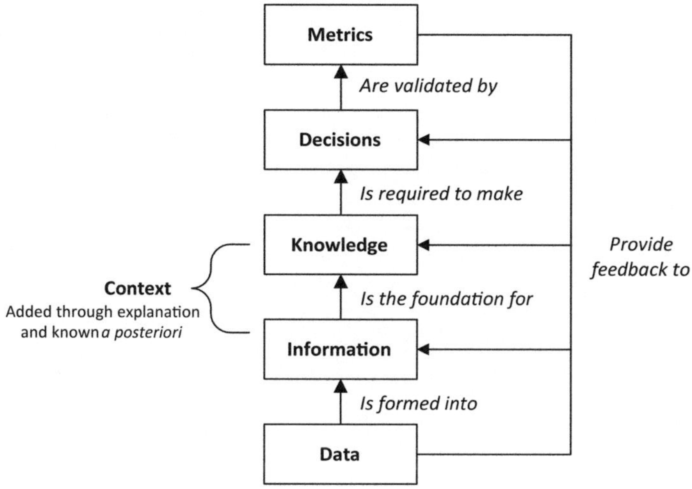

Fig. 15.2 Relationship between data, information, knowledge, decisions, and metrics (DIKDM) (from Chap. 9)

different explanations according to current context" (Brézillon, 1999, p. 57). Furthermore, because context is dynamic (i.e., the situational information is changing over time), it can only be represented a posteriori. Three points may be made using this structure:

1. Data do not relate to any specific context.
2. Information, on the other hand, does relate to a specific context. Information is formed by selecting, organising and summarising data to be meaningful and useful within a specific context.
3. Knowledge is more generic, relevant to many contexts and longer lasting. (French, 2013, p. 553)

Boisot (1998) refers to the contextual understanding as a function of the perceptual and conceptual filters that permit the agent or knowledge source to create relationships among the events or data sources, as depicted in Fig. 15.3. It is from these relationships that we are able to generate inferences and formulate decisions.

In Fig. 15.4 there are four separate processes, each of which are described in Table 15.4.

In summary, the observation of empirical phenomena is the single, essential process responsible for the processing of empirical data into information, and information into knowledge. Armed with this knowledge, users are able to develop inferences and make decisions based upon factual, real-world occurrences. In the next section, we will discuss how technological systems and the observer's ability to perceive, comprehend, and to project data, information, and knowledge in support of subsequent decisions may be modeled.

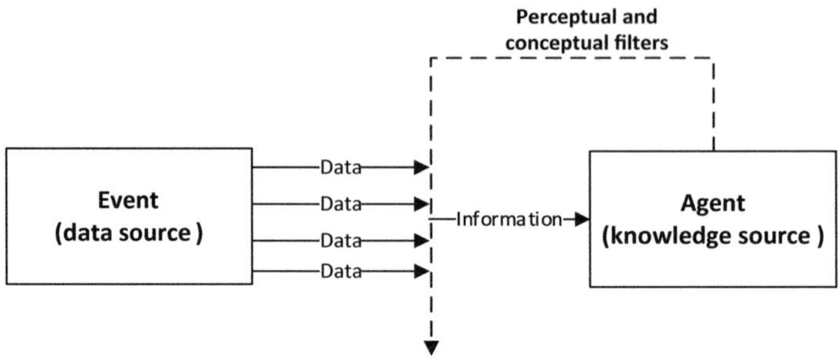

Fig. 15.3 Perceptual and conceptual filters in knowledge generation (Boisot, 1998)

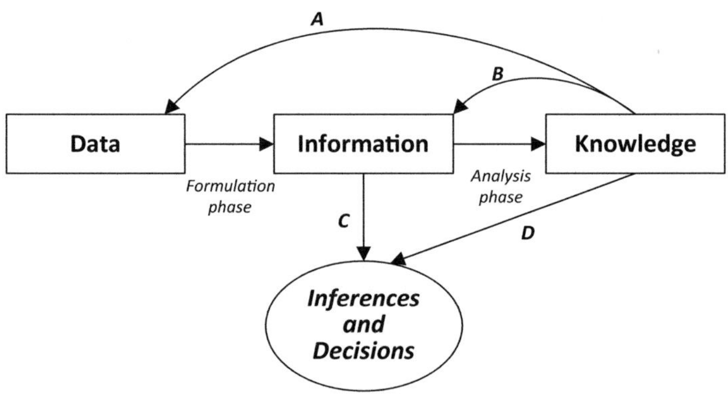

Fig. 15.4 DIK and the making of inferences and decisions (French, 2013)

Table 15.4 Use of data, information and knowledge in forming inferences, forecasts, and decisions

Arrow	Description of process
A	Arrow A indicates these uses of knowledge in the formulation phase and the detailed organization of the data in the analysis phase
B	Arrow B indicates that in order to recognize such patterns and form new knowledge, it requires insight and higher-level knowledge
C	Arrow C indicates that the decision makers are informed by the outputs of the formulation and analysis phases
D	Arrow D indicates that in order to understand the information and to be able to use it to improve their thinking, the users need to be conscious of their knowledge of the analytic methods used and how these have helped them in the past

15.4 Observation and Situated Cognition

The theory-laden observation process must involve consideration of both technological and human elements, and can be thought of as residing within a larger construct. A model to describe this observation process is the *Dynamic Model of*

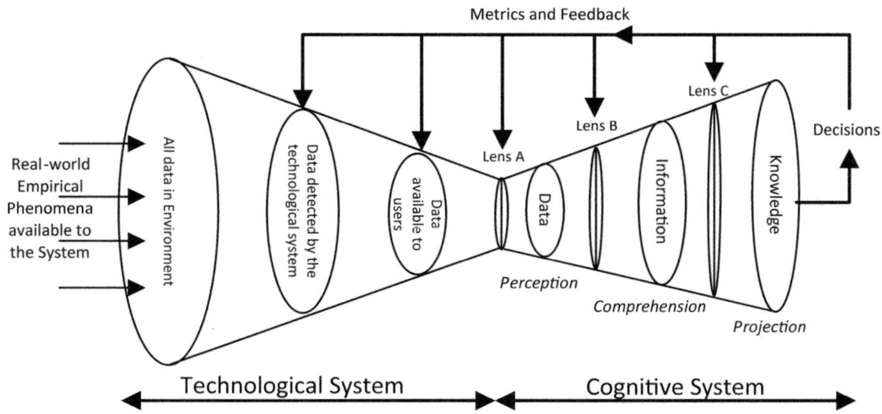

Fig. 15.5 The dynamic model of situated cognition (Shattuck & Miller, 2006)

Situated Cognition (DMSC), which captures both the human and technological components of systems in a single model that depicts how observation is influenced by a variety of agents (Miller and Shattuck, 2005; Shattuck & Miller, 2006). Figure 15.5 is our interpretation of the DMSC. Note that we have included the DIKDM structure, as the terminology used by Miller & Shattuck is not entirely consistent with Coombs and what we have used throughout this book.

15.4.1 Technological System in the DMSC

Our model of the DMSC in Fig. 15.5 processes phenomena from left to right. In the first oval, the technological system has the opportunity to observe all of the real-world, i.e., empirical phenomena present in the environment. This process is a function of the environment and of conditions external to the technological system. In the second oval, the system detects some smaller sub-set of the available data from the environment. The system is incapable of processing all of the empirical data from the environment due to the system's capability and its operational state, which may change based upon the condition of the system over time. In the third oval, the system's users have an opportunity to detect what the system has both detected and presented for human processing. The proportion of the data processed is, once again, less than that of the previous oval and is a function of the system's capabilities, its operational state, and how it was configured to present data for human interpretation (Phase 1 in Fig. 15.1 and Table 15.1).

15.4.2 Cognitive System in the DMSC

The first lens in the cognitive side of the DMSC represents the first human encounter with the data presented by the system. The processes conducted in the first lens requires human selection (i.e., of what data to observe from that presented by the system) and is termed *perception* (the subjective Phase 1 in Fig. 15.1 and Table 15.1). All of the users involved in the conduct of such interpretations must be keenly aware that this is rarely *truth*, but is instead some subset of the real-world empirical phenomena that exist in the environment. The second lens is representative of the process where the user transforms the perceived data into information in what is termed *comprehension* (the interpretive Phase 2 in Fig. 15.1 and Table 15.1). Comprehension requires human recognition and processing of a variety of raw data into more focused information. This information is passed to the third and final lens where the user processes the information and subsequently merges it with the ever-changing, time-variant context in what is termed *projection* (the inference Phase 3 in Fig. 15.1 and Table 15.1). The inference(s) developed by the user during projection are based on the inferential model(s) used to analyze information and current context. The inferences produced form the basis for decision making.

15.4.3 Cybernetic Nature of the DMSC

The final element of the DMSC depicted in Fig. 15.5 is the inclusion of a cybernetic feedback mechanism. The feedback mechanism permits the results of decisions to be evaluated and corrective actions developed and implemented. Corrective actions may include adjustments to the technological system, changes to the cognitive system, improvements in the inferential models, and subsequent training of users of the system.

15.5 Measurement and Observation

Good science is based upon four generally accepted criteria that ensure quality: (1) truth value, (2) applicability, (3) consistency and (4) neutrality (Lincoln & Guba, 1985). The third criterion, consistency, addresses the conformity in the application and generation of knowledge and establishes guidelines for ensuring stability during generation (i.e., design and technique), of new knowledge. The ability to accurately repeat observations, independent of the original observer, is an essential element. The requirement for independent reproducibility ensures that observations by different observers are comparable. Because the physiological capacity for input perception in humans is subjective and qualitative (i.e., the five senses perform differently from human to human), this makes them difficult to record and hence, to compare.

The concept of measurement evolved to permit different human observers to record and compare observations made at different times and places. Measurement consists of using observation to compare the real-world phenomena being measured to an established standard which can be reliably reproduced for use by multiple, independent observers. Measurement's goal is to reduce an observation to a discrete measure which can be recorded and used as the basis for comparison with other measures.

Qualities of criterion such as reproducibility may be invoked through the use of formal methods and measurement. However, the nagging issue and difficulties generated by the presence of theory-laden observation must be addressed by an understanding of how bias is introduced into the process. This leads to the next section which will discuss the mitigation of bias as an element of personal beliefs during observation.

15.6 Bias and Heuristics in Observation

Our ability to observe is affected, both negatively and positively, by our own biases and heuristics. First, we discuss bias, defined as:

> Any process at any stage of inference which tends to produce results or conclusions that differ systematically from the truth. (Sackett, 1979, p. 60)

Bias may be introduced during each and every stage and phase depicted in Fig. 15.1. As a result, the observer must ensure that the process depicted in Fig. 15.1 and related in Table 15.1 provides reasonable controls that mitigate bias as much as possible.

> The difficulties generated for scientific inquiry by unconscious bias and tacit value orientations are rarely overcome by devout resolutions to eliminate bias. They are usually overcome, often only gradually, through self-corrective mechanisms of science as a social enterprise. (Nagel, 1961, p. 489)

Part of understanding how to mitigate human bias requires knowledge of the source and major types of unconscious bias. Because all human beings have unintentional cognitive biases that affect their decision making, knowledge of the types of bias may help improve their detection and elimination. Cognitive biases include behaviors that are labeled *heuristics*. Table 15.5 lists a variety of definitions for the term heuristic.

The unintentional biases and heuristics that operate at the subconscious level are the most difficult to prevent. The sections that follow will provide a short discussion of major heuristics and how to mitigate their effect.

Table 15.5 Definitions for heuristic

Definition	Sources
A heuristic is a procedure for achieving a result which does not consist simply in applying certain general rules which are guaranteed to lead to the result in question	(Proudfoot & Lacey, 2010, p. 165)
A rule or solution adopted to reduce the complexity of computational tasks, thereby reducing demands on resources such as time, memory, and attention	(Audi, 1999, p. 379)
Heuristics are 'rules of thumb' that are used to find solutions to problems quickly	(Kynn, 2008, p. 242)
Heuristics are often described as judgmental shortcuts that generally get us where we need to go- and quickly- but at the cost of occasionally sending us off course	(Gilovich & Savitsky, 1996, p. 35)

15.6.1 Availability Heuristic

The availability heuristic refers to the practice of applying probabilistic evidence to an available piece of information from one's own set of experiences (Tversky & Kahneman, 1973, 1974). That is to say, humans estimate the likelihood of an event based on a similar event that they can remember, which is by definition, from a biased and unrepresentative sample in their memory. Further, since newer events provide greater saliency in one's mind, newer events influence an individual's reasoning to a larger degree than do older events. Additionally, events with unusual characteristics stand out more in one's mind (e.g., you don't remember the hundreds of times you went to a given restaurant, but you definitely remember the time you got food poisoning). Furthermore, humans may be biased based on the actual efficacy of the retrieval mechanism that they utilized to obtain the experience from memory. Depending on who is asking the question, for example, an individual may consciously or unconsciously block memories. Or, if they are very upset or tired, their retrieval process might not deliver as much information as it would in another situation. In order to mitigate this problem, observers should employ techniques that take into account how their experiences bias the data they retrieve about a particular set of observations.

15.6.2 Representativeness Heuristic

The representativeness heuristic refers to the phenomenon that occurs when individuals assume commonalities between objects and estimate probabilities accordingly (Tversky & Kahneman, 1974). The determination of similarity between objects is typically performed by comparing their known attributes. Individuals compute a running tally of matches versus mismatches and then estimate whether or not the item fits a category based on the total. Once the item is categorized, automatic category-based judgments are made about the member item. Using this type of analysis has its issues. To combat this bias, observers must use base rates (i.e., unconditional, or prior, probabilities) to compare the underlying category probability versus the specific scenario. Then, the base rate can be adjusted to accurately reflect the specific scenario's characteristics (i.e., its conditional factors).

It should be noted that the availability and representativeness heuristics are often confused, but they are not the same phenomenon. With the availability heuristic, individual instances are retrieved and a judgment concerning the frequency of the item is made based on the item's saliency (i.e., relative importance) and ease of information retrieval. Alternately, the representativeness heuristic involves retrieving information about generic concepts and then a similarity match is made between the item in question and a proposed category. The category association, along with goodness-of-match or degree of similarity, produces confidence or a frequency estimate.

The most important element to take away from this heuristic is that although observers may rely on representativeness to make judgments, they are likely to judge wrongly because the fact that something is more representative does not actually make it more likely to occur.

15.6.3 Conjunction Fallacy

Another bias that individuals may be prone to is the conjunction fallacy (Tversky & Kahneman, 1983). This is the idea that some very specific condition is more probable than a single general one. Tversky and Kahneman (1983) illustrate this phenomenon regarding an individuals' particular assumptions and conclusions with the following scenario. Linda is 31, single, outspoken and very bright. She majored in philosophy. As a student, she was deeply concerned with issues of discrimination and social justice and also participated in antinuclear demonstrations. Is she more likely to be (a) a bank teller, or (b) a bank teller and active in the feminist movement?

The overwhelming majority of survey respondents answered (b), despite the fact that (b) is more restrictive (and therefore less probable) than (a). People report the more complicated scenario as being *more likely* or that it *made more sense*. The conjunction fallacy is counteracted by analyzing individual event probabilities and then combining them.

The reader should be aware of some issues surrounding this heuristic. First, the wording of the "Linda problem" has been thought to directly influence the outcomes and as a result has been studied and criticized more than other types of demonstration of this heuristic (Gigerenzer, 1996; Hertwig & Gigerenzer, 1999). Secondly, an experimental study has shown a correlation between a decrease in the conjunction fallacy in those that had higher test scores on a cognitive reflection test. The conclusion was that "biases are significantly more pronounced for individuals with low cognitive abilities" (Oechssler, Roider, & Schmitz, 2009, p. 150). Third, it has also been shown that the conjunction fallacy becomes less prevalent when subjects are allowed to consult with other subjects (Charness, Karni, & Levin, 2010). For a more complete understanding of this heuristic the reader may wish to review each of these three issues in the articles referenced in this paragraph.

15.6.4 Anchoring and Adjustment Heuristic

Another bias is the anchoring and adjustment heuristic (Tversky & Kahneman, 1973). Humans establish anchors as starting points for their judgments and then base subsequent observations on the initial value that was provided to them. In other words, values developed in the beginning of the individual's learning process or when an individual made similar judgments will be given higher weights than values developed later and will thereby serve as *anchors* for future analysis. Anchors tend to create a bias in future information that will be sought and incorporated into one's later analysis. The status quo is a powerful anchor. It is often easier for individuals to take an existing value and adjust it to their specifications. The anchoring and adjustment effect can be both beneficial or detrimental. Its effects may be combated by independently generating values prior to conducting observations of the real-world value that is the subject of the evaluation and subsequent judgment.

15.6.5 Recognition Heuristic

The recognition heuristic refers to the heuristic by which an individual selects an alternative theory, observation, or conclusion that is the most familiar to them (Goldstein & Gigerenzer, 1999). While this approach may seem to be a fundamentally unsound approach to decision making, Goldstein and Gigerenzer (1999) discovered through their experiments that this approach often outperforms more rigorous approaches to decision making. It can be useful for *on-the-fly* decision making in inconsequential scenarios such as choosing a pair of shoes that you are already familiar with. This approach, however, has both positive and negative effects and should probably be avoided while conducting empirical observations.

15.6.6 Confirmation Bias

One other possible bias involved in thinking is confirmation bias. Hypothesis creation and testing represent an important part of an expert's toolbox. An expert may be exposed to accusations of handling his or her developed hypothesis in such a way that he or she is biased to confirm them. This bias means that data is searched for, interpreted and recalled in such a way that it consistently hampers the possibility that the expert's familiar or *go-to* hypothesis will be discarded. In effect this form of bias may falsely promote acceptance of the expert's *go-to* hypothesis. In this case, the issue is not the application of deceptive strategies to distort data or subsequent conclusions, but is rather the application, and continuous implementation of information processing that occur more or less inadvertently in a non-systematic way. In the common vernacular, this may be likened to *jumping to conclusions prematurely*.

To see confirmation bias in action, let's review a problem described by Oswald and Grosjean (2004). Let us imagine that the following task is presented to us:

> I made up a rule for the construction of sequences of numbers. For instance, the three numbers "2-4-6" satisfy this rule. To find out what the rule is, you may construct other sets of three numbers to test your assumption about the rule I have in mind. I gave you one set of three already, and for every three numbers you come up with, I will give you feedback as to whether it satisfies my rule or not. If you are sure you have the solution, you may stop testing, and tell me what you believe the rule to be. (Oswald & Grosjean, 2004, p. 79)

When presented with this task, how would you start? Which sets of numbers would you draw to test the rule? What is the most efficient way of confirming such a rule? Popper (2002) suggests that the general mistake involves attempting to confirm a hypothesis rather than seeking to falsify it. To demonstrate this, for instance, when a person is typically given the sequence of three numbers *2-4-6*, the individual almost automatically assembles a hypothesis about the rule (e.g., *as a sequence of even numbers*). Then, one tries to test the rule by suggesting different sets of numbers satisfying this assumed rule. (e.g., *8-10-12*, *14-16-18*, and *22-24-26*). A positive feedback is given by these examples (i.e., *this set corresponds to given rule*). After multiple iterations of similar testing, one may feel confident about the hypothesis they've developed, and might stop searching for the problem's solution. They may already be thinking that the correct rule has been identified. Unfortunately, the reality is that their conclusion would be incorrect. As an explanation, the rule specified that the sequences were *any set of three increasing numbers*. The original hypothesis initially formulated was only a subset of all possible sets of three numbers satisfying the rule. The testing strategy initially employed eventually advanced towards a misleading confirmation of the resulting hypothesis. As suggested earlier, the fallacy was the tendency to confirm instead of to falsify. Seeking to falsify the hypothesis of the *first rule that came to mind* would have required fewer rounds of testing. This is a common error in hypothesis testing.

It is impossible to prove that a hypothesis is true scientifically; we can only say that it has not yet been disproven. Thus, attempts to disprove a hypothesis (and formulate it accordingly to encourage such an analysis) are much more fruitful in obtaining meaningful results. In summary, without any knowledge of the truth, whenever people search, interpret or remember information in a way that the corroboration of a hypothesis becomes likely, somewhat like a *self-fulfilling prophecy*, they show confirmation bias.

15.7 Summary

Observation is the central method in which we engage with the real-world. As such, observation is our source for factual data which is assembled into information, and processed into knowledge where we may use it as a source for the generation of inferences and the formulation of decisions.

This chapter has shown how the human observer impacts each and every observation. We have discussed that, in order to make good, repeatable decisions, we must use formal processes. We also address the use of measurement, and provide an account of a number of biases and account for heuristics in an effort to suppress their inadvertent use in decision making. The chapter includes a model and associated four phase process for observation. A formal model that recognizes the relationship between technological systems for observation and the cognitive system of human processing was discussed and proposed as a means for making sense in real-world situations involving decision making.

After reading this chapter, the reader should:

1. Understand the four phase process for proper observation; and
2. Be able to avoid bias in conducting observation.

References

Adams, K. M., & Hester, P. T. (2012). Errors in systems approaches. *International Journal of System of Systems Engineering, 3*(3/4), 233–242.

Audi, R. (Ed.). (1999). *Cambridge dictionary of philosophy*. New York: Cambridge University Press.

Boisot, M. H. (1998). *Knowledge assets: Securing competitive advantage in the information economy*. New York: Oxford University Press.

Charness, G., Karni, E., & Levin, D. (2010). On the conjunction fallacy in probability judgment: New experimental evidence regarding Linda. *Games and Economic Behavior, 68*(2), 551–556. doi:10.1016/j.geb.2009.09.003.

Coombs, C. H. (1964). *A theory of data*. New York: John Wiley & Sons.

Davenport, T. H., & Prusak, L. (2000). *Working knowledge: How organizations manage what they know*. Boston: Harvard Business School Press.

Ellis, W. D. (Ed.). (1997). *A source book of gestalt psychology*. Gouldsboro, ME: The Gestalt Journal Press.

Ellson, D. G. (1963). The scientists' criterion of true observation. *Philosophy of Science, 30*(1), 41–52. doi:10.1086/287910.

Endsley, M. R. (1995). A taxonomy of situation awareness errors. In R. Fuller, N. Johnston, & N. McDonald (Eds.), *Human factors in aviation operations* (pp. 287–292). Aldershot, England: Avery Aviation Ashgate Publishing Ltd.

Feyerabend, P. K. (1962). Explanation, reduction, and empiricism. In H. Feigl & G. Maxwell (Eds.), *Minnesota studies in philosophy of science* (Vol. III—Scientific explanation, space and time, pp. 28–97). Minneapolis: University of Minnesota Press.

French, S. (2013). Cynefin, statistics and decision analysis. *Journal of the Operational Research Society, 64*(4), 547–561. doi:10.1057/jors.2012.23.

Gigerenzer, G. (1996). On narrow norms and vague heuristics: A reply to Kahneman and Tversky. *Psychological Review, 103*(3), 592–596. doi:10.1037/0033-295X.103.3.592.

Gilovich, T., & Savitsky, K. (1996). Like goes with like: The role of representativeness in erroneous and pseudo-scientific beliefs. *Skeptical Inquirer, 20*(2), 30–40. doi:10.1017/CBO9780511808098.036.

Godfrey-Smith, P. (2003). *An introduction to the philosophy of science: Theory and reality*. Chicago: University of Chicago Press.

Goldstein, D. G., & Gigerenzer, G. (1999). The recognition heuristic: How ignorance makes us smart. In G. Gigerenzer & P. M. Todd (Eds.), *Simple heuristics that make us smart* (pp. 37–58). Oxford: Oxford University Press.

Hertwig, R., & Gigerenzer, G. (1999). The 'Conjunction Fallacy' revisited: How intelligent inferences look like reasoning errors. *Journal of Behavioral Decision Making, 12*(4), 275–305. doi:10.1002/(sici)1099-0771(199912)12:4<275:aid-bdm323>3.3.co;2-d.

Klein, G. A., Calderwood, R., & Clinton-Cirocco, A. (1986). Rapid decision making on the fireground. *Proceedings of the Human Factors and Ergonomics Society 30th Annual Meeting* (pp. 576–580). Santa Monica, CA: Human Factors Society.

Kuhn, T. S. (1996). *The structure of scientific revolutions*. Chicago: University of Chicago Press.

Kynn, M. (2008). The 'Heuristics and Biases' bias in expert elicitation. *Journal of the Royal Statistical Society: Series A (Statistics in Society), 171*(1), 239–264. doi:10.2307/30130739.

Lincoln, Y. S., & Guba, E. G. (1985). *Naturalistic inquiry*. Newbury Park, CA: Sage Publications.

Miller, N. L., & Shattuck, L. G. (2005). Applying a dynamic model of situated cognition to the investigation of mishaps. *Proceedings of the Human Factors and Ergonomics Society 49th Annual Meeting, 49*(3), 219–223. doi:10.1177/154193120504900302.

Nagel, E. (1961). *The structure of science: Problems in the logic of scientific explanation*. New York: Harcourt, Brace & World.

Oechssler, J., Roider, A., & Schmitz, P. W. (2009). Cognitive abilities and behavioral biases. *Journal of Economic Behavior & Organization, 72*(1), 147–152. doi:10.1016/j.jebo.2009.04.018.

Oswald, M. E., & Grosjean, S. (2004). Confirmation bias. In R. F. Pohl (Ed.), *Cognitive illusions: A handbook on fallacies and biases in thinking, judgment and memory* (pp. 79–96). New York: Psychology Press.

Popper, K. (2002). *The logic of scientific discovery*. New York: Routledge.

Proudfoot, M., & Lacey, A. R. (2010). *The Routledge dictionary of philosophy* (4th ed.). Abingdon: Routledge.

Sackett, D. L. (1979). Bias in analytic research. *Journal of Chronic Diseases, 32*(1–2), 51–63.

Shattuck, L. G., & Miller, N. L. (2006). Extending naturalistic decision making to complex organizations: A dynamic model of situated cognition. *Organization Studies, 27*(7), 989–1009.

Tversky, A., & Kahneman, D. (1973). Availability: A heuristic for judging frequency and probability. *Cognitive Psychology, 5*(2), 207–232.

Tversky, A., & Kahneman, D. (1974). Judgment under uncertainty: Heuristics and biases. *Science, 185*(4157), 1124–1131. doi:10.1126/science.185.4157.112.

Tversky, A., & Kahneman, D. (1983). Extension versus intuitive reasoning: The conjunction fallacy in probability judgment. *Psychological Review, 90*(4), 293–315.

Chapter 16
Systemic Learning

Abstract If you have followed this book from the beginning, you will note that we have provided you with a systemic frame of reference (Part I), exposed you to the *Who*, *What*, *Why*, *When*, *How*, and *Where* of systemic thinking (Part II), and have discussed the course of action selection and implementation (Chaps. 12–14 in Part III). This chapter will address learning and the individual, group, organizational, and inter-organizational aspects of learning in an effort to further enforce the thoughts on observation from the previous chapter.

16.1 Introduction

Learning is the *raison d'être* of the feedback element in the systemic decision making multimethodology. Every viable organism and organization includes a reflexive response or behavior that permits it to change its behavior based upon its experience. This ability to rationally change, termed learning, is what permits the organism or organization to remain viable. The most successful organisms and organizations not only remain viable, but also grow as a result of the feedback and consequent learning that the organism or organization incorporates within its memory. In advanced life forms and organizations, the feedback mechanisms and learning modes are formalized and permit the inclusion of more complex behaviors. In the sections that follow, we will address learning as the act that permits organizations to detect and recognize errors, analyze the errors, and adapt their behaviors, in a process of organizational learning. We propose the following definition:

> Systemic learning is the ability to detect and correct error.

Because all complex problems require groups of individuals to properly address them, learning has been approached from an organizational perspective. Organizational learning is a multifaceted field of study with branches in four of the

P.T. Hester and K.M. Adams, *Systemic Decision Making*, Topics in Safety, Risk, Reliability and Quality 33, DOI 10.1007/978-3-319-54672-8_16

social sciences: (1) psychology; (2) economics and business; (3) education; and (4) sociology (OECD, 2007). The location of the emerging discipline within the scientific community is not consistent and tends to vary between research institutions, and colleges and universities. The sections that follow will expose you to the evolution of the major theories associated with organizational learning, how organizational learning is an established organizational process, and how systemic learning is a means for measuring learning.

16.2 Learning Theory

The sections that follow will trace the development of organizational learning theories from their early roots in the 1950s to their current form. The principal proponents that supplied significant theoretical insights include Gregory Bateson, the cybernetics movement, and the current thought leaders Donald Schön and Chris Argyris. Each of these will be briefly discussed in order to provide a foundation to understand the importance of learning theory and systemic learning.

16.2.1 Gregory Bateson and Early Learning Theory

Gregory Bateson [1904–1980] was an anthropologist and behavioral scientist who was introduced to cybernetics through his wife, the well-known cultural anthropologist, Margaret Mead [1901–1978] during the Macy Conferences of the 1950s. His introduction to the Macy conferences started a long relationship with many of the leading cyberneticists, who included both Norbert Wiener and John von Neumann (Levy & Rappaport, 1982; Visser, 2003). Bateson was able to apply the ideas from cybernetics to his own work, which resulted in his theorizing that there are two classes of learning, *proto-learning* and *deutero-learning*.

> Bateson distinguishes between two levels of learning, proto- and deutero-learning. These levels of learning are simultaneous. The term deutero-learning describes the context in which (proto-) learning processes occur. You "learn" not only what you are supposed to learn (in a common sense understanding); so, for example, riding a bike, learning a language, or repairing a car – these processes are all proto-learning. At the same time you are learning this, you are also learning something about the world and something about how things occur. You develop habits. This is, at least partly, a result of deutero-learning. (Lutterer, 2012, p. 939)

Bateson went on to develop a more extensive learning theory where he distinguishes between five levels of learning as described in Table 16.1.

Table 16.1 Bateson's five levels of learning (Bateson, 1972)

Phase	Description
0—Zero Learning	"Characterized by *specificity of response*, which-right or wrong-is not subject to correction" (Bateson, 1972, p. 293)
1—Learning I	"*Is change in specificity of response* by correction of errors of choice within a set of alternatives" (Bateson, 1972, p. 293)
2—Learning II	"*Is change in the process of Learning I*, e.g., a corrective change in the set of alternatives from which choice is made, or it is a change in how the sequence of experience is punctuated" (Bateson, 1972, p. 293)
3—Learning III	"*Is change in the process of Learning II*, e.g., a corrective change in the system of sets of alternatives from which choice is made. (We shall see later that to demand, this level of performance of some men and some mammals is sometimes pathogenic)" (Bateson, 1972, p. 293)
4—Learning IV	"*Learning IV would be change in Learning III*, but probably does not occur in any adult living organism on this earth. Evolutionary process has, however, created organisms whose ontogeny brings them to Level III. The combination of phylogenesis with ontogenesis, in fact, achieves Level IV" (Bateson, 1972, p. 293)

Table 16.1 is one of the first attempts to characterize learning as a process that includes the notion of *learning to learn*. Bateson's ideas on learning formed the foundation for the work in organizational learning described in the final section.

16.2.2 Cybernetics and Learning Theory

Learning in an organization is a function of the desired behaviors and the actual outcomes that occur within an organization's processes. The ability to learn is what permits an organization to remain viable in an ever-changing external environment. When we apply our simple definition of systemic learning to an organization, we generate the following definition:

> Systemic organizational learning is the ability of an organization to detect and correct error.

As such, organizational learning is the means through which an organization detects error, and in an effort to correct the error (i.e., to satisfy or improve an established outcome measure), it modifies or changes the process which produced the detected error. The process for detection of the error is feedback, the central element of cybernetics. Figure 16.1 is a depiction of process for the creation of a product or service and the cybernetic feedback that permits detection of error in the product or service delivered to a customer.

Once an error has been detected, the organization must take action to correct the error as well as the processes, associated policies, and procedures that contributed to the error's commission.

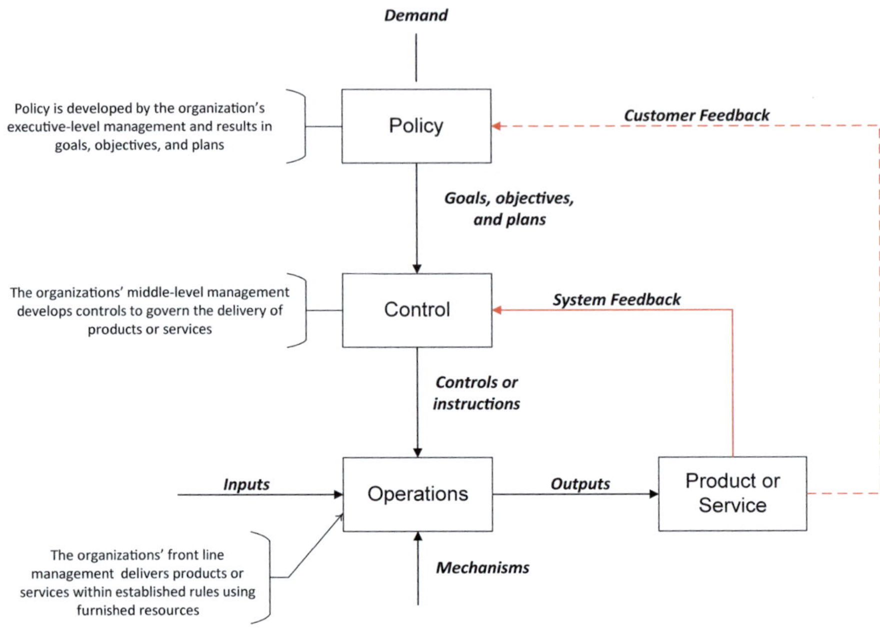

Fig. 16.1 Cybernetic feedback process

16.2.3 Chris Argyris, Donald Schön, and Learning Theory

Imagine the process in Fig. 16.1 is running for a long period of time and producing a large number of products or services. During this period, the process periodically detects the presence of error, reports the error through consequent feedback, and ultimately corrects the error. After some time, the organization's staff adapts—the production processes are adjusted, improved materials and reliable supply sources are established, and equipment operations are tweaked; the process is *in-control* and yielding designed results. The staff's adaptation is the result of learning.

Chris Argyris and Donald Schön followed up on Bateson's work and initiated research in the field of organizational learning. Their research established formal concepts for what they termed single-loop learning and double-loop learning (Argyris & Schön, 1974, 1978).

> Learning is defined as occurring under two conditions. First, learning occurs when an organization achieves what it intended; that is, there is a match between its design for action and the actuality or outcome. Second, learning occurs when a mismatch between intentions and outcomes is identified and it is corrected; that is, a mismatch is turned into a match. (Argyris, 1999, p. 67)

As described in the earlier paragraph, it is the organization's members who adapt and learn, focusing the learning on the individuals within the organization. Figure 16.2 is a depiction of the two types of learning that occur in organizations.

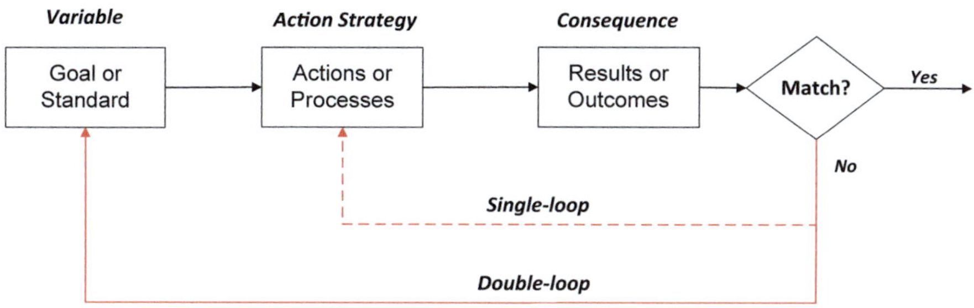

Fig. 16.2 Single-loop and double-loop learning

- **Single-loop learning**. "Single-loop learning occurs when a mismatch is detected and corrected without changing the underlying values and status quo that govern the behaviors" (Argyris, 2003, p. 1178).
- **Double-loop learning**. "Double-loop learning occurs when a mismatch is detected and corrected by first changing the underlying values and other features of the status quo" (Argyris, 2003, pp. 1178, 1179).

Argyris (1999) provides an excellent description of both single-loop learning and double-loop learning:

> Whenever an error is detected and corrected without questioning or altering the underlying values of the system (be it individual, group, intergroup, or organizational or interorganizational), the learning is single-loop. The term is borrowed from electrical engineering or cybernetics where, for example, a thermostat is defined as a single-loop learner. The thermostat is programmed to detect states of "too cold" or "too hot," and to correct the situation by turning the heat on or off. If the thermostat asked itself such questions as why it was set at 68 degrees, or why it was programmed as it was, then it would be a double-loop learner. (Argyris, 1999, p. 68)

It should now be clear from the depiction in Fig. 16.2, the definitions, and Argyris' description that single-loop learning and double-loop learning are essential elements of control in the management of organizational processes. The next section will frame learning using a construct that permits its mathematical representation.

16.3 Relating Performance to First-order, Second-order, and Deutero-Learning

The concepts associated with learning levels continue to be the subject of ongoing discussions (Visser, 2007). However, we can establish that "deutero-learning is not a third level but is the process of learning about either of the other two forms of learning" (Easterby-Smith & Lyles, 2003, p. 51). Figure 16.3 provides the framework for this understanding based on concepts of performance introduced in Chap. 13.

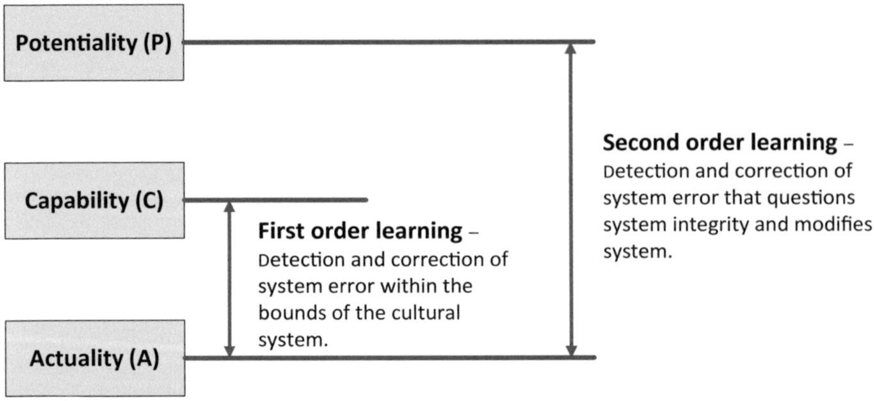

Fig. 16.3 Constructs for first, second, and deutero-learning (Adams & Bradley, 2012, p. 18)

- **First-order Learning**: Detection and correction of system error within the bounds of the current system.
- **Second-order Learning**: Detection and correction of system error that questions system integrity and modifies system.
- **Deutero-Learning**: Learning when, why, and how to do first- and second-order learning. Emphasis is on design and execution of the learning system.

16.4 Learning in Organizations

We have established that organizational learning is a systemic process where learning is defined as *the ability of an organization to detect and correct error.* Organizationally, learning is a behavioral process that permits organizations to detect and recognize errors, analyze the errors, and adapt their behaviors in an effort to remain viable. Viability is a function of the organization's relationship with the environment, which is defined as "a set of elements and their relevant properties, which elements are not part of the system but a change in any of which can produce a change in the state of the system" (Ackoff, 1971, pp. 662, 663). For most organizations, these environmental elements include competitors (both known and unknown), government regulators, weather, financial markets, suppliers, buyers, and technological innovation. Organizations must develop strategies to address each of these environmental elements in order to maintain competitive advantage and ultimately, remain viable. The three sections that follow will show how successful strategies for developing and maintaining a competitive advantage rely upon organizational learning.

16.4.1 Strategy and Competitive Advantage

The development of a formal strategy for success is an essential element of every viable organization. Harvard professor and strategy expert Michael Porter (1991) has developed a *chain of causality* for businesses that links environmental circumstances and firm actions to market success. The chain of causality and associated determinants are depicted in Fig. 16.4.

Figure 16.4 clearly shows that the ability to develop and sustain competitive advantage is a major causal determinant in organizational success. The next section will discuss the relationship between organizational learning and competitive advantage.

16.4.2 Competitive Advantage and Organizational Learning

In order to survive in a dynamic, interconnected, worldwide, and fast-moving technological marketplace, organizations must control a number of the environmental elements previously addressed in order to develop and maintain competitive advantage. Principal among these is technological innovation. Sustained technological innovation requires organizations to "create superior knowledge management capabilities, and thereby foster ongoing innovation" (Rebernik & Širec, 2007, p. 407). Figure 16.5 depicts the relationships between (1) competitive advantage;

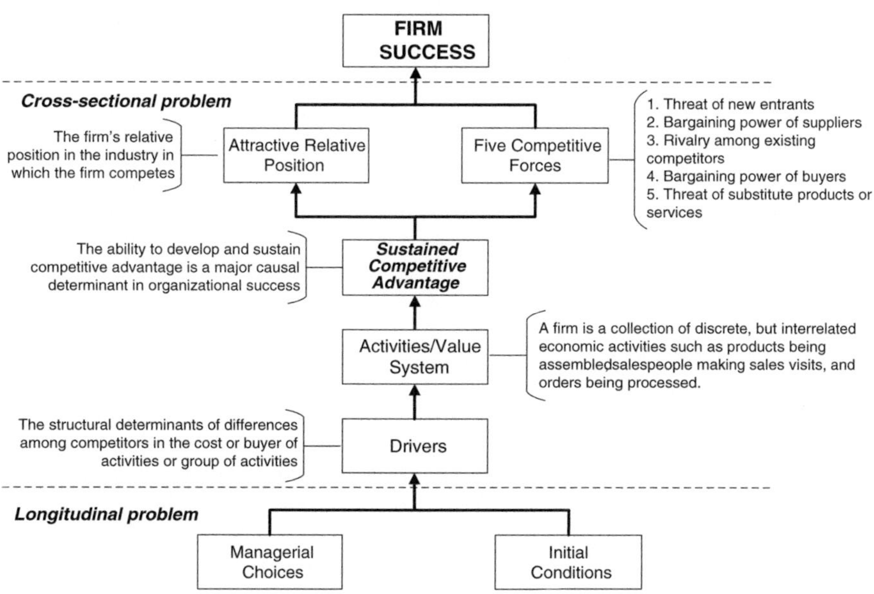

Fig. 16.4 Porter's determinants of success (adapted from Porter, 1991)

(2) continuous innovation; (3) new knowledge; and (4) learning and unlearning at the individual, organizational, and environmental levels. In order to successfully implement the attributes shown in Fig. 16.5, an organization must adopt a formal learning management plan which addresses: (1) building the organizational culture for learning; (2) instituting a formal knowledge management plan that formally addresses knowledge assets efficiently; and (3) facilitates effective knowledge application as a step in becoming a learning organization.

The elements of the learning management plan ensure that the organization can foster both explicit and tacit knowledge resident in its knowledge workers in order to retain a truly sustainable competitive advantage.

> Much of our knowledge is only the basis for a transient competitive advantage as our competitors reverse-engineer our products, copy our best practices and develop parallel (or superior) technologies. In contrast, tacit knowledge and superb knowledge management capabilities can form the basis of a relatively inimitable competitive advantage. Tacit knowledge can be spread within a firm but will be very difficult for other firms to copy. Superior knowledge management capabilities are the basis for the rapid acquisition and spread of new knowledge and, therefore, foster continuous innovation and improvement. (Rebernik & Širec, 2007, p. 416)

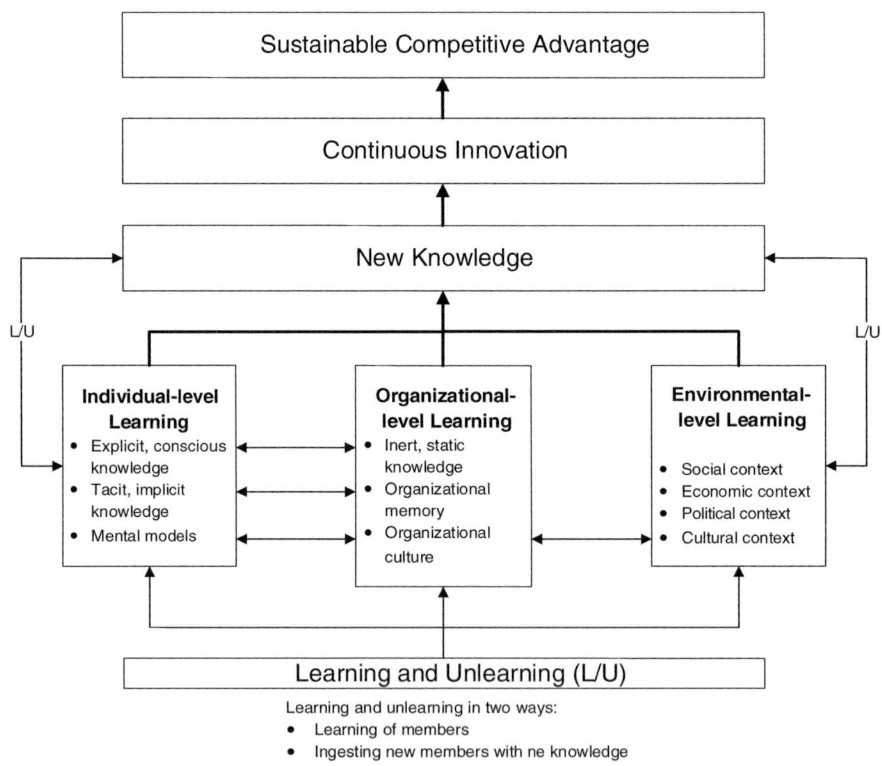

Fig. 16.5 A typology of attributes needed to create new knowledge (adapted from Rebernik & Širec, 2007)

16.4.3 Leaders and the Learning Organization

MIT professor Edgar Schein (1992) states:

> In fact, one could argue that the only thing of real importance that leaders do is to create and manage culture and that the unique talent of leaders is their ability to understand and work with culture. If one wishes to distinguish leadership from management or administration, one can argue that leaders create and change cultures, while managers and administrators live within them. (p. 5)

Possibly the most critical element in establishing a learning organization is the development and sustainment of an organizational culture for learning. Schein has ten essential characteristics and an associated question and scale, presented in Table 16.2, that are relevant to the capacity of an organization to learn.

Each of the questions associated with the characteristics in Table 16.2 serves to examine a critical dimension of the culture required to maintain a productive and innovative learning environment. The leaders in the organization that invoke new paradigms must ensure that they manage the environment or context in which work is done, rather than controlling the workers themselves (Stewart, 1997; Sveiby, 1997). The next section will discuss the new roles and challenges facing workers in a learning organization.

16.4.4 Workers in the Learning Organization

The principal worker in a learning organization is termed a *knowledge worker*. In an earlier text, we proposed the following definition for a knowledge worker: "A human being that possesses the required knowledge, skills, and abilities to perform a specific function in support of a desired output or outcome in the achievement of a purposeful objective or goal" (Hester & Adams, 2014, p. 175). As such, a knowledge worker normally has highly specialized knowledge, skills, and abilities that add value to the organizations goals and objectives. In Table 16.3, we contrast five characteristics of the traditional worker with the knowledge worker.

Knowledge workers are typically working with the data-information-knowledge-decisions-metrics relationships depicted in Figs. 9.4 and 9.5 in Chap. 9. Because the knowledge, skills, abilities, and context within which they work are so different for knowledge workers, leaders who manage the learning organization are faced with some unique challenges.

16.4.5 Leadership Challenges in the Learning Organization

Renowned management consultant and professor Peter Drucker [1909–2005] popularized the idea of knowledge workers as far back as (1959). He discussed this further in an important article *The new society of organizations* (1992), stating:

Table 16.2 Characteristics of a learning culture (Schein, 1992)

Characteristic	Question
1. Organization-environment relationship	✓ *Is the leadership environment viewed as being manageable?* • Position on a sliding scale directly affects ability to learn
2. Nature of human activity	✓ *Are workers viewed by leaders to be proactive problem solvers and learners?* • Position on a sliding scale directly affects ability to learn relative to changes in the external environment
3. Nature of reality and truth	✓ *Do the leaders share the belief that solutions to problems are derived from truth and that truth can be found anywhere?* • Position on a sliding scale directly affects ability to learn relative to new problems being posed from the external environment
4. Nature of human nature	✓ *Do the leaders have faith in people and believe that they want to learn, improve, and remain viable?* • Knowledge, skills, and abilities are widely distributed and require leaders to be more dependent on others
4. Nature of human relationships	✓ *Do the leaders foster individualism or groupism?* • Complex interdependent solutions that require learning favor the groupist kind of organization
5. Nature of time	✓ *What is the leadership time horizon?* • A short range vision requires no learning behaviors. A near to far future perspective requires adaptation and learning
6. Information and communications	✓ *How does leadership handle information and communications?* • Multichannel communications where everyone is connected to everyone else permit robust learning and knowledge transfer
7. Subcultural uniformity versus diversity	✓ *How does leadership view diversity of resources?* • A robust, diverse organization is more likely to be able to react to a turbulent environment and cope with unpredicted events
8. Task versus relationship orientation	✓ *How does leadership view relationships versus tasks?* • In a complex turbulent environment where technological interdependence is high, relationship achieves the level of trust and communications that make joint problem solving possible
9. Linear versus systemic field logic	✓ *How does leadership view problems?* • The ability to think systemically, to analyze joint causal effects, by abandoning a linear causal model permits complex mental models to be used in joint problem solving

In this society, knowledge is the primary resource for individuals and for the economy overall. Land, labor, and capital-the economist's traditional factors of production- do not disappear, but they become secondary. Knowledge by itself produces nothing. It can become productive only when it is integrated into a task. And that is why the knowledge society is also a society of organizations: the purpose and function of every organization, business and nonbusiness alike, is the integration of specialized knowledges into a common task. (p. 96)

Table 16.3 The ideal-types of traditional work and knowledge work (Pyöriä, 2005, p. 124)

Characteristic	Traditional worker	Knowledge worker
Education	Requires some formal education and on-the-job training	Requires extensive formal education and continuous on-the-job training
Skills	Strictly defined skills	Transferrable skills
Nature of work	High level of standardization, involves working with physical matter either directly or indirectly through electronic interfaces (e.g., control of production processes)	Low level of standardization, involves working with abstract knowledge and symbols (e.g., design and planning of production processes)
Organization	Ranges from bureaucracy to teams, fixed roles and positions, knowledge as a secondary production factor	Ranges from professional bureaucracies to self-managing teams, job and task circulation, knowledge as a primary production factor
Medium of work	Physical materials and/or people	Symbols and/or people

Drucker was concerned that the managers trained to manage the traditional worker would be ill-prepared to manage the new class of knowledge workers beginning to take over organizations as we moved away from an industrial age to an information age. He specified six specific challenges:

1. Knowledge worker productivity demands that we ask the question: "What is the task?"
2. It demands that we impose the responsibility for their productivity on the individual knowledge workers themselves. Knowledge workers have to manage themselves. They have to have autonomy.
3. Continuing innovation has to be part of the work, the task and the responsibility of knowledge workers.
4. Knowledge work requires continuous learning on the part of the knowledge worker, but equally continuous teaching on the part of the knowledge worker.
5. Productivity of the knowledge worker is not—at least not primarily—a matter of the quantity of output. Quality is at least as important.
6. Finally, knowledge worker productivity requires that the knowledge worker is both seen and treated as an "asset" rather than a "cost". It requires that knowledge workers want to work for the organization in preference to all other opportunities. (Drucker, 1999, p. 142)

Knowledge workers are the mechanisms through which a learning organization gathers data, forms data into information, processes information into knowledge, and uses this knowledge to create and sustain competitive advantage in the marketplace. Leaders must not only address the learning culture, but also they must also ensure that both (1) a formal knowledge management plan and (2) modern information technology assets are in place to enable knowledge workers to store, transfer, and access knowledge within the entire organization.

Knowledge is the organization's premier asset and must be treated as such. This premier asset includes *tacit knowledge* that is embedded in the individual knowledge workers and *explicit knowledge* that may be embedded in the organization's information systems and databases. Knowledge workers flourish in a properly ran learning organization.

16.5 Avoiding the Type VI Error

So far this chapter we have discussed learning in all its forms—from the lens of an individual, group, and organization. But we have yet to touch on the role of learning in avoiding the Type VI error. Recall that the Type VI error is characterized as an unsubstantiated inference. It is most famously described by Holland (1986) as, "Correlation does not imply causation…" (p. 945). But, how, exactly, *do* we establish causality? Utts (2014) offers the following advice:

> Given the number of possible explanations for the relationship between two variables, how do we ever establish that there actually is a causal connection? It isn't easy. Ideally, in establishing a causal connection, we would change nothing in the environment except the suspected causal variable and then measure the result on the suspected outcome variable.
>
> The only legitimate way to establish a causal connection statistically is through the use of randomized experiments. (pp. 239, 240)

Utts (2014, p. 240) provides four conditions for evidence of a possible causal connection: (1) a reasonable explanation of cause and effect, (2) the connection happens under varying conditions, (3) potential confounding variables are ruled out, and (4) the presence of a "dose-response" relationship. Each condition is expanded on below.

There is a reasonable explanation of cause and effect. Utts (2014) discusses the correlation between the number of pages in a hardcover book and the price of the book. It doesn't seem reasonable that higher prices translate necessarily to more pages, but the converse seems reasonable. That is, more pages likely result in a higher price. Thus, this is a reasonable explanation for how an increase in the number of pages could *cause* an increase in the book's price.

The connection happens under varying conditions. If many observational studies under varying conditions also show the same link between two variables, evidence for causality is strengthened.

Potential confounding variables are ruled out. Confounding variables, extraneous variables that correlate with both a dependent variable and an independent variable, must be ruled out in order to strengthen the casual belief between two variables.

There is a "dose-response" relationship. An increase in the dose of an explanatory (independent) variable results in an increased magnitude of response from the dependent variable.

It is clear that it is not a simple undertaking to establish causality. Further, we may be prone to many causal fallacies, as summarized by Damer (2012):

> There may be a confusion between a necessary condition and a sufficient condition (confusion of a necessary with a sufficient condition), or the causal factors in a situation may be too few to account for the effect in question (causal oversimplification). Some faulty causal analyses claim that because something happened right after another event, it was caused by that prior event (post hoc fallacy), while others confuse an effect with a cause (confusion of cause and effect) or fail to recognize that there is a third or common event or situation that provides a better causal account for two different events erroneously thought to be causally related (neglect of a common cause). Finally, a faulty causal analysis may lead one to draw an unwarranted conclusion that a series of events leading to an inevitable and undesirable end follows from a single identified event (domino fallacy) or that is possible to make reliable predictions about a chance event based on the past performance of similar chance events (gambler's fallacy). (p. 188)

So, ultimately, how are the establishment of causality and learning linked? The answer is through theory. We have set out in this text to guide the reader toward the development of a sophisticated representation of a mess under study and this model represents our theory. Argyris and Schön (1974, p. 5) elaborate:

> Theories are vehicles for explanation, prediction, or control. An explanatory theory explains events by setting forth propositions from which these events may be inferred, a predictive theory sets forth propositions from which inferences about future events may be made, and a theory of control describes the conditions under which events of a certain kind may be made to occur (Argyris & Schön, 1974, p. 5).

The theory (model) that we develop about our mess helps us explain it, predict it, and control it. Part of this control involves improvement of our system's performance through learning. But how?

In reality, the best we can do is experiment with, and learn about, our system. This is not ideal, and not always possible, but it is the *real solution* to our causality (and understanding) problem. If necessary, we update our theory (model) to reflect the results of our observations if they are inconsistent. We will never have a full set of data at our disposal for a mess due to the *darkness principle*, but nonetheless, we must continue to revise our understanding of our problems and improve our

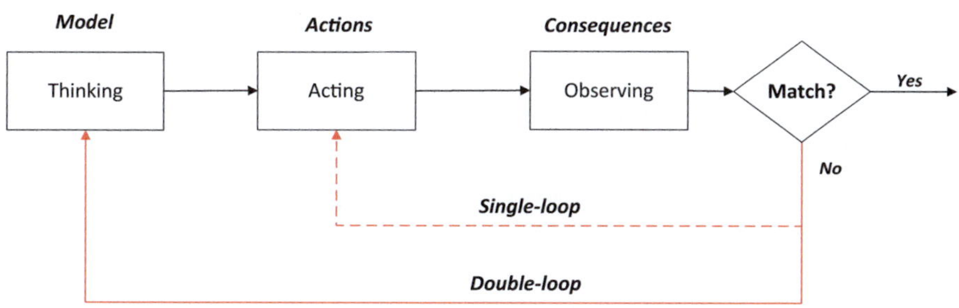

Fig. 16.6 Single-loop and double-loop learning applied to the systemic decision making process

system's model (and accompanying performance) as best as possible. After all, that's why we modeled (i.e., created a purposeful abstraction of reality) in the first place.

Ultimately, our learning process should follow the process shown in Fig. 16.6, which shows single-loop learning and double-loop learning applied to the systemic decision making process discussed throughout this text. In this case, double-loop learning leads to a modification of our situation's model (i.e., a return to the thinking stage), whereas single-loop learning leads to a different action (i.e., a return to the acting stage). This fundamental model should guide our learning throughout the systemic decision making process.

16.6 Summary

This chapter addressed learning and the individual, group, organizational, and inter-organizational aspects of learning. Learning is the *raison d'être* of the feedback element in the systemic decision making multimethodology and it is a fundamental requirement for maintaining viability. This chapter discussed the basics of learning theory, learning in organizations, and how to properly implement systemic learning.

After reading this chapter, the reader should:

1. Be able to explain single- and double-loop learning; and
2. Understand how to use learning to improve our system's performance.

References

Ackoff, R. L. (1971). Toward a system of systems concepts. *Management Science, 17*(11), 661–671. doi:10.1287/mnsc.17.11.661.

Adams, K. M., & Bradley, J. M. (2012). *Productivity, performance and availability and intermediate level maintenance*. Retrieved from Norfolk, VA.

Argyris, C. (1999). *On organizational learning* (2nd ed.). Maldem MA: Blackwell Publishers Inc.

Argyris, C. (2003). A life full of learning. *Organization Studies, 24*(7), 1178–1192. doi:10.1177/0170840603024700 9.

Argyris, C., & Schön, D. A. (1974). *Theory in practice: Increasing professional effectiveness*. San Francisco, CA: Jossey-Bass.

Argyris, C., & Schön, D. A. (1978). *Organizational learning: A theory of action perspective*. Reading, MA: Addison-Wesley.

Bateson, G. (1972). *Steps to an ecology of mind: A revolutionary approach to man's understanding of himself*. San Francisco: Chandler Publishing.

Damer, T. (2012). *Attacking faulty reasoning* (7th ed.). Boston, MA: Cengage Learning.

Drucker, P. F. (1959). *The landmarks of tomorrow: A report on the new "Post-Modern" world*. New York: Harper and Row.

Drucker, P. F. (1992). The new society of organizations. *Harvard Business Review, 70*(5), 95–105.

Drucker, P. F. (1999). *Management challenges for the 21st century*. New York: Harper Collins.

Easterby-Smith, M., & Lyles, M. (2003). Re-reading "organizational learning": Selective memory, forgetting, and adaptation. *The Academy of Management Executive, 17*(2), 51–55. doi:10.5465/AME.2003.10025192.

Hester, P. T., & Adams, K. M. (2014). *Systemic thinking—Fundamentals for understanding problems and messes*. New York: Springer.

Holland, P. W. (1986). Statistics and causal inference. *Journal of the American Statistical Association, 81*(396), 945–960.

Levy, R. I., & Rappaport, R. (1982). Gregory Bateson, 1904–1980. *American Anthropologist, 84* (2), 379–384. doi:10.1525/aa.1982.84.2.02a00100.

Lutterer, W. (2012). Deutero-learning. In N. M. Seel (Ed.), *Encyclopedia of the sciences of learning* (pp. 939–941). Boston, MA: Springer, US.

OECD. (2007). *Revised field of science and technology (FOS) classification in the Frascati Manual*. Retrieved from Paris.

Porter, M. E. (1991). Towards a dynamic theory of strategy. *Strategic Management Journal, 12* (52), 95–117. doi:10.1002/smj.4250121008.

Pyöriä, P. (2005). The concept of knowledge work revisited. *Journal of Knowledge Management, 9*(3), 116–127. doi:10.1108/13673270510602818.

Rebernik, M., & Širec, K. (2007). Fostering innovation by unlearning tacit knowledge. *Kybernetes, 36*(3/4), 406–419. doi:10.1108/03684920710747039.

Schein, E. H. (1992). *Organizational culture and leadership* (2nd ed.). San Francisco: Jossey-Bass.

Stewart, T. A. (1997). *Intellectual capital: The new wealth of organizations*. New York: Doubleday.

Sveiby, K. E. (1997). *The new organizational wealth: Managing and measuring knowledge-based assets*. San Francisco: Berrett-Koehler.

Utts, J. (2014). *Seeing through statistics*. Boston, MA: Cengage Learning.

Visser, M. (2003). Gregory Bateson on deutero-learning and double bind: A brief conceptual history. *Journal of History of the Behavioral Sciences, 39*(3), 269–278. doi:10.1002/jhbs.10112.

Visser, M. (2007). Deutero-learning in organizations: A review and reformulation. *The Academy of Management Review, 32*(2), 659–667. doi:10.5465/AMR.2007.24351883.

Chapter 17
Ford Pinto Case Study

Abstract This text has discussed a real estate-focused case study throughout in disparate pieces. The aim of this chapter is to provide a cradle-to-grave case study which illustrate the entire multimethodology developed in the text. The case study focuses on the Ford Pinto and the National Highway Transportation Safety Administration.

17.1 Introduction

This chapter discusses a complete case study based on the Ford Pinto and its relationship with the US auto industry and federal government in the early 1970s. In the chapter, we will utilize the entire multimethodology developed in the text in the assessment of this case. The aim of the chapter is to demonstrate the utility of systemic decision making multimethodology on a practical (albeit historical) problem. A snapshot of the process discussed during this chapter is provided in Fig. 17.1. The figure originally appeared as Fig. 3.7 in Chap. 3.

This chapter will discuss problem structuring, thinking (to include all six perspectives for each of the constituent problems), and acting.

17.2 Problem Structuring

In 1970, the National Highway Transportation Safety Administration (NHTSA) was established with the following mission: "Save lives, prevent injuries and reduce economic costs due to road traffic crashes, through education, research, safety standards and enforcement activity."[1] The NHTSA, in January 1970, proposed a set of strict fuel-system integrity regulations. Concurrently, international competition for compact car sales had heated up in the USA. Competitors such as Volkswagen and Toyota had moved into the US market.

[1]http://www.nhtsa.gov/About+NHTSA/NHTSA's+Core+Values.

P.T. Hester and K.M. Adams, *Systemic Decision Making*, Topics in Safety, Risk, Reliability and Quality 33, DOI 10.1007/978-3-319-54672-8_17

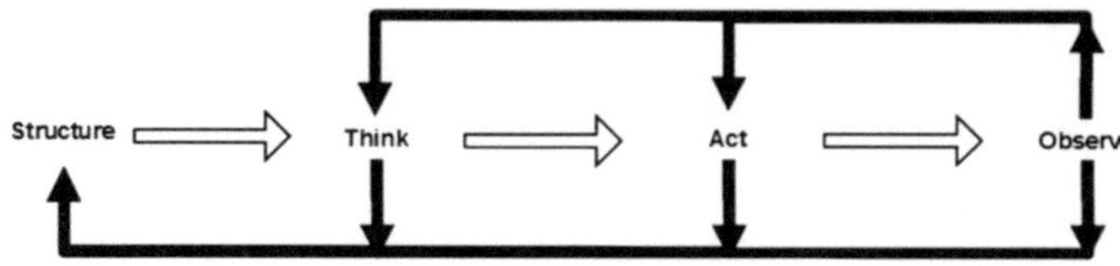

Fig. 17.1 Systemic decision making process (from Chap. 3)

Stricter requirements (which Ford met), along with increased competition, turned up the heat on Ford. In investigating the new NHTSA standards, Ford discovered a potential issue with respect to the gas tank of its Ford Pinto (a new subcompact car) and rear impacts. Its cost-benefit analysis showed a financial benefit in making payouts for injury and death due to rear impacts, as compared with the costs accompanying an auto recall. Ford lobbied for pushing this standard off and eventually, in part due to highly publicized injuries and deaths as a result of their negligence, the Pinto was recalled.

To investigate this mess, we have identified the following two problems for further analysis (with the problem owner in parentheses):

1. (Ford) In the early 1970s, due to an influx of international competition and a renewed national focus on safety, Ford struggled to maintain its market share.
2. (NHTSA) In the early 1970s, the NHTSA was looking to standardize safety guidelines for vehicles in the USA

We will begin by analyzing Ford's problem (Problem 1).

17.3 Problem 1: Ford Problem

The following subsections address the systemic thinking perspectives as they pertain to the first problem, for which Ford is the owner:

> In the early 1970s, due to an influx of international competition and a renewed national focus on safety, Ford struggled to maintain its market share.

17.3.1 Who Perspective

The following subsections discuss the *who* perspective analysis for this problem.

17.3.1.1 Stakeholder Brainstorming

Brainstorming stakeholders for the Ford problem yield the following stakeholders and their associated wants:

1. Ford wants to maximize its market share
2. NHTSA wants safe vehicle operation
3. Foreign competitors want to maximize their market share
4. Domestic competitors want to maximize their market share
5. Customers want consumer confidence.

It should be noted that foreign and domestic competitors are being considered separately as they are subject to distinct laws and regulations. While many more individuals and groups could be added into the analysis, it is thought that the initial stakeholder analysis should include, at a minimum, these five entities and their associated desires.

17.3.1.2 Stakeholder Classification

Table 17.1 shows evaluations of the attributes and class for each of the stakeholders identified in the previous section. They have been sorted according in decreasing order of prominence.

Clearly, the most prominent stakeholder is Ford. Ford's prominence is obvious in its own problem. The next tier of prominence is the NHTSA and customers. The NHTSA's prominence comes from its power to influence legislation as well as its legitimacy as a federal agency, while customers' prominence comes from their purchasing power, as well as their legitimate interest in a safe, affordable vehicle. Finally, foreign and domestic competitors are both powerful as they represent direct competition to Ford, but they are not legitimate from Ford's perspective as they wish to see Ford fail, and they are not urgent with respect to Ford's problem as it doesn't affect them directly.

Table 17.1 Stakeholder classification

Stakeholder	Stakeholder attribute			Prominence
	Power	Legitimacy	Urgency	
Ford	1	1	1	1.0
NHTSA	1	1	0	0.67
Customers	1	1	0	0.67
Foreign competitors	1	0	0	0.33
Domestic competitors	1	0	0	0.33

17.3.1.3 Stakeholder Attitude Evaluation

Table 17.2 shows evaluations of the potential for threat and potential for cooperation for each of the stakeholders identified in the previous subsection. These two parameters provide an identification of the attitude of each stakeholder. They have been sorted in decreasing order of support according to their assigned stakeholder attitude.

Only Ford is supportive in this scenario. The NHTSA is indifferent and offers neither threat nor cooperation potential. Customers have both the potential for threat and cooperation due to their consumer purchases. Foreign and domestic competitors have full potential for threat and no potential for cooperation due to their business relationship with Ford.

17.3.1.4 Stakeholder Objective Mapping

With classification and attitude defined in the previous two sections, Fig. 17.2 shows a stakeholder objective map, including the influence (direction and magnitude) for all identified stakeholders involved in the problem. The thicker the line, the stronger the causal influence.

17.3.1.5 Stakeholder Engagement Priority

In order to calculate the stakeholder engagement priority for all the stakeholders in the Ford problem, we need to calculate k_i^{in}, k_i^{out}, $s_i^{*-\text{in}}$, $s_i^{*-\text{out}}$, *Popularity*, and *Activity*, in accordance with equations found in Chap. 6. These results are shown in Table 17.3.

We then sort the stakeholders by activity first (in descending order), and then, by popularity (in ascending order). Table 17.4. illustrates the order in which stakeholders should be engaged in support of this effort.

It is clear that the NHTSA should be prioritized for this effort. This is due to their large influence and connectivity; however, influencing them may be problematic. Next up for engagement is the domestic competitors and then the foreign competitors. Domestic competitors are a higher priority as they have a greater influence

Table 17.2 Stakeholder attitude evaluation

Stakeholder	Potential for threat	Potential for cooperation	Support
Ford	0	1	1
NHTSA	0	0	0
Customers	1	1	0
Foreign competitors	1	0	−1
Domestic competitors	1	0	−1

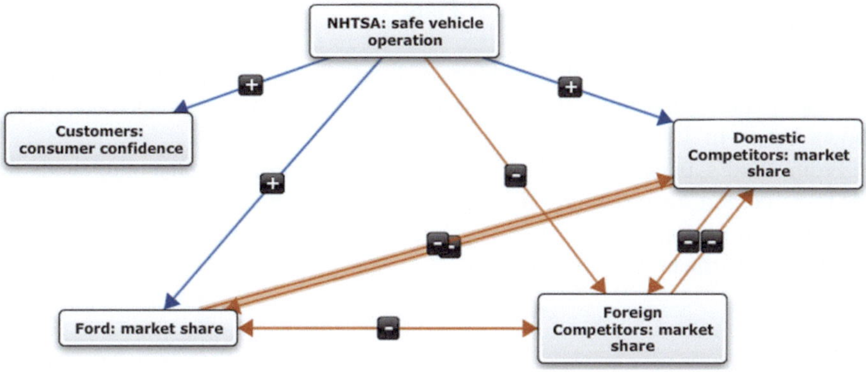

Fig. 17.2 Stakeholder objective map

Table 17.3 Network characteristics

Stakeholder	k_i^{in}	k_i^{out}	$s_i^{*-\text{in}}$	$s_i^{*-\text{out}}$	Activity	Popularity
NHTSA: safe vehicle operation	0	4	0	1	2	0
Ford: market share	3	2	1	0.75	1.22	1.73
Domestic competitors: market share	3	2	1	0.75	1.22	1.73
Foreign competitors: market share	3	2	0.75	0.5	1	1.5
Customers: consumer confidence	1	0	0.25	0	0	0.5

Table 17.4 Stakeholder prioritization

Stakeholder	Activity	Popularity	Engagement priority
NHTSA: safe vehicle operation	2	0	1
Domestic competitors: market share	1.22	1.73	2
Foreign competitors: market share	1	1.5	3
Customers: consumer confidence	0	0.5	4
Ford: market share	1.22	1.73	n/a

on the US market. Last are the customers as they have influence in the problem, but it is minimal. This is somewhat surprising but is likely due to their lack of a direct influence on Ford itself, but rather on the industry in general. Ford is removed from engagement consideration as they are the problem owner and, as a result, their engagement is automatic.

17.3.1.6 Stakeholder Management Plan

The final step in analyzing this problem as it pertains to the who perspective is to develop a stakeholder management plan. Ford's stakeholder management plan is shown below in Table 17.5.

From Ford's perspective, their focus is to monitor and collaborate with the NHTSA, defend against domestic and foreign competitors, and monitor and collaborate with customers as best as possible. These all seem like reasonable strategies for engagement moving forward.

17.3.2 What Perspective

The following subsections discuss the *what* perspective analysis for this problem.

17.3.2.1 Articulate Objectives

In terms of Ford, it must be concerned with both short- and long-term company performance. So, we can define one fundamental objective as *Maximize profit*. However, there are numerous mechanisms for maximizing profit that could be problematic in the long run (i.e., unethical and/or illegal), so they should consider these concerns as well. Thus, we can add an objective of *Minimize risk*. Now, we can organize our objectives.

17.3.2.2 Fundamental Objectives Hierarchy

Organizing our two fundamental objectives into a hierarchy yields Fig. 17.3, showing a further decomposition of our objectives. Profit is broken down into

Table 17.5 Ford's stakeholder management plan

Stakeholder name	Wants	Prominence	Support	Priority of engagement	Strategy
NHTSA	Safe vehicle operation	0.67	0	1	Monitor/ Collaborate
Domestic competitors	Market share	0.33	−1	2	Defend
Foreign competitors	Market share	0.33	−1	3	Defend
Customers	Consumer confidence	0.67	0	4	Monitor/ Collaborate
Ford	Market share	1.0	1	n/a	Involve

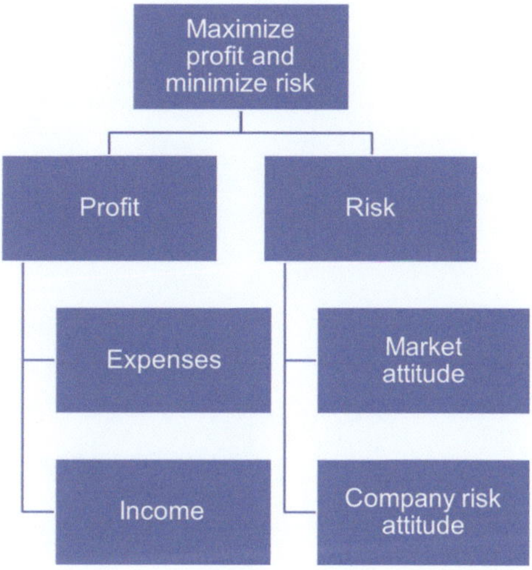

Fig. 17.3 Fundamental objectives hierarchy

expenses and income, which represent a standard way of evaluating a company's balance sheet (i.e., profit = income − expenses). Risk can be decomposed into market attitude (how do consumers feel about risk) and company risk attitude (how the company feels internally about risk).

17.3.2.3 Means-Ends Network

The means-ends network shows our understanding of the means necessary to produce our desired ends (i.e., our fundamental objectives). Using the same two fundamental objectives as before, we can create the network shown in Fig. 17.4. Both profit and risk are means to achieve the ends of maximize profit and minimize risk. Additionally, profit influences risk and vice versa. A high number of cars sold, high profit per car, and low number of recalls lead to high profit. The number of recalls, number of insurance payouts (as a result of injuries and/or deaths due to vehicle malfunctions), and the average cost of insurance payouts are means which help to influence the end of risk.

17.3.2.4 FCM Update

Armed with our means-ends network, we can now integrate it into our existing FCM. Our modified FCM reflects the revision of Ford's objective to incorporate both short- and long-term concerns. This revised scenario depiction is shown in Fig.17.5.

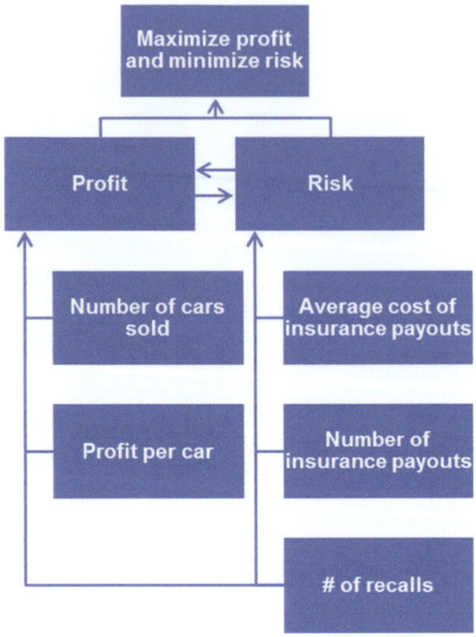

Fig. 17.4 Means-ends network

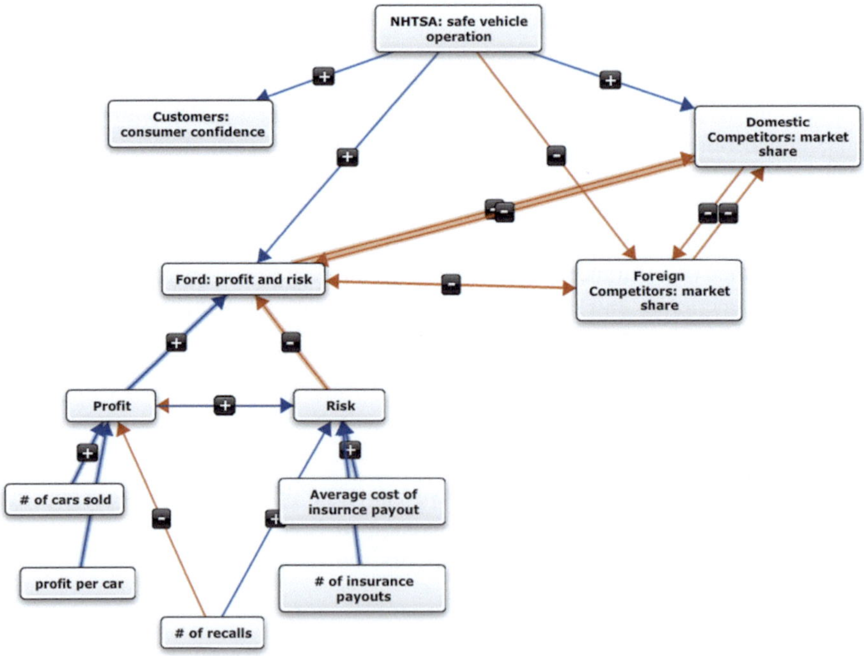

Fig. 17.5 Updated FCM with means-end network incorporated

17.3.3 Why Perspective

The following subsections discuss the *why* perspective analysis for this problem.

17.3.3.1 Motivation/Feedback Analysis

If we are interested primarily in the *profit and risk* concept (Ford's objective), then we can make a few observations:

- Ford's *profit and risk* concept is involved in a vicious circle with both foreign and domestic competitors. This is not a surprise; in a capitalistic endeavor, this should be expected.
- It seems as though Ford should have a relationship with the NHTSA. As Ford's profit increases and it seeks to minimize risk, safe vehicle operation will increase. So, a link is necessary to reflect this change.
- Also, an increase in consumer confidence should lead to an increase in Ford's profit, which leads to an increase in the risk that Ford is willing to take, which decreases consumer confidence. This is accurate but unfortunate.
- Additionally, as the # of recalls goes up, consumer confidence decreases. This link does not currently exist but should.

17.3.3.2 FCM Update

Our feedback analysis in the previous subsection indicates the need for a causal link between *profit and risk* and *safe vehicle operation*, as well as a link from *# of recalls* to *consumer confidence*. These changes are reflected in Fig. 17.6.

17.3.3.3 Proposed Changes During Act Stage

No new proposed changes to the FCM are required.

17.3.4 Where Perspective

The following subsections discuss the *where* perspective analysis for this problem.

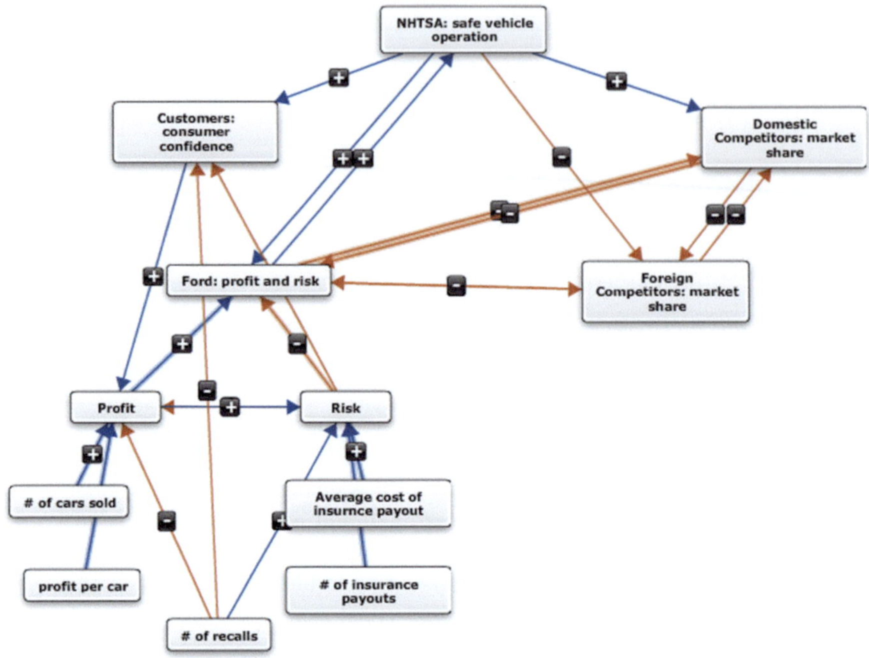

Fig. 17.6 Updated FCM with feedback analysis incorporated

17.3.4.1 Boundary Articulation

Based on analysis of critically heuristic boundary issues of our problem using guidelines found in Ulrich (2000), we can generate the boundary critique shown in Table 17.6.

Assessment of our boundary yields a few insights. The NHTSA is seen as having too much power and knowledge. Ford would rather an independent entity, and customers lead a drive toward more safety (if it is warranted). Ford also sees a number of legitimate perspectives beyond its own, including the NHTSA, itself, its competitors, and customers.

Table 17.6 Boundary critique

Boundary issue	What is	What ought to be
Sources of motivation	Profit	Profit; risk
Sources of power	NHTSA	Customers
Sources of knowledge	NHTSA	Independent entity
Sources of legitimation	Ford	Ford/NHTSA/ Competitors/ Customers

17.3.4.2 Context

We can assess the contextual elements of our problem, including its abstraction (circumstances, factors, and conditions) and culture (values and patterns) as shown in Table 17.7.

This problem's context has several competing elements at play. Driving the problem toward resolution is the patterns of safety-related autonomy (i.e., individuals should be given the freedom to decide on their vehicle purchases), American culture and bravado pushing for American purchases by Americans, and customer loyalty to Ford. Further, market share for Ford is strong, but declining (incorporating the factor of Ford's evaluation of cost/benefit of recalls). Working against Ford as restraining forces are the circumstance that they must comply with regulatory implications and laws (including safety recalls), the conditions of the emergence of the NHTSA as a safety watchdog, and foreign competition. There is also a value that passengers should be safe in cars (also leading to potential safety recalls).

17.3.4.3 Force Field Diagram

Combining elements from the previous two subsections yields Table 17.8.

Ultimately, this problem is one of the Ford's difficulties in adapting to changing market conditions. It sees its decline in market share and blames others (including its competitors and the NHTSA), when, in reality, it should embrace the opportunity to serve as an industry leader and safety champion.

17.3.4.4 Updated FCM

Many elements are missing in the FCM as a result of boundary and context analysis. They include safety-related autonomy, NHTSA emergence as a safety watchdog, American cultural bravado, customer loyalty, and safety recalls. These concepts, as well as their connections, are reflected in Fig. 17.7.

Table 17.7 Context articulation

Category	Element
Circumstance	Need to comply with regulatory implications and laws (including recalls)
Factor	Cost/benefit ratio of recalls
Condition	Emergence of the NHTSA as a safety watchdog
	Foreign competition
Value	Passengers should be safe in cars
Pattern	Safety-related autonomy
	American culture and bravado
	Customer loyalty

Table 17.8 Force field diagram

Driving force	Strength as-is	Strength ought-to-be	Problem	Strength ought-to-be	Strength as-is	Restraining force
Safety-related autonomy	0.5	0.5	*Ford sees its decreasing market share as a sign of its decline. This is further exacerbated by the presence of the NHTSA as a safety watchdog. (Ford should embrace the challenge of safety-centric, perhaps even serving in a champion role, and focus on its investments, rather than its market share, as its competitors were facing a similar squeeze from foreign competition)*	−0.5	−1	NHTSA's emergence as a safety watchdog
Market share	1	0.5		−0.25	−0.5	Foreign competition
American cultural bravado	0.5	0.5		−0.25	−0.25	Safety recalls
Customer loyalty	0.5	0.5				

17.3.4.5 Proposed Ought-to-Be Changes

Ford may wish to focus more on safety at their own insistence rather than because they are forced to do so by the NHTSA. While this is a difficult financial decision in the short term, it is hoped that it might lead to positive, long-term consequences for Ford.

17.3.5 How Perspective

The following subsections discuss the *how* perspective analysis for this problem.

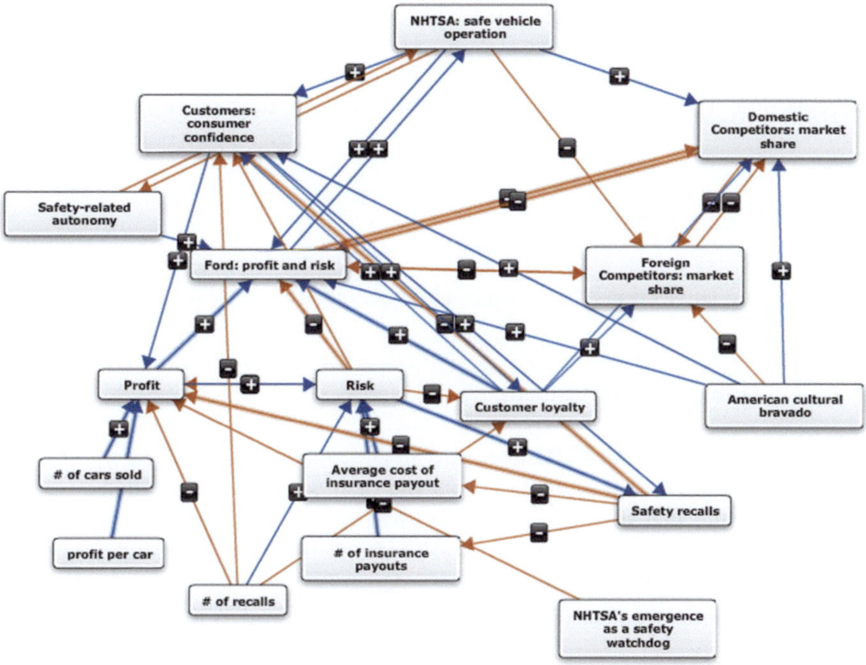

Fig. 17.7 Updated FCM with boundary and context incorporated

17.3.5.1 Cynefin Analysis

Ford's problem seems relatively well ordered. It appears to be in the complicated domain, however, as there are some major uncertainties stemming from a lack of information regarding what regulations will come from the NHTSA and what consumer demands are.

17.3.5.2 Mechanism Selection

Ford should employ manpower and KSAs to commission a study by the marketing department to ensure Ford's products are in keeping with consumer demands, and information to find out more about the NHTSA's intentions. These changes can be captured in our FCM as *clarity of intentions* to represent uncertainty regarding the NHTSA and a *consumer demand study* to represent a study that Ford can commission to find out more information regarding consumer demands to use to both shape its future strategic direction as well as inform its discussions with NHTSA.

17.3.5.3 Updated FCM

Reflecting the inclusion of the *clarity of intentions* and *consumer demand study* concepts, Fig. 17.8 shows the updated FCM based on mechanism analysis.

17.3.6 When Perspective

The following subsections discuss the *when* perspective analysis for this problem.

17.3.6.1 Time Scale Assessment

First, we must analyze our FCM to ensure all concept transitions occur on the same time scale. We can list all of our concepts and their accompanying time horizon for change to ensure that they change at the same rate. This information is found in Table 17.9. Note that proposed changes indicate whether the total magnitude should be increased (+) or decreased (−). An indication of two or more plus or minus values indicates that a stronger temporal adjustment is necessary.

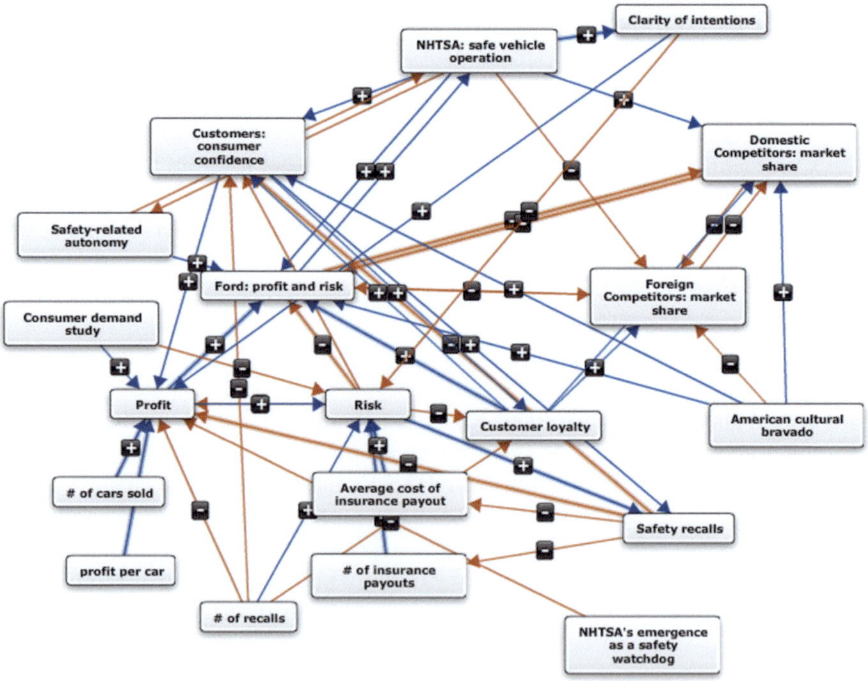

Fig. 17.8 Updated FCM with mechanisms incorporated

Table 17.9 Assessment of concept time horizons

Concept	Time period for change	Proposed change
Ford: profit and risk	Monthly	None
Domestic competitors: market share	Monthly	None
Foreign competitors: market share	Monthly	None
NHTSA: safe vehicle operation	Monthly	None
Customers: consumer confidence	Monthly	None
Profit	Monthly	None
Safety recalls	Monthly	None
Customer loyalty	Monthly	None
Safety-related autonomy	Monthly	None
Clarity of intentions	Monthly	None
profit per car	Monthly	None
# of cars sold	Monthly	None
Risk	Monthly	None
# of recalls	Monthly	None
# of insurance payouts	Monthly	None
Average cost of insurance payout	Monthly	None
NHTSA's emergence as a safety watchdog	Monthly	None
American cultural bravado	Monthly	None
Consumer demand study	Monthly	None

Although we could make an argument that many of these concepts could be updated more or less frequently than monthly, we can make a safe assumption that they could all be considered to operate on the same (or similar) time scale. Thus, no changes are needed to our FCM.

17.3.6.2 Intervention Timing

Armed with our completed FCM, we must work our way through the decision flowchart in Chap. 10. Starting with element 1, we can definitively conclude that the benefit remaining in the problem (as it pertains to profit and risk) certainly outweighs the cost of intervention. So, $\max(B/C) \geq 1$. Next, we must ask whether or not our problem is stable (Element 3). This requires us to consider initial values for our concepts as the status quo. In this case, we believe that *NHTSA's emergence as a safety watchdog* is at +1, reflecting conditions at the time (early 1970s), while all remaining concepts are taken to be 0 (absent any further information). The results of this analysis are shown in Fig. 17.9.

Although it may be unclear from the figure, analysis of the data reveals that the scenario exhibits periodic behavior. Thus, it is complex. In this case, we move to Step 4b, *act to increase understanding*. This represents the conclusion of the

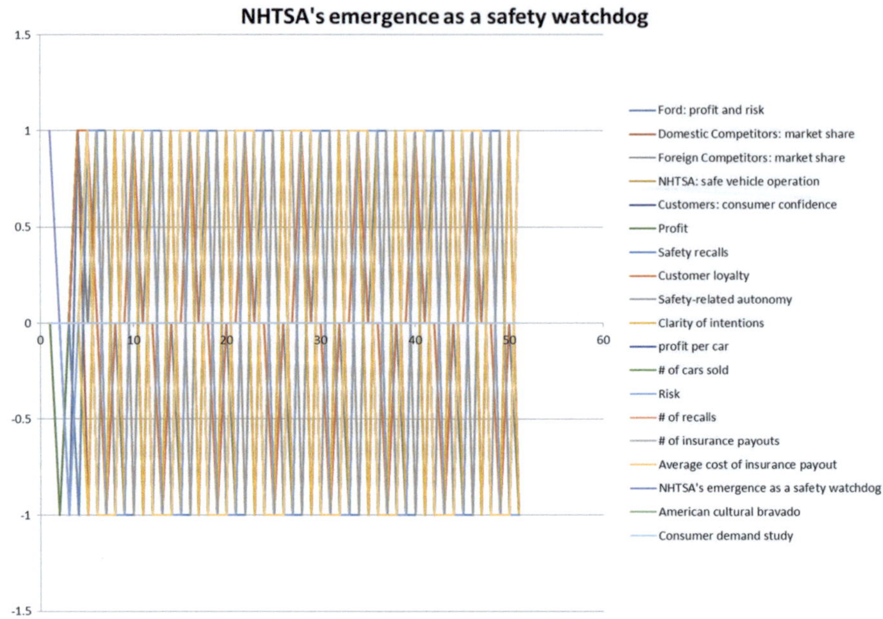

Fig. 17.9 Stability analysis of Ford's problem

perspective-driven analysis of this problem (i.e., the thinking phase). At this point, we can transition to the second problem analysis, that of the NHTSA.

17.4 Problem 2: NHTSA Problem

The following subsections address the systemic thinking perspectives as they pertain to the second problem, for which the NHTSA is the owner:

In the early 1970s, the NHTSA was looking to standardize safety guidelines for vehicles in the USA.

17.4.1 Who Perspective

The following subsections discuss the *who* perspective analysis for this problem.

17.4.1.1 Stakeholder Brainstorming

Brainstorming stakeholders for the NHTSA problem yield the following stakeholders and their associated wants:

1. Ford wants minimal government intervention
2. NHTSA wants safe vehicle operation
3. Domestic competitors want minimal government intervention
4. Customers want consumer confidence.

It should be noted that foreign competitors are not considered in this problem as they were with Ford's problem as they are not under the control of the NHTSA. While many more individuals and groups could be added into the analysis, it was thought that the initial stakeholder analysis should include, at a minimum, these four entities and their associated desires.

17.4.1.2 Stakeholder Classification

Table 17.10 shows evaluations of the attributes and class for each of the stakeholders identified in the previous section. They have been sorted according in decreasing order of prominence.

Clearly, the most prominent stakeholder is the NHTSA. The NHTSA's prominence is obvious in its own problem. The next tier of prominence is Ford and other domestic competitors. Their prominence comes from a somewhat powerful stance (i.e., they could all band together and refute the NHTSA's recommendations) and urgency due to perceived pressure from the NHTSA. Finally, customers are legitimacy, but neither powerful nor urgent as it pertains to the NHTSA (given the social and political landscape operating at the time).

17.4.1.3 Stakeholder Attitude Evaluation

Table 17.11 shows evaluations of the potential for threat and potential for cooperation for each of the stakeholders identified in the previous subsection. These two parameters provide an identification of the attitude of each stakeholder. They have

Table 17.10 Stakeholder classification

Stakeholder	Stakeholder attribute			Prominence
	Power	Legitimacy	Urgency	
NHTSA	1	1	1	1.0
Ford	0.5	0	1	0.5
Domestic competitors	0.5	0	1	0.5
Customers	0	1	0	0.33

Table 17.11 Stakeholder attitude evaluation

Stakeholder	Potential for threat	Potential for cooperation	Support
NHTSA	0	1	1
Customers	0	0	0
Ford	1	0	−1
Domestic competitors	1	0	−1

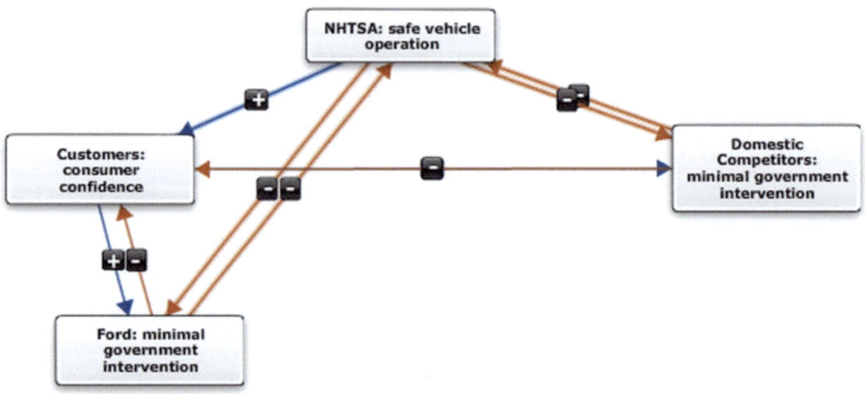

Fig. 17.10 Stakeholder objective map

been sorted in decreasing order of support according to their assigned stakeholder attitude.

Only the NHTSA is supportive in this scenario. Customers are indifferent and offer neither threat nor cooperation potential. Ford and domestic competitors have full potential for threat and no potential for cooperation due to their view of the NHTSA as a threat to their business model.

17.4.1.4 Stakeholder Objective Mapping

With classification and attitude defined in the previous two sections, Fig. 17.10 shows a stakeholder objective map, including the influence (direction and magnitude) for all identified stakeholders involved in the problem. The thicker the line, the stronger the causal influence.

17.4.1.5 Stakeholder Engagement Priority

In order to calculate the stakeholder engagement priority for all the stakeholders in NHTSA problem, we need to calculate k_i^{in}, k_i^{out}, s_i^{*-in}, s_i^{*-out}, *Popularity*, and

Table 17.12 Network characteristics

Stakeholder	k_i^{in}	k_i^{out}	$s_i^{*-\text{in}}$	$s_i^{*-\text{out}}$	Activity	Popularity
NHTSA: safe vehicle operation	2	3	1	1.5	2.12	1.41
Ford: minimal government intervention	2	2	0.75	0.75	1.22	1.22
Domestic competitors: minimal government intervention	2	2	0.75	0.75	1.22	1.22
Customers: consumer confidence	3	2	1	0.5	1.00	1.73

Table 17.13 Stakeholder prioritization

Stakeholder	Activity	Popularity	Engagement priority
Ford: minimal government intervention	1.22	1.22	1
Domestic competitors: minimal government intervention	1.22	1.22	1
Customers: consumer confidence	1.00	1.73	2
NHTSA: safe vehicle operation	2.12	1.41	n/a

Activity, in accordance with equations found in Chap. 6. These results are shown in Table 17.12.

We then sort the stakeholders by activity first (in descending order) and then by popularity (in ascending order). Table 17.13 illustrates the order in which stakeholders should be engaged in support of this effort.

It is clear that Ford and domestic competitors should be prioritized for this effort. This is due to their power when acting collectively and their opposition to the NHTSA's effort. Secondarily, customers should be engaged, but only after the auto manufacturers have been accounted for. The NHTSA is removed from engagement consideration as they are the problem owner and, as a result, their engagement is automatic.

17.4.1.6 Stakeholder Management Plan

The final step in analyzing this problem as it pertains to the who perspective is to develop a stakeholder management plan. The NHTSA's stakeholder management plan is shown below in Table 17.14.

From the NHTSA's perspective, their primary focus is to defend against Ford and other domestic competitors due to their lack of support for the NHTSA's mission, and monitor and collaborate with customers to gauge and sway public opinion as best as possible. These all seem like reasonable strategies for engagement moving forward.

Table 17.14 NHTSA stakeholder management plan

Stakeholder name	Wants	Prominence	Support	Priority of engagement	Strategy
Ford	Minimal government intervention	0.5	−1	1	Defend
Domestic competitors	Minimal government intervention	0.5	−1	1	Defend
Customers	Consumer confidence	0.33	0	2	Monitor/Collaborate
NHTSA	Safe vehicle operation	1.0	1	n/a	Involve

17.4.2 What Perspective

The following subsections discuss the *what* perspective analysis for this problem.

17.4.2.1 Articulate Objectives

In terms of the NHTSA, it must be concerned with minimization of risk but not at the expense of unnecessary recalls. This will ensure they fulfill their mission at a minimal inconvenience to customers (i.e., society at large) and to the auto manufacturers. Minimal inconvenience is important as the NHTSA seeks to establish legitimacy as a newly formed agency. So, we can establish two fundamental objectives for the agency: (1) *minimize risk* and (2) *minimize unnecessary recalls*. Now, we must organize these objectives.

17.4.2.2 Fundamental Objectives Hierarchy

Organizing our two fundamental objectives into a hierarchy yields Fig. 17.11, showing a further decomposition of our objectives. Risk is broken down into deaths and injuries. Unnecessary recalls is decomposed into unnecessary expenses for manufacturers (as a result of recalls), loss of NHTSA legitimacy (as a result of being seen as *crying wolf*), and eroded consumer confidence (as a combination of the previous two factors).

17.4.2.3 Means-Ends Network

The means-ends network shows our understanding of the means necessary to produce our desired ends (i.e., our fundamental objectives). Using the same two

Fig. 17.11 Fundamental objectives hierarchy

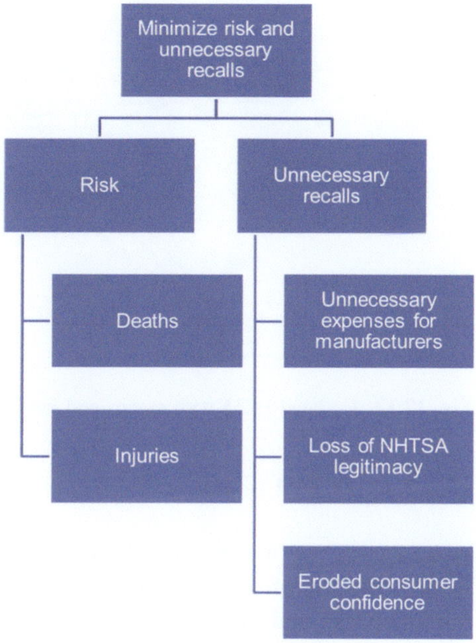

Fig. 17.12 Means-ends network

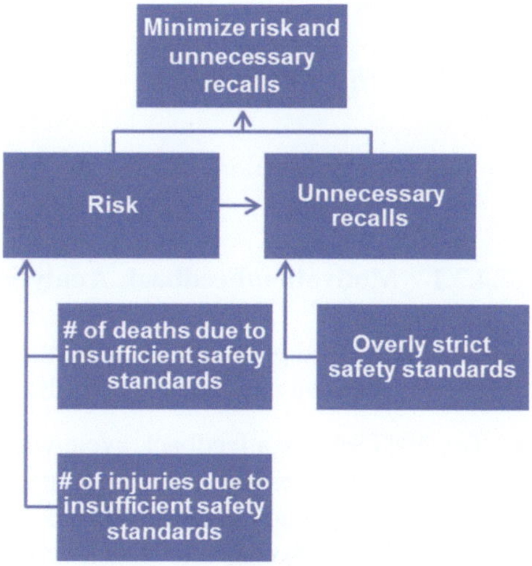

fundamental objectives as before, we can create the network shown in Fig. 17.12. First, risk is a means to achieve the end of unnecessary recalls. That is, if we choose to eliminate risk, we must recall all vehicles (while obviously doing so unnecessarily in many instances). Minimization of risk can be achieved by minimized by minimizing the number of deaths and injuries due to insufficient safety standards. Overly strict safety standards can increase the number of unnecessary recalls.

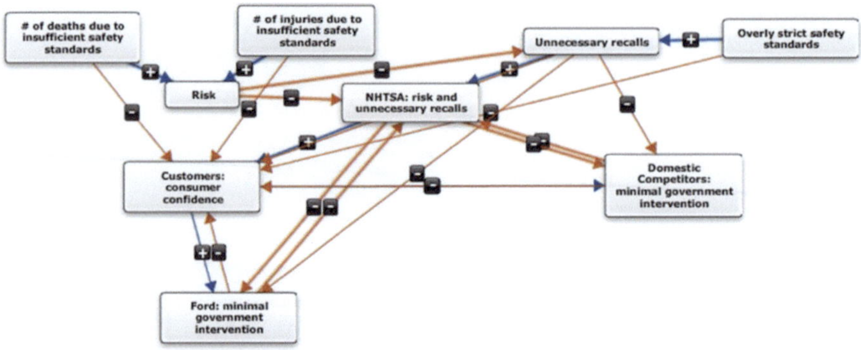

Fig. 17.13 Updated FCM with means-end network incorporated

17.4.2.4 FCM Update

Armed with our means-ends network, we can now integrate it into our existing FCM. Our modified FCM reflects the revision of the NHTSA's objectives to minimize risk and unnecessary recalls, as well as the means identified in the previous subsection. This revised scenario depiction is shown in Fig. 17.13.

17.4.3 Why Perspective

The following subsections discuss the *why* perspective analysis for this problem.

17.4.3.1 Motivation/Feedback Analysis

If we are interested primarily in the *risk and unnecessary recalls* concept (NHTSA's objective), then we can make a few observations:

- The NHTSA has a feedback cycle with both Ford and domestic competitors; as risk and unnecessary recalls go up, minimal government intervention goes down, which leads to a rise in risk and recalls, and so on.
- The NHTSA also has a feedback cycle with customers. As risk and recalls goes up, consumer confidence goes up—this is incorrect. It should go down. This reduction will then lead to a decrease in minimal government intervention (i.e., more intervention). This is correct.

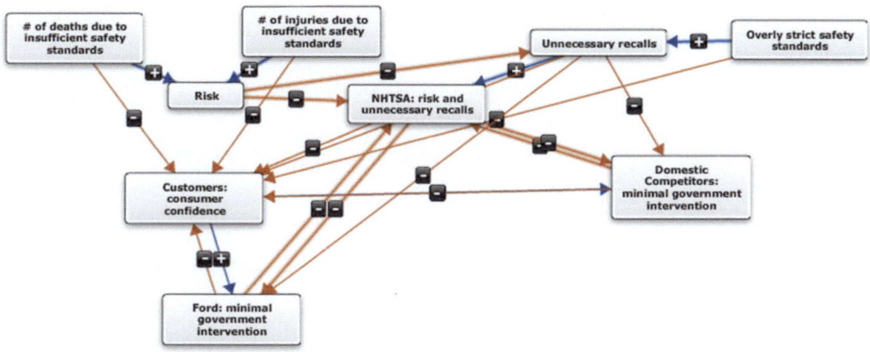

Fig. 17.14 Updated FCM with feedback analysis incorporated

17.4.3.2 FCM Update

Our feedback analysis in the previous subsection indicates the need for a correction in the direction of the causal link between *risk and unnecessary recalls* and *consumer confidence*. This change is reflected in Fig. 17.14.

17.4.3.3 Proposed Changes During Act Stage

No new proposed changes to the FCM are required.

17.4.4 Where Perspective

The following subsections discuss the *where* perspective analysis for this problem.

17.4.4.1 Boundary Articulation

Based on analysis of critically heuristic boundary issues of our problem using guidelines found in Ulrich (2000), we can generate the boundary critique shown in Table 17.15.

Assessment of our boundary yields a few insights. The NHTSA currently has all the power, knowledge, and legitimation as it pertains to the problem. This is dangerous if the NHTSA truly wishes to have buy-in from others (a must for successful operation). Thus, it recognizes the need to involve others, including the public at large and, potentially, an agency such as a professional society for independent verification of any recall mandates it may put forward. Further, the

Table 17.15 Boundary critique

Boundary issue	What is	What ought to be
Sources of motivation	Safety	Safety while not inhibiting business
Sources of power	NHTSA	NHTSA/Public
Sources of knowledge	NHTSA	NHTSA/Independent agency (i.e., professional society)
Sources of legitimation	NHTSA	Public

NHTSA is currently motivated entirely by safety, but it must not forget business-related concerns in an effort to establish itself.

17.4.4.2 Context

We can assess the contextual elements of our problem, including its abstraction (circumstances, factors, and conditions) and culture (values and patterns) as shown in Table 17.16.

Analysis of this problem's context reveals several competing elements. Driving the problem toward resolution is the circumstance of the NHTSA's ability to establish regulations, as well as a value that the government should ensure that passengers are safe in cars. Restraining the NHTSA is the condition that it must justify its existence as a new agency, the pattern that personal freedoms should not impinge on public safety (leading to a lack of public support), and the factor of risk/benefit concerns of recalls (exacerbated by increased foreign competition, leading to pressure on domestic manufacturers).

17.4.4.3 Force Field Diagram

Combining elements from the previous two subsections yields Table 17.17.

Ultimately, this problem is one of the NHTSA's difficulties in resolving what it sees as its charge from the Federal Government of ensuring automotive safety for the American public with the opposition it receives from auto manufacturers. The NHTSA should (and does) prefer to work with automotive manufacturers to

Table 17.16 Context articulation

Category	Element
Circumstance	Ability to establish regulations
Factor	Risk/benefit ratio (Increasing pressure on domestic manufacturers)
Condition	Need to justify the establishment of the NHTSA
Value	Government should ensure that passengers are safe in cars (safety-centric culture)
Pattern	Personal freedoms should not impinge on public safety (lack of public support)

Table 17.17 Force field diagram

Driving force	Strength as-is	Strength ought-to-be	Problem	Strength ought-to-be	Strength as-is	Restraining force
Emergence of safety-centric culture	0.25	1.0	*NHTSA sees that it has been tasked by the Federal Government with ensuring automotive safety for the American public, but it faces opposition from car manufacturers as it affects their bottom line. (NHTSA would prefer to work in concert with automotive manufacturers to ensure safety standards are met)*	−0.5	−1	Need to justify the NHTSA's existence
Ability to establish regulations	0.5	1.0		−0.5	−0.5	Increased foreign competition leading to pressure on domestic manufacturers
				−0.25	−0.25	Lack of public support

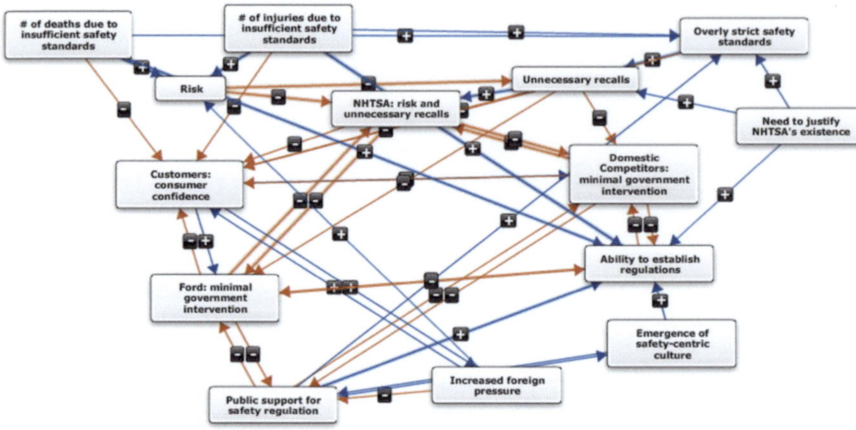

Fig. 17.15 Updated FCM with boundary and context incorporated

ensure safety standards and appropriate and adhered to. This requires a collaborative, rather than adversarial, relationship.

17.4.4.4 Updated FCM

Many elements are missing in the FCM as a result of boundary and context analysis. They include public support for safety regulations, increased foreign pressure, emergence of safety-centric culture, ability to establish regulations, and need to justify NHTSA's existence. These concepts, as well as their connections, are reflected in Fig. 17.15.

17.4.4.5 Proposed Ought-to-Be Changes

NHTSA must figure out a way to garner public support for its efforts, whether it's directly from customers or through automotive manufacturers as an intermediary.

17.4.5 How Perspective

The following subsections discuss the *how* perspective analysis for this problem.

17.4.5.1 Cynefin Analysis

The NHTSA's problem is fairly unordered. It is new as the agency has been recently established, and it is unclear exactly what to do next, and thus, there is a great deal of uncertainty. It faces pressure from consumers, automotive manufacturers, and the Federal Government at large. While it may be in a brief *honeymoon phase*, it will soon need to produce results to justify its existence.

17.4.5.2 Mechanism Selection

The NHTSA should employ information in an effort to learn more about the public and manufacturers to avoid committing a Type III error by instituting too many regulations without justification. This can be achieved by undertaking two actions: (1) *survey consumer attitudes regarding safety* and (2) *survey auto manufacturer attitudes regarding safety*. Both of these mechanisms should be captured as concepts in our FCM.

17.4.5.3 Updated FCM

Reflecting the inclusion of the two survey concepts discussed in the previous subsection, Fig. 17.16 shows the updated FCM based on mechanism analysis.

17.4.6 When Perspective

The following subsections discuss the *when* perspective analysis for this problem.

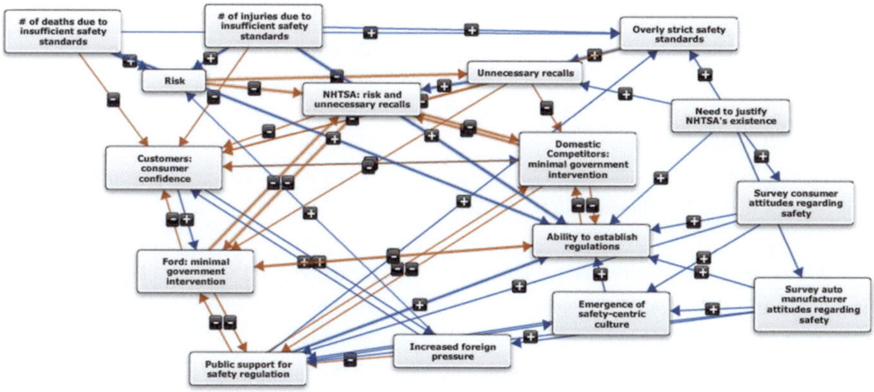

Fig. 17.16 Updated FCM with mechanisms incorporated

17.4.6.1 Time Scale Assessment

First, we must analyze our FCM to ensure all concept transitions occur on the same time scale. We can list all of our concepts and their accompanying time horizon for change to ensure that they change at the same rate. This information is found in Table 17.18. Note that proposed changes indicate whether the total magnitude should be increased (+) or decreased (−). An indication of two or more plus or minus values indicates a stronger temporal adjustment is necessary.

Just like with the Ford problem before, although we could make an argument that many of these concepts could be updated more or less frequently than monthly, we can make a safe assumption that they could all be considered to operate on the same (or similar) time scale. Thus, no changes are needed to our FCM.

17.4.6.2 Intervention Timing

Armed with our completed FCM, we must work our way through the decision flowchart in Chap. 10. Starting with element 1, we can definitively conclude that the benefit remaining in the problem (as it pertains to profit and risk) certainly outweighs the cost of intervention. So, $\max(B/C) \geq 1$. Next, we must ask whether or not our problem is stable (Element 3). This requires us to consider initial values for our concepts as the status quo. In this case, we believe that the environment

Table 17.18 Assessment of concept time horizons

Concept	Time period for change	Proposed change
Ford: minimal government intervention	Monthly	None
NHTSA: risk and unnecessary recalls	Monthly	None
Customers: consumer confidence	Monthly	None
Public support for safety regulation	Monthly	None
Ability to establish regulations	Monthly	None
Domestic competitors: minimal government intervention	Monthly	None
Increased foreign pressure	Monthly	None
Overly strict safety standards	Monthly	None
Unnecessary recalls	Monthly	None
# of injuries due to insufficient safety standards	Monthly	None
Risk	Monthly	None
# of deaths due to insufficient safety standards	Monthly	None
Emergence of safety-centric culture	Monthly	None
Need to justify NHTSA's existence	Monthly	None
Survey consumer attitudes regarding safety	Monthly	None
Survey auto manufacturer attitudes regarding safety	Monthly	None

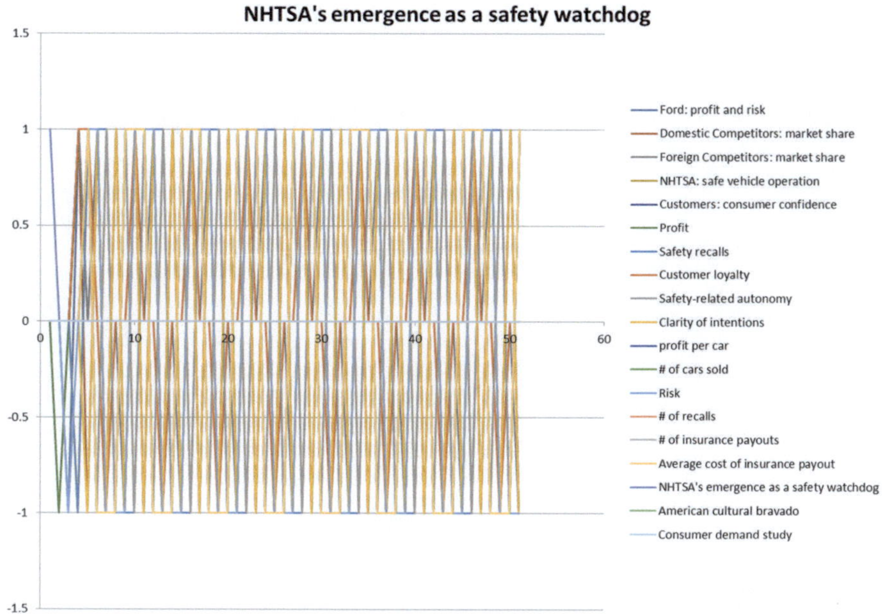

Fig. 17.17 Stability analysis of NHTSA's problem

indicates that *Ford: minimal government intervention* and *Domestic competitors: minimal government intervention* should be set to +1, indicating the prevailing thought at the time that the NHTSA and other government agencies should not interfere in business. All other concepts are taken to be 0 (absent any further information). The results of this analysis are shown in Fig. 17.17.

Although it may be unclear from the figure, analysis of the data reveals that the scenario exhibits periodic behavior. Thus, it is complex. In this case, we move to Step 4b, *act to increase understanding*. This represents the conclusion of the perspective-driven analysis of this problem (i.e., the thinking phase).

17.5 Ford Pinto Mess

We begin our mess-level analysis by first generating a "what-is" meta-perspective of our mess. It is important to check for any conflicts, spelling errors, etc. to avoid duplicate concept entries. None were found, and thus, we were able to generate a depiction of our what-is meta-perspective as shown in Fig. 17.18.

Following the steps for action discussed in Chap. 13, we can now calculate activity and popularity for our mess concepts and determine a priority for them. The top twenty concepts (in terms of order of engagement priority), as well as an assessment of their feasibility, are shown in Table 17.19.

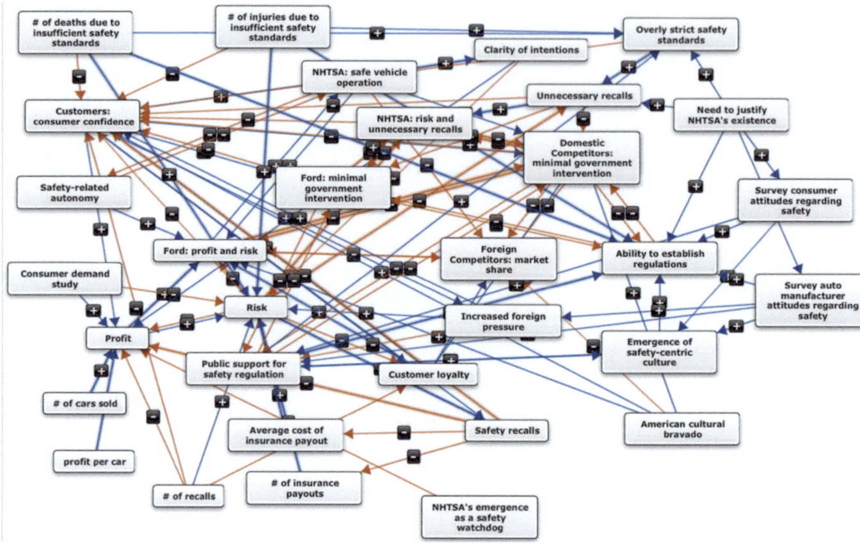

Fig. 17.18 Ford Pinto mess what-is meta-perspective

Table 17.19 Assessment of concept influence feasibility Ford Pinto mess

Priority	Concept	Activity	Popularity	Feasibility
1	Risk	4.39	5.41	N
2	NHTSA: safe vehicle operation	3.24	1.00	N
3	Customers: consumer confidence	3.00	7.25	N
4	Public support for safety regulation	2.74	3.00	N
5	Need to justify NHTSA's existence	2.50	0.00	N
6	# of injuries due to insufficient safety standards	2.45	0.00	N
7	# of deaths due to insufficient safety standards	2.45	0.00	N
8	Safety recalls	2.45	1.22	Y
9	Customer loyalty	2.24	1.50	N
10	Unnecessary recalls	2.24	1.94	Y
11	Ford: minimal government intervention	2.24	2.74	N
12	Domestic competitors: minimal government intervention	2.24	2.74	N
13	# of recalls	2.00	0.00	Y
14	American cultural bravado	2.00	0.00	N
15	Survey auto manufacturer attitudes regarding safety	2.00	0.50	N
16	NHTSA: risk and unnecessary recalls	1.94	2.83	N
17	Ford: profit and risk	1.73	4.90	N
18	Survey consumer attitudes regarding safety	1.50	0.50	Y
19	Increased foreign pressure	1.50	1.00	N
20	Overly strict safety standards	1.22	2.00	Y

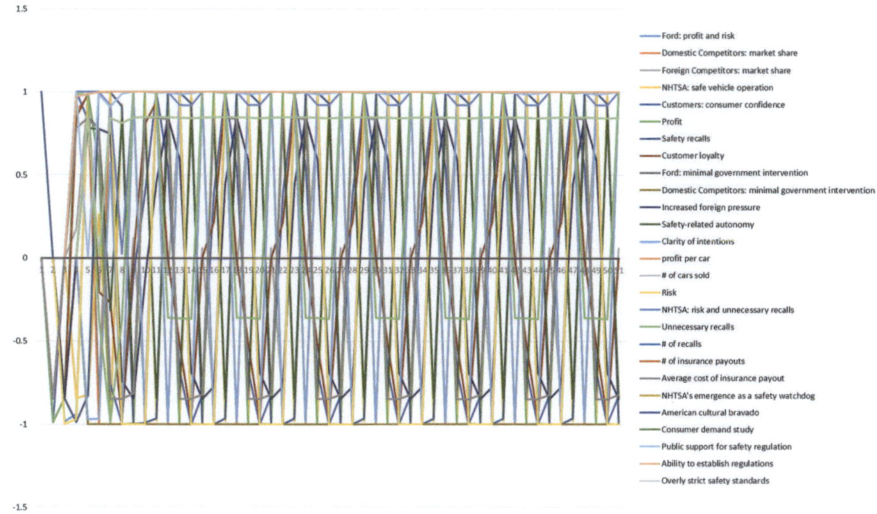

Fig. 17.19 *Safety recalls* set to +1 initially

Assessing the top twenty concepts, we see there are only five feasible concepts for influence. *Safety recalls* is directly influenced by the NHTSA, as are *unnecessary recalls* and *# of recalls*. Further, the NHTSA can directly *survey consumer attitudes regarding safety*, and they have the ability to reduce *overly strict safety standards*. Interestingly, none of the highest priority, feasible concepts are under Ford's jurisdiction. Further thought into the mess provides an explanation for this. Ford's actions are far more focused on it, whereas the NHTSA can directly affect resolution of its own objectives, as well as those of domestic (including Ford) and foreign competitors and the NHTSA.

We can now explore scenarios to achieve the objectives of the two problems. *Safety recalls* is the highest ranked feasible element for intervention. Under the current situation, it could be reasoned that the NHTSA would wish to increase recalls in order to increase automobile safety. Thus, we can initially set *safety recalls* to +1. Results of this scenario are shown in Fig. 17.19.

This scenario exhibits periodic, complex behavior. This is not desirable and obviously very turbulent as the figure shows. Alternatively, the NHTSA can make a long-term commitment to increasing *safety recalls* (thus clamping this concept at +1) in the hope for scenario stabilization. This scenario is shown in Fig. 17.20.

This new scenario continues to exhibit periodic, complex behavior. It seems reasonable that we may need to invoke changes in more than one concept to produce stable behavior. Increasing *safety recalls* makes sense; however, in an effort to avoid damaging the automotive industry, we must also curb *unnecessary recalls*, the second highest priority, feasible concept. As a result, setting *safety recalls* to +1 and *unnecessary recalls* to −1 initially is a scenario worth exploring. Results of this scenario are shown in Fig. 17.21.

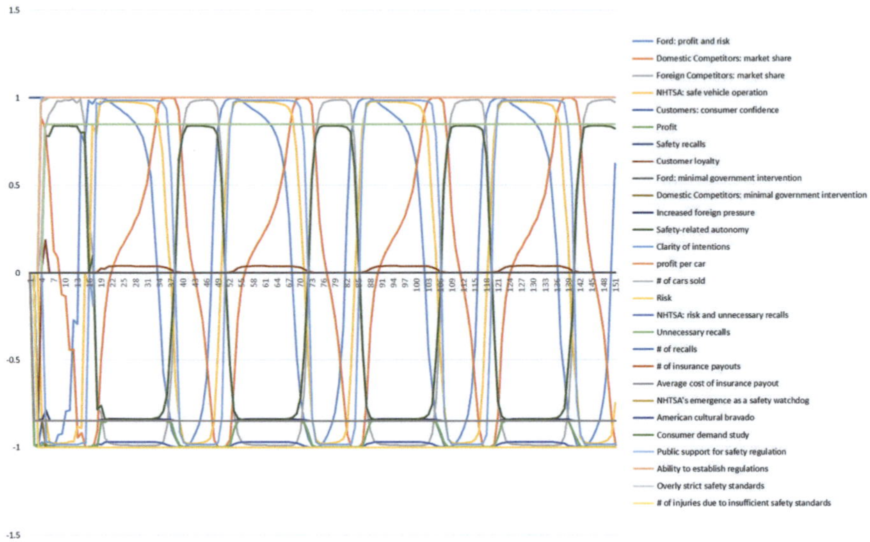

Fig. 17.20 *Safety recalls* clamped to +1

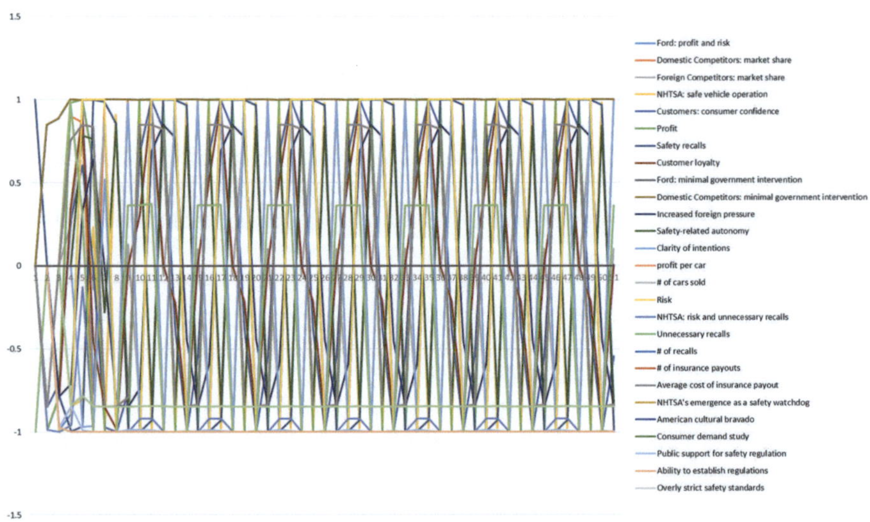

Fig. 17.21 *Safety recalls* set to +1 and *unnecessary recalls* set to −1, both initially

Once again, we have complex, periodic behavior. This remains unacceptable. The next scenario we can investigate is to clamp both *safety recalls* to +1 and *unnecessary recalls* to −1, making a long-term commitment on the part of the NHTSA to increase safety, yet not do so via unnecessary recalls. This scenario is shown in Fig. 17.22.

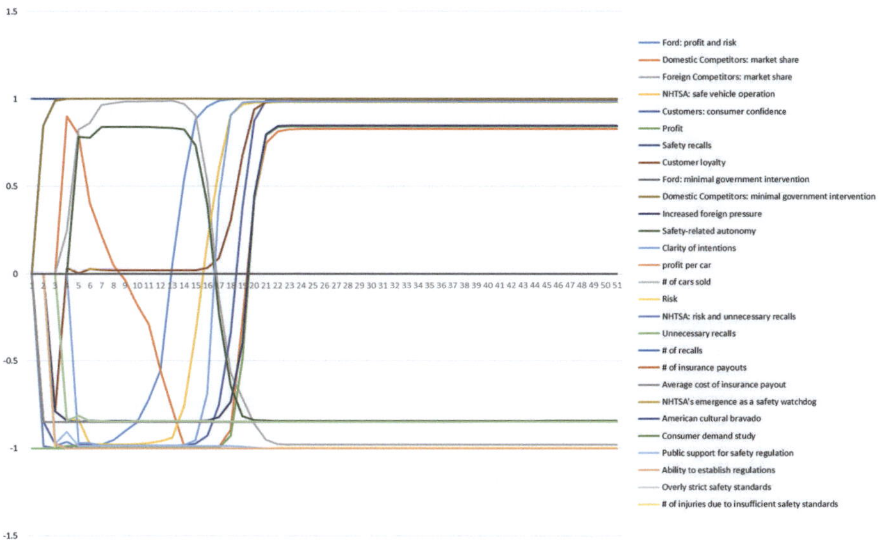

Fig. 17.22 *Safety recalls* clamped to +1 and *unnecessary recalls* clamped to −1

Finally, we have arrived at a stable, albeit complicated, scenario. This scenario stabilizes after about 22 time periods and results in the most value for both Ford (maximum profit, minimum risk) and the NHTSA (minimum risk, minimal unnecessary recalls). A summary of the four scenarios that were investigated is shown in Table 17.20.

Ultimately, the best course of action appears to be clamping safety recalls to +1 and unnecessary safety recalls to −1. If we believe this strategy is sustainable, then we cannot do better from a performance perspective as performance is at unity (potentiality/actuality = 1/1 = 1). This situation represents one in which we understand that an increase in safety recalls is likely an inevitability, but the NHTSA will do their best to minimize unnecessary safety recalls, thereby gaining the public's trust and ensuring the auto industry can remain viable. We could explore other alternatives in an effort to improve the potentiality of our situation (especially, if we didn't believe an increase in safety recalls was an inevitability),

Table 17.20 Summary of mess scenario exploration

Potential course of action	Feasible?	Performance (desirability)
Safety recalls set to +1 initially	Yes	Unstable performance
Safety recalls clamped to +1	Yes	Unstable performance
Safety recalls set to +1 and *unnecessary recalls* set to −1, both initially	Yes	Unstable performance
Safety recalls clamped to +1 and *unnecessary recalls* clamped to −1	Yes	Best performance on both fundamental concepts

although a perusal of the four scenarios discussed in this chapter shows the difficult in getting this scenario to stabilize.

One additional detail bears mentioning. Recall that the time period we specified was monthly, meaning that this stabilization will take nearly two years. This will take patience on the part of the NHTSA, Federal Government, Ford and domestic competitors, and customers. This may be a tall order, but it does not appear that a better course of action is readily apparent.

17.6 Conclusions

This chapter worked through a comprehensive case study focused on the Ford Pinto and the establishment of the National Highway Transportation Safety Administration utilizing the multimethodology developed throughout the text.

After reading this chapter, the reader should:

1. Understand the utility of the book's developed multimethodology on a practical mess from *cradle to grave*.

Reference

Ulrich, W. (2000). Reflective practice in the civil society: The contribution of critically systemic thinking. *Reflective Practice, 1*(2), 247–268.

Chapter 18
Conclusion

Abstract Well, we have finished a long journey through a wide variety of topics we felt would be useful to you in understanding the landscape associated with complex decision making. Our take on decision making utilizes systemic thinking as the prime motivation for the decision making methods and techniques we have described in the previous 17 chapters. However, this is not simple decision making, but decision making associated with ill-structured, wicked situations where multiple problems are presented as messes. Hence, the title of the book: *Systemic Decision Making: Fundamentals for Addressing Problems and Messes*. Hopefully, each of you has been able to take away the following key points.

18.1 Part I: A Frame of Reference for Systemic Thinking

Part I provided the foundation we felt was required to think systemically in five separate chapters. Chapter 1 is where we introduced the Think–Act–Observe (TAO) approach for improving our understanding about a complex problem. We also provided a formal taxonomy for errors that humans are prone to making when addressing problems. Chapter 2 emphasized the integration of technical, organizational, political, and human perspectives during the analysis of the problem. We emphasized that systems age messes are much grander and more complex than their machine age problem predecessors. As a result, we point out that our experience indicates that most systems age problem domains have deeply rooted or philosophical divergence which add to the difficulty in developing a mutually agreeable problem formulation. An essential quotation from Mitroff and Linstone (1993) summarizes the view of this chapter:

> …"everything interacts with everything," that all branches of inquiry depend fundamentally on one another, and that the widest possible array of disciplines, professions, and branches of knowledge — capturing distinctly different paradigms of thought — must be consciously brought to bear on the problem. (p. 91)

Chapter 3 introduces systems approaches and discussed their shortfalls in addressing the complex problems found in a systems age. We introduced a

P.T. Hester and K.M. Adams, *Systemic Decision Making*, Topics in Safety, Risk, Reliability and Quality 33, DOI 10.1007/978-3-319-54672-8_18

methodology for systemic thinking and contrasted it with traditional systematic thinking with the goal of putting the approach into practice in a manner which will garner increased understanding for systems age messes. Chapter 4 introduced our notion of *systems theory* as the underlying theoretical foundation for understanding systems. The extensive integration and transformation of existing information (i.e., scientific propositions from a variety of disciplines) in systems theory, along with its seven axioms and attendant principles, is what provides the improved understanding necessary to design and manage modern systems and their associated problems and messes. The first section concluded with an introduction to complex systems modeling methods. Specifically, a comparison of available methods was presented, which outlined the appropriateness of fuzzy cognitive mapping for understanding complex systems. Finally, a framework was presented for the development and use of a fuzzy cognitive map as a mechanism for thinking about, acting on, and observing messes.

18.2 Part II: Thinking Systemically

Part II introduced you to a formal concept and methodology for thinking systemically. The methodology addressed the who, what, where, how, and when associated with problems and decision making in six separate chapters. Chapter 6 introduced the concept of stakeholders, those humans and organizational entities that exist at the center of all systems problems and who serve as the principal contributors to the solution of these problems. We developed a six-step approach to stakeholder analysis and management which included the following: (1) identification of stakeholders; (2) classification of these stakeholders; (3) assessment of their attitude; (4) calculation of their engagement priority; (5) mapping them in relation to one another; and (6) developing a plan for managing them. It also included an implicit 7th step, carrying out the management plan in a formal stakeholder management and communication plan. This comprehensive technique serves as an important discriminator enabling systems practitioners to deal with stakeholders in an effective manner. Chapter 7 reviewed the anatomy of a problem and discussed the importance of objectives. The organization of objectives into both a fundamental objective hierarchy and means-ends network was emphasized as an important consideration required to answer the *what* question of systemic thinking. Chapter 8 provided the background for *why* by investigating a wide variety of theories associated with motivation as the incentive, the stimulus, and the inspiration for continued involvement. We provided a cybernetic model with clear feedback loops that ensures continued performance by ensuring goals remain synchronized with the individual and situational characteristics that form the context of the messes and constituent problems. This cybernetic model provides a congruent, current, and logical framework for achieving goals and objectives developed to address the elements of messes and associated problems. In Chap. 9, we provided a foundation for understanding the *where* in a mess or problem. The importance of the circumstances, factors, conditions, values, and patterns that surround messes and problems (i.e., the context) was presented. In addition, problem *boundaries*, the representations we use that provide lines of demarcation between messes and problems and the surrounding environment, were presented along with a framework for operationalizing the process of assessing problem boundary and context issues. Chapter 10 addressed the *how* question as it relates to the attainment of specific, purposeful goals. Nine physical, human, and abstract mechanisms were identified as the means for moving a mess from a current state toward a desired state. Specific focus was placed on abstract mechanisms, namely methods and information, because of their nonintuitive nature and their importance in achieving increased understanding in complex problem domains. The processes used to develop knowledge were addressed using three separate models of knowledge generation. The third of these models, the Cynefin framework, was discussed as an approach by which to achieve increased understanding in the five domains found in complex systems. The section was finalized in Chap. 11 which discussed the *when* question of systemic thinking. In order to determine the appropriate time for intervention in our mess, we developed an approach to assess the *maturity* and *stability* of our mess. The maturity discussion focused on life-cycle concerns and on evaluating the cost-to-benefit ratio of mess intervention, while our stability perspective focused on a discussion of system evolution and self-organization. The resulting six-element framework served as a guide for individuals interested in determining timing issues as they pertain to increasing understanding about their mess.

18.3 Part III: Acting Systemically

Part III addresses how the practitioner will implement the systemic decision making multimethodology on the mess. Chapter 12 introduced two meta-perspectives, the *what is* and the *what ought-to-be*, in an effort to make sense at the mess level of our problem-level analyses. This guidance provided a framework for systemic understanding of our mess and a general guideline for undertaking a case study using the multimethodology presented in Chaps. 6 through 11 of Part II. Chapter 13 introduced the basics of decision analysis and provided a structured approach to determine an appropriate course of action for our mess that included concerns such as robustness and optimality. Chapter 14 discussed the very real potential for human error as an inevitable part of decision making. The ability to classify, manage, and prevent human error helps practitioners in their understanding with respect to the implementation of decisions.

18.4 Part IV: Observing Systemically

The final part of the book, Part IV, addresses the practitioner's ability to observe messes and their attendant problems systemically. Consistent application of the behaviors in this section will ensure that practitioners use valid observations as the source for their interpretations and follow-on decisions when dealing with problems and messes. Chapter 15 specifies that observation is the central method in which we engage with the real world. As such, observation is the source for factual data which are assembled into information and processed into knowledge where we may use it in the generation of inferences and the formulation of decisions. Practitioners are advised that the human observer impacts each and every observation and that in order to make good, repeatable decisions a formal process must be invoked. In conjunction, the use of measurement, and an ability to account for biases and heuristics, in an effort to suppress their inadvertent use in decision making, must be accounted for. We include a model and associated four-phase process for observation. This chapter concludes with a powerful model that recognizes the relationship between technological systems for observation and the cognitive system of human processing as a means for making sense in real-world situations involving decision making. Chapter 16 addresses learning and the individual, group, organizational, and inter-organizational aspects of learning. We emphasize that learning is the *raison d'être* of the feedback element in a systemic approach. Every viable organism and organization includes a reflexive response or behavior that permits it to change its behavior based upon its experience. This ability to rationally change, termed learning, is what permits the organism or organization to remain viable. This chapter also addresses that learning is the act that permits organizations to detect and recognize errors, analyze the errors, and adapt their behaviors, in a process of organizational learning. Chapter 17 concludes Part IV with a comprehensive case study focused on the Ford Pinto and the establishment of the National Highway Transportation Safety Administration utilizing the multimethodology presented throughout the text. The goal in this chapter is to present a *cradle-to-grave* demonstration of the approach outlined in the text.

18.5 Summary

Fundamentally, we need a novel way to *understand* and *address* ill-structured, wicked, complex problems and the larger messes. In order to this successfully, we believe that practitioners must think, act, and observe *systemically*, hence the title of this book. It is the hope of the authors that after reading this book, readers will gain an appreciation for a novel way of thinking and reasoning about complex problems that encourages increased understanding. We have presented this material in a manner that did not require the reader to be either an engineer or a scientist. Indeed, most of the real-world, complex problems vexing us are not restricted to any one discipline or area of practice, but cross boundaries indiscriminately. We have endeavored to show that making decisions in a systemic fashion requires the inclusion of the widest number of perspectives possible, in an effort to improve understanding to the maximum extent feasible.

As always, we alone are responsible for the notions, ideas, and information presented in this book and would appreciate constructive criticism and feedback.

Reference

Mitroff, I. I., & Linstone, H. A. (1993). *The unbounded mind: Breaking the chains of traditional business thinking*. New York: Oxford University Press.

Appendix A
Real Estate Problem 2

This appendix discusses the perspective-centric (i.e., who, what, etc.) analysis of Problem 2 of the real estate mess discussed throughout the book.

A.1 Problem Description

This problem is derived from the example discussed briefly in Hester, Bradley, and Adams (2012). In this example, a local real estate developer sought to rezone portions of an upscale, single family home residential neighborhood. The intended rezoning needed to take into account the values of important stakeholders (e.g., neighbors, local government) in order to ensure project success. This problem is being discussed from the perspective of the local communities, who are concerned with maintaining their property values, irrespective of any development concerns.

A.2 Who Perspective

The following subsections discuss the *who* perspective analysis for this problem.

A.2.1 Stakeholder Brainstorming

Brainstorming stakeholders for the local communities' problem yields the following stakeholders and their associated wants:

1. Nine local communities *want* to maximize their property values.
2. The real estate developer *wants* to maximize ROI of the development.
3. Potential homeowners *want* to maximize happiness.

P.T. Hester and K.M. Adams, *Systemic Decision Making*, Topics in Safety, Risk, Reliability and Quality 33, DOI 10.1007/978-3-319-54672-8

4. Banks *want* to maximize revenue.
5. Federal Reserve *wants* to maximize economic activity.

While many more individuals and groups could be added to the analysis, it was thought that the initial stakeholder analysis should include, at a minimum, these five entities and their associated desires.

A.2.2 Stakeholder Classification

Table A.1 shows evaluations of the attributes and class for each of the stakeholders identified in the previous section. They have been sorted according to decreasing order of prominence.

Clearly, the two most prominent stakeholders are the nine local communities and the Federal Reserve. The local communities' prominence is obvious; less obvious is the Federal Reserve. Their prominence comes from an interest in maintaining economic activity, a goal that they are seen as powerful, urgent, and mostly legitimate in achieving. Next, potential homeowners are powerful (they choose whether or not to move in), legitimate (their finances are on the line), but less urgent as they home multiple choices for homeownership. Next, the real estate developer, as it pertains to this problem, is somewhat powerful (they have some influence over the property values of the communities, but not all), no legitimacy (in the eyes of the communities, outside developers should not wield authority), and urgency as they own the land and need to see it developed. Finally, banks are powerful as they can choose to loan money to potential homeowners, current homeowners, and the real estate developer, but they are not legitimate as it pertains to this problem as they will seek to advance their own financial interests irrespective of the effect on the communities, and they are not urgent as there is no large need to be involved in this situation due to the presence of other customers.

Table A.1 Stakeholder classification

Stakeholder	Stakeholder attribute			Prominence
	Power	Legitimacy	Urgency	
Nine local communities	0.75	1	1	0.92
Federal Reserve	1	0.75	1	0.92
Potential homeowners	1	1	0.5	0.83
The real estate developer	0.5	0	1	0.5
Banks	1	0	0	0.33

A.2.3 Stakeholder Attitude Evaluation

Table A.2 shows evaluations of the potential for threat and potential for cooperation for each of the stakeholders identified in the previous section. These two parameters provide an identification of the attitude of each stakeholder. They have been sorted in decreasing order of support according to their assigned stakeholder attitude.

Both the nine local communities and potential homeowners are seen as supportive of this effort. The nine local communities' support is obvious, while potential homeowners are supportive of maximized property values as they wish to make an investment in a desirable neighborhood, as indicated by strong financial returns. The Federal Reserve and banks could pose a threat, but they could also cooperate if they side with the communities and not the developer. Finally, the real estate developer has strong potential for threat due to the impending development, and some potential for cooperation in the eyes of the local communities.

A.2.4 Stakeholder Objective Mapping

With classification and attitude defined in the previous two sections, Fig. A.1 shows a stakeholder objective map, including the influence (direction and magnitude) for

Table A.2 Stakeholder attitude evaluation

Stakeholder	Potential for threat	Potential for cooperation	Support
Nine local communities	1	1	1
Potential homeowners	0	1	1
Federal Reserve	1	1	0
Banks	1	1	0
The real estate developer	1	0.5	−0.5

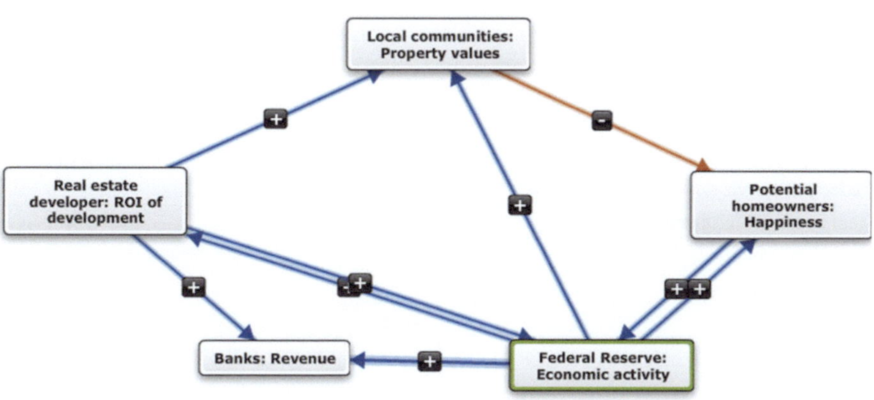

Fig. A.1 Stakeholder objective map

all identified stakeholders involved in the problem. The thicker the line, the stronger the causal influence.

A.2.5 Stakeholder Engagement Priority

In order to calculate the stakeholder engagement priority for all the stakeholders in the local communities' problem, we need to calculate k_i^{in}, k_i^{out}, $s_i^{*-\text{in}}$, $s_i^{*-\text{out}}$, *Popularity*, and *Activity*, in accordance with equations found in Chap.6. These results are shown in Table A.3.

We then sort the stakeholders by activity first (in descending order) and then by popularity (in ascending order). Table A.4 illustrates the order in which stakeholders should be engaged in support of this effort.

It is clear that the Federal Reserve should be prioritized for this effort. This is due to their large influence and connectivity; however, influencing them may be problematic. Thus, we can move to the second stakeholder, the real estate developer. Working with the developer makes sense as they have a strong influence on the financial return of the local communities. At the lowest end of the priority spectrum is the banks. They have no outgoing arcs, so they cannot affect change beyond themselves; thus, they should be given a low priority in terms of their engagement.

Table A.3 Network characteristics

Stakeholder	k_i^{in}	k_i^{out}	$s_i^{*-\text{in}}$	$s_i^{*-\text{out}}$	Activity	Popularity
Local communities	2	1	0.5	0.25	0.5	1
The real estate developer	1	3	0.25	0.75	1.5	0.5
Potential homeowners	2	1	0.5	0.25	0.5	1
Banks	2	0	0.75	0	0	1.22
Federal Reserve	2	4	0.5	1.25	2.24	1

Table A.4 Stakeholder prioritization

Stakeholder	Activity	Popularity	Engagement priority
Federal Reserve	2.24	1	1
Real estate developer	1.5	0.5	2
Local communities	0.5	1	3
Potential homeowners	0.5	1	4
Banks	0	1.22	5

A.2.6 Stakeholder Management Plan

The final step in analyzing this problem is to develop a stakeholder management plan. An example stakeholder management plan is shown below in Table A.5. As this stakeholder assessment is being performed by the local communities, their priority of engagement is a nonissue. They are inherently a part of the stakeholder management process. Thus, although they are both prominent and supportive, they are moved to the bottom of the list.

From the local communities' perspective, their focus is to monitor the banks and Federal Reserve, defend against the real estate developer, and involve potential homeowners as best as possible. These all seem like reasonable strategies for engagement moving forward.

A.3 What Perspective

The following subsections discuss the *what* perspective analysis for this problem.

A.3.1 Articulate Objectives

In terms of the communities, they are concerned with maintaining their investment. So we can define our first fundamental objective as *Maximize property values*, in accordance with the previous perspective's analysis. They might be willing to accept a short-term reduction in property values for a longer-term improvement in quality of life (i.e., a new community swimming pool would improve their life but be disruptive in the meantime). Thus, we should add one objective, *Maximize quality of life*. Now, we can organize our objectives.

Table A.5 Example stakeholder management plan

Stakeholder name	Wants	Prominence	Support	Priority of engagement	Strategy
Federal Reserve	Economic activity	0.92	0	1	Monitor
The real estate developer	ROI of development	0.5	−0.5	2	Defend
Potential homeowners	Happiness	0.83	1	3	Involve
Banks	Revenue	0.33	0	4	Monitor
Nine local communities	Property values	0.92	1	n/a	Involve

A.3.2 Fundamental Objectives Hierarchy

Organizing our two fundamental objectives into a hierarchy yields Fig. A.2, showing a further decomposition of our objectives. Property values are broken down into land value (heavily influenced by surrounding developments and the neighborhood) and home value (influenced by both the neighborhood and individual home maintenance). Quality of life is broken down into noise and safety. The communities' quality of life would be negatively affected by significant construction noise or crime.

A.3.3 Means-Ends Network

The means-end network shows our understanding of the means necessary to produce our desired ends (i.e., our fundamental objectives). Using the same two fundamental objectives as before, we can create the network shown in Fig. A.3. Both property values and quality of life are means to achieve the end of maximize property values and quality of life. Additionally, property values help achieve high quality of life and vice versa. A strong economy (e.g., high GDP, low unemployment), home and neighborhood maintenance (e.g., mowed lawns, freshly painted exteriors), and good schools (e.g., high test scores) contribute to high property values, as well as quality of life. In addition to these factors, community services (e.g., parks and recreational opportunities), laws and regulations, and a neighborhood watch program all improve quality of life.

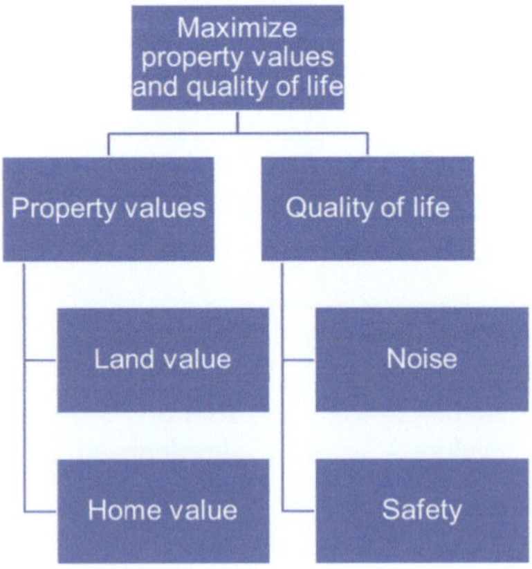

Fig. A.2 Fundamental objectives hierarchy

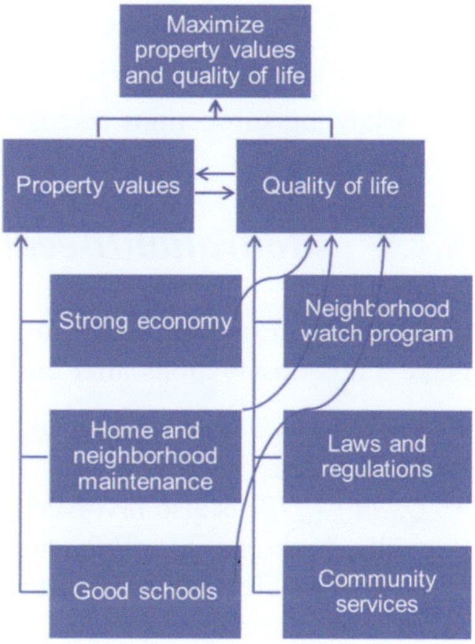

Fig. A.3 Means-ends network

A.3.4 FCM Update

Armed with our means-ends network, we can now integrate it into our existing FCM. Our modified FCM reflects the combination of happiness and quality of life as one concept to reduce clutter. This revised scenario depiction is shown in Fig. A.4.

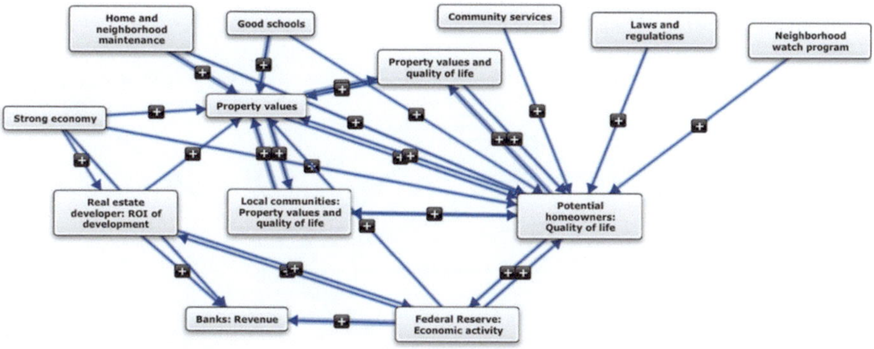

Fig. A.4 Updated FCM with means-end network incorporated

A.4 Why Perspective

The following subsections discuss the *why* perspective analysis for this problem.

A.4.1 Motivation/Feedback Analysis

If we are interested primarily in *property values* and *quality of life*, then we can make a few observations after analysis of our FCM and its feedback mechanisms:

- *Property values* and *quality of life* are involved in a virtuous circle. As each goes up, the other improves.
- *Quality of life* is also involved in a virtuous circle with *economic activity*. Those with a high *quality of life* choose to encourage *economic activity*. Higher *economic activity* tends to lead to a higher *quality of life*.
- As it is currently depicted, *property values* and *economic activity* are related only unidirectionally (*economic activity* causes an increase in *property values*, but not the other way around). This seems problematic, and thus, the FCM should be modified to reflect the perceived relationship of a virtuous circle.

A.4.2 FCM Update

Our feedback analysis in the previous subsection indicates the need for a causal link from *property values* to *economic activity*. This update is reflected in Fig. A.5.

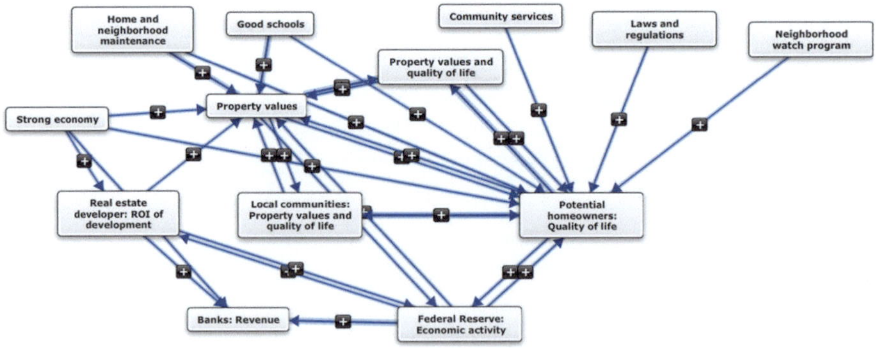

Fig. A.5 Updated FCM with feedback analysis incorporated

A.4.3 *Proposed Changes During Act Stage*

No new proposed changes to the FCM are required.

A.5 Where Perspective

The following subsections discuss the *where* perspective analysis for this problem.

A.5.1 *Boundary Articulation*

Based on analysis of critically heuristic boundary issues of our problem using guidelines found in Ulrich (2000), we can generate the boundary critique shown in Table A.6.

Assessment of our boundary yields a few insights. Banks have too much perceived power in the eyes of the local communities. They would rather trust an independent agency such as the Federal Reserve, rather than the banks. Further, they do not want the developer to have power, although it is currently perceived that they have it.

A.5.2 *Context*

We can assess the contextual elements of our problem, including its abstraction (circumstances, factors, and conditions) and culture (values and patterns) as shown in Table A.7.

This problem's context has several competing elements at play. Driving the problem toward resolution is the value of "The American dream," a belief held by

Table A.6 Boundary critique

Boundary issue	*What is*	*What ought-to-be*
Sources of motivation	Property values and quality of life	Property values and quality of life
Sources of power	Communities, potential homeowners, banks, Federal Reserve, Developer	Communities, potential homeowners, Federal Reserve
Sources of knowledge	Banks, Federal Reserve	Federal Reserve
Sources of legitimation	Communities, potential homeowners	Communities, potential homeowners

Table A.7 Context articulation

Category	Element
Circumstance	Current laws regulating residential development
Factor	Homeowner budgets
Condition	Housing bubble status, mortgage crisis
Value	"The American dream"
Pattern	Business-friendly environment

many that an individual can work full time to earn enough to buy a home, and the circumstance of current laws regulating residential development which prevent the real estate developer (and others) from doing whatever it wants with its land parcel. On the other hand, there are several restraining forces involved, including the factor that homeowner budgets are limited in a tough economic climate, conditions which include recovery from a housing bubble burst and mortgage crisis, both of which may reduce consumer confidence, and a pattern in the USA of a capitalist-driven, business-friendly environment in most cities.

A.5.3 Force Field Diagram

Combining elements from the previous two subsections yields Table A.8.

Ultimately, this problem involves a modern day David versus Goliath scenario. The local communities see themselves as the "little guy" battling the corporate giants. While this may be true, it is the reality in which they operate and, thus, they ought to work cooperatively with the real estate developer and local banks to arrive at a satisfactory solution to their situation.

A.5.4 Updated FCM

Many elements are missing in the FCM as a result of boundary and context analysis. They include a business-friendly environment, housing bubble status/mortgage crisis, homeowner budgets, laws and regulations, and "The American Dream." These concepts, as well as their connections, are reflected in Fig. A.6.

A.5.5 Proposed Ought-to-Be Changes

At this point, no new proposed changes are necessary for later analysis.

Table A.8 Force field diagram

Driving force	Strength as-is	Strength ought-to-be	Problem	Strength ought-to-be	Strength as-is	Restraining force
Current laws regulating residential development	0.5	0.5	*The nine local communities see local developers and banks as having too much power in their property values. (They should understand that these are the "rules of the game" and work within this structure).*	−0.5	−1	Business-friendly environment
"The American dream"	0.5	1		−0.25	−0.5	Homeowner budgets
				−0.25	−0.25	Housing bubble status, mortgage crisis

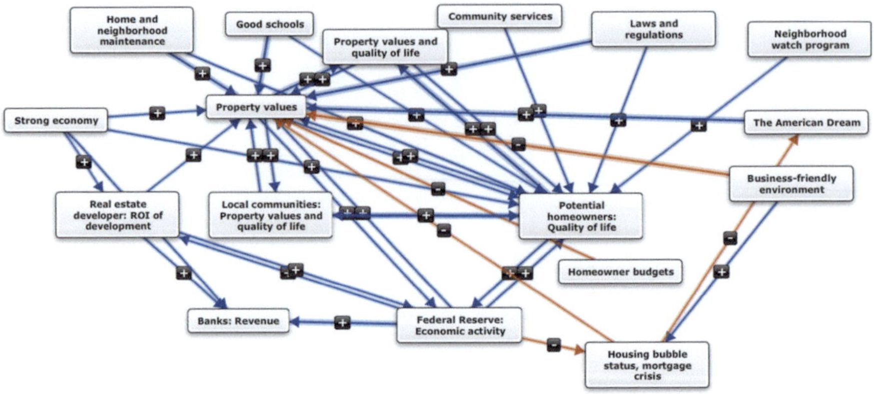

Fig. A.6 Updated FCM with boundary and context incorporated

A.6 How Perspective

The following subsections discuss the *how* perspective analysis for this problem.

A.6.1 Cynefin Analysis

The *property value* and *quality of life* problem seems relatively well ordered. Relationships are understood and there is acknowledgment on the behalf of the local communities as to where they stand (waiting for action from the banks, Federal Reserve, and real estate developer). It is in the complicated domain, however, as there are some major uncertainties stemming from a lack of information regarding what the overall market will do.

A.6.2 Mechanism Selection

We should employ a number of mechanisms to our problem. We can expend our time, money, materiel, equipment, KSAs, and manpower to maintain our neighborhood property values and increase quality of life overall. As appropriate, we should try to stay informed regarding the economy as a whole. Given the status of the economy, we may have no choice but to wait (invoking time). As the problem unfolds, the scenario may change. We should capture *community improvements* as a concept (the result of our mechanisms) in our FCM.

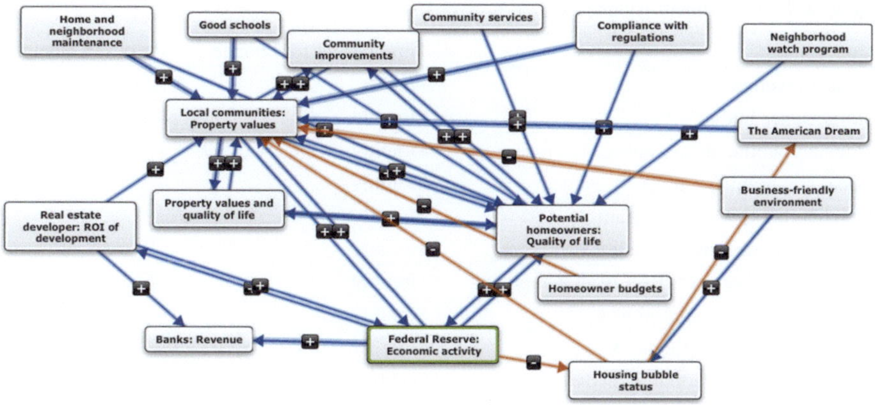

Fig. A.7 Updated FCM with mechanisms incorporated

A.6.3 Updated FCM

Reflecting the inclusion of the *community improvements* concept, Fig. A.7 shows the updated FCM based on mechanism analysis.

A.7 When Perspective

The following subsections discuss the *when* perspective analysis for this problem.

A.7.1 Timescale Assessment

First, we must analyze our FCM to ensure that all concept transitions occur on the same timescale. We can list all of our concepts and their accompanying time horizon for change to ensure that they change at the same rate. This information is found in Table A.9. Note that proposed changes indicate whether the total magnitude should be increased (+) or decreased (−). An indication of two or more plus or minus values indicates a stronger temporal adjustment is necessary.

We then adjust our causal weights using the information found in Table A.9.

Table A.9 Assessment of concept time horizons

Concept	Time period for change	Proposed change
Banks: revenue	Weekly	None
Homeowner budgets	Monthly	–
Housing bubble status, mortgage crisis	Monthly	–
Local communities: property values	Monthly	–
Neighborhood watch program	Monthly	–
Potential homeowners: quality of life	Monthly	–
Real estate developer: ROI of development	Monthly	–
Business-friendly environment	Monthly	–
Compliance with regulations	Monthly	–
Federal Reserve: economic activity	Quarterly	–
Community services	Yearly	–
Good schools	Yearly	–
Home and neighborhood maintenance	Yearly	–
Community improvements	Yearly	–
Property values and quality of life	Yearly	–
The American Dream	Yearly	–

A.7.2 Intervention Timing

Armed with our modified (for timing) FCM, we must work our way through the decision flowchart in Chap.10. Starting with element 1, we can definitively conclude that the benefit remaining in the problem (as it pertains to revenue maximization and quality of life) certainly outweighs the cost of intervention. So, max (B/C) $\geq$ 1. Next, we must ask whether or not our problem is stable (Element 3). This requires us to consider initial values for our concepts as the status quo. In this case, we believe *Developer ROI* is at −0.25, a reflection of a slight ROI loss for the developer due to carrying costs and a lack of economic activity on their parcel of land, while all remaining concepts are taken to be 0 (absent any further information). The results of this analysis are shown in Fig. A.8.

Clearly, the scenario is stable, but complicated. In this case, we move to Step 4a, *Act to achieve objectives*. Thus, overall we can conclude that the problem has sufficient time to act and is stable enough to warrant action to resolve its objectives. This represents the conclusion of the perspective-driven analysis of this problem (i.e., the thinking phase).

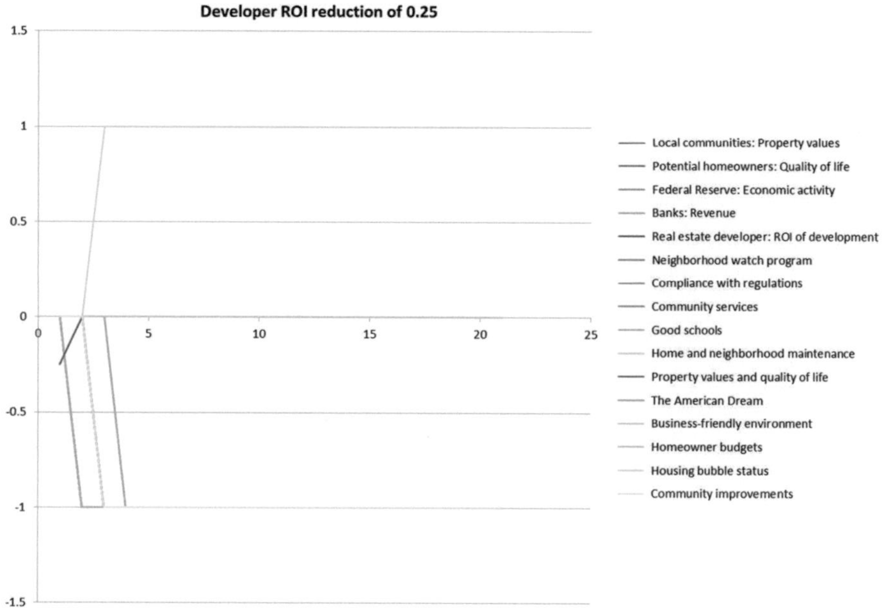

Fig. A.8 Stability analysis of local communities' problem

A.8 Conclusions

This appendix provided analysis of the second problem in the real estate mess discussed throughout this mess. Armed with this analysis, as well as those conducted in *thinking* chapters, the two problems can be integrated into a mess-level articulation, using the techniques outlined in Chap. 12.

References

Hester, P. T., Bradley, J. M., & Adams, K. M. (2012). Stakeholders in systems problems. *International Journal of System of Systems Engineering, 3*(3/4), 225–232.

Ulrich, W. (2000). Reflective practice in the civil society: the contribution of critically systemic thinking. *Reflective Practice, 1*(2), 247–268.

Index

P.T. Hester and K.M. Adams, *Systemic Decision Making*, Topics in Safety, Risk, Reliability and Quality 33, DOI 10.1007/978-3-319-54672-8

D

E

R

S

MIX
Papier aus verantwortungsvollen Quellen
Paper from responsible sources
FSC® C105338

If you have any concerns about our products,
you can contact us on
ProductSafety@springernature.com

In case Publisher is established outside the EU,
the EU authorized representative is:
Springer Nature Customer Service Center GmbH
Europaplatz 3, 69115 Heidelberg, Germany

Printed by Libri Plureos GmbH
in Hamburg, Germany